CLIMATE CHANGE AND ITS CAUSES, EFFECTS AND PREDICTION SERIES

GLOBAL CHANGE AND FORESTRY: ECONOMIC AND POLICY IMPACTS AND RESPONSES

CLIMATE CHANGE AND ITS CAUSES, EFFECTS AND PREDICTION SERIES

Global Climate Change
Harace B. Karling (Editor)
2001. ISBN: 1-56072-999-6

Global Climate Change Revisited
Harace B. Karling (Editor)
2007. ISBN: 1-59454-039-X

Climate Change Research Progress
Lawrence N. Peretz (Editor)
2008. ISBN: 1-60021-998-5

Climate Change: Financial Risks
United States Government Accountability Office
2008. ISBN: 978-1-60456-488-4

Post-Kyoto: Designing the Next International Climate Change Protocol
Matthew Clarke
2008. ISBN: 978-1-60456-840-0

Post-Kyoto: Designing the Next International Climate Change Protocol
Matthew Clarke
2008. ISBN: 978-1-61668-101-2 (Online book)

The Effects of Climate Changeon Agriculture, Land Resources, Biodiversity in the United States
Peter Backlund, Anthony Janeto and David Schimel
2009. ISBN: 978-1-60456-989-6

Economics of Policy Optionsto Address Climate Change
Gregory N. Bartos
2009. ISBN: 978-1-60692-116-6

Climate Variability, Modeling Tools and Agricultural Decision-Making
Angel Utset (Editor)
2009. ISBN: 978-1-60692-703-8

Climate Variability, Modeling Tools and Agricultural Decision-Making
Angel Utset (Editor)
2009. ISBN: 978-1-60876-791-5 (Online Book)

Emissions Trading: Lessons Learned from the European Union and Kyoto Protocol Climate Change Programs
Ervin Nagy and Gisella Varga (Editors)
2009. ISBN: 978-1-60741-194-9

Constructing Climate Change Legislation: Background and Issues
Gerald P. Overhauser (Editor)
2009. ISBN: 978-1-60692-986-5

Designing Greenhouse Gas Reduction and Regulatory Systems
Sonja Enden (Editor)
2009. ISBM: 978-1-60741-195-6

Global Climate Change: International Perspectives and Responses
Elias D'Angelo (Editor)
2009. ISBN: 978-1-60741-233-5

Disputing Global Warming
Anton Horvath and Boris Molnar (Editors)
2009. ISBN: 978-1-60741-235-9

Disputing Global Warming
Anton Horvath and Boris Molnar (Editors)
2009. ISBN: 978-1-60876-503-4 (Online book)

Focus on Climate Change and Health
Viroj Wiwanitkit
2009. ISBN: 978-1-60741-247-2

Land Use and Climate Change
Suresh C. Rai
2009. ISBN: 978-1-60741-362-2

Lightning in the Tropics: From a Source of Fire to a Monitoring System of Climatic Changes
Osmar Pinto, Jr.
2009. ISBN: 978-1-60741-764-4

Global Change and Forestry: Economic and Policy Impacts and Responses
Jianbang Gan, Stephen Grado and Ian A. Munn (Editors)
2010. ISBN: 978-1-60876-262-0

CLIMATE CHANGE AND ITS CAUSES, EFFECTS AND PREDICTION SERIES

GLOBAL CHANGE AND FORESTRY: ECONOMIC AND POLICY IMPACTS AND RESPONSES

JIANBANG GAN,
STEPHEN GRADO
AND
IAN. A. MUNN
EDITORS

Nova Science Publishers, Inc.
New York

For permission to use material from this book please contact us:
Telephone 631-231-7269; Fax 631-231-8175
Web Site: http://www.novapublishers.com

LIBRARY OF CONGRESS CATALOGING-IN-PUBLICATION DATA

Global change and forestry : economic and policy impacts and responses / editors, Jianbang Gan, Stephen Grado, and Ian A. Munn.
p. cm.
Includes index.
ISBN 978-1-60876-262-0 (hardcover)
1. Forest management. 2. Forests and forestry--Economic aspects. I. Gan, Jianbang. II. Grado, Stephen C. III. Munn, Ian A.
SD431.G56 2009
333.75--dc22

2009035590

Published by Nova Science Publishers, Inc. ✢ New York

CONTENTS

PREFACE

Global change that impacts forestry is multi-faceted, and includes changes in climate, markets and trade, and human and natural interfaces, among others. These changes have had, and will continue to have, effects on forest resources and the forest products industry. Understanding the economic and social impacts of these changes and exploring policy responses can help us better meet future challenges and take advantage of opportunities these changes will bring about.

This book is a collection of current research work on economic and social impacts of, and policy responses to, global change mainly in regard to the southern United States. This work primarily represents papers presented at the 2007 Southern Forest Economics Workshop in San Antonio, Texas. Several papers have been published in journals and are reprinted here with permission. The book covers a variety of changes including climate change, land use change, globalization, and other forces as they relate to forestry. These papers address broad aspects of impacts and responses, including those related to markets, forest resource management, forest landowners, and consumers. Though the book focuses on the southern United States--an important player in the domestic and global forest products markets and forest ecosystems, the issues addressed here well reflect those facing many other parts of the world. Hence, the research approaches and findings presented here have useful implications for forest resource management and forest products markets beyond the region.

The book is organized in three parts. Part I, with seven papers, focuses on climate change, land use change, and disturbance mitigation. Chapter 1 describes and analyzes carbon markets, forest carbon accounting, and implications of carbon credits for forest management. Chapter 2 assesses the impact of climate change on forest-based recreation. Chapters 3, 4, and 5 present various modeling approaches and assessments of forest land use change. Chapter 6 addresses tropical deforestation in China, another type of competing land use. Part I concludes with Chapter 7, a case study of invasive species control, an increasingly challenging forest management issue throughout the world.

There are seven papers in Part II, which encompasses the theme of forest products markets and trade. Chapter 8 starts with an overview of the United States pulpwood market with projected future trends under different scenarios including increased paper recycling and competing uses of woody biomass for energy production. This is followed by Chapters 9, 10, and 11 that address, respectively, various aspects of recent developments in forest biomass/bioenergy, ranging from mill/refinery location and economic impacts to forest management for bioenergy production. Chapter 12 reviews the evolution of hardwood timber markets since the 1970s. The value chain of wood products in the southeastern U.S. is analyzed in Chapter 13. Last, Chapter 14 compares the softwood production response by the forest products industry and family landowners to price changes.

Forest landownership and the demand for forest ecological services have changed in the U.S. South. The first five chapters in Part III address family forest landowners, the dominant group of forest landowners in the region. Chapter 15 addresses changing patterns of small-scale family forests, followed by the behavior of these landowners in forest management in Chapter 16, their awareness and adoption of forest regeneration programs in Chapter 17, their promptness in forest regeneration in Chapter 18, and tax incentives for reforestation on their lands in Chapter 19. Recreation is the second most economically important use of forest resources in the southern U.S. Chapter 20 depicts the economic impact of birding festivals, Chapter 21 the spending patterns of park visitors, and Chapter 22 the market value of hunting leases.

We would like to thank all the chapter authors and reviewers for their hard work and patience. Without them, this book would have been impossible. We are also very grateful for the professional assistance from Nova Science Publishers.

Jianbang Gan
Stephen C. Grado
Ian A. Munn

PART I. CLIMATE CHANGE, LAND USE CHANGE, AND DISTURBANCE MITIGATION

In: Global Change and Forestry
Editors: J. Gan, S. Grado and I. A. Munn
ISBN: 978-1-60876-262-0

Chapter 1

FORESTRY AND CARBON: MARKET, MEASUREMENT, AND MANAGEMENT

***Jianbang Gan*[*,1] and Bruce A. McCarl*[2]**

[1] Department of Ecosystem Science and Management,
Texas A&M University, Texas, USA

[2] Department of Agricultural Economics,
Texas A&M University, Texas, USA

ABSTRACT

We review and analyze markets for greenhouse gases, measurements of forestry carbon sinks, and implications of carbon credits for forest management. Currently forestry carbon offsets are transacted mainly on voluntary markets and to a lesser extent under the Clean Development Mechanism on regulated markets. The carbon price is much lower, yet more volatile on voluntary markets than on regulated carbon markets, but they are highly correlated. There are several tools available for measuring forestry carbon pools including afforestation/reforestation, avoided deforestation, sustainably managed forests, wood products, and fuel substitution; yet improvements and standardization of these tools are needed to meet the requirements for registries on carbon markets. The implications of carbon credits on forest management are multifaceted. At the ecosystem level, inclusion of carbon values will lead to more standing inventory and likely complement timber production in the long run. Given the joint provision of multiple forest benefits and the interlink among different forestry carbon pools, systematic approaches are imperative in developing forest GHG mitigation strategies as well as designing forest carbon account methods.

Keywords: Greenhouse gases; Forestry offset; Emission trading; Carbon accounting; Forest management.

---

* E-mail address: j-gan@tamu.edu, phone: (979) 862-4392, fax: (979) 845-6049, Department of Ecosystem Science and Management, 2138 TAMU, Texas A&M University, College Station, TX 77843-2138.

1. Introduction

Forestry plays an important role in the global carbon cycle. Globally forests store 86% of above-ground terrestrial carbon and 73% of soil carbon (Sejdo 1993). A forest can be either a carbon source or a sink, depending upon human and/or natural forces. Growth of trees absorbs carbon via photosynthesis and stores it in wood and soil; while deforestation and forest fires release carbon. Reductions in deforestation and forest fires can alleviate carbon emissions while tree planting enhances sequestration. As such, forests, if properly managed, can contribute to global greenhouse gas (GHG) emission mitigation, an option for combating GHG net emissions and in turn climate change. Kauppi et al. (2001) estimates that the global forest sector has an average biophysical mitigation potential of 5,380 MtCO2 per year until 2050 accounting for about 46% of the biological mitigation potential on all lands (IPCC 2000). Additionally, harvested wood products can store carbon; and substitutions of wood products for metals and plastics and forest biofuels for fossil fuels can also help reduce GHG emissions.

Many studies have been conducted on forest carbon sinks and sources including work on carbon accounting, cost estimation, and forest management impacts and responses, among others (Eggleston et al. 2006, Easterling et al. 2007, Nabuurs et al. 2007, Malmsheimer et al. 2008). Meanwhile carbon markets have emerged and evolved though still at their infant stage. We survey the literature in these interrelated aspects encompassing forestry and carbon, review the status of forestry in carbon markets, and synthesize existing findings coupled with further analysis.

The rest of the chapter is organized as follows. The next section will give an overview of existing carbon markets. The third section will review forest carbon sinks, accounting methods, and costs. The fourth section will highlight impacts on forest management as well as forest management responses. This chapter will conclude with a summary and synthesis along with recommendations.

2. Carbon Markets

2.1. Market Classification and Eligibility of Forestry Offsets

The development of carbon trading markets represents a market based approach to mitigating GHG emissions. Currently there are two major types of carbon markets: voluntary and compliance (regulated) markets. Carbon offset trades transacted in a voluntary market are not mandated by regulation. The Chicago Climate Exchange (CCX) (http://www.chicago climatex.com/), created in 2003 as the world's first marketplace for carbon trading, is an example of a voluntary, rules based GHG trading system that is legally binding for participating entities. Another variant of voluntary markets is the over-the-counter (OTC) market that is based on bilateral deals instead of exchanges.

On the other hand, compliance (regulated) markets are operated under certain mandatory GHG reduction regimes such as the Kyoto Protocol (UNFCCC 1997) and the European Union's Emissions Trading Scheme (EU ETS). Under the Kyoto Protocol, Annex I Parties can use three mechanisms including the Joint Implementation (JI), Clean Development

Mechanism (CDM) and trading of Assigned Amount Units (AAUs) to fulfill their commitments to reducing GHG emissions. The EU ETS was launched on January 1, 2005 to help EU countries to meet their Kyoto commitments and beyond.

Although there are a variety of potential forestry carbon sinks including afforestation, reforestation, avoided deforestation, sustainably managed forests, forest biofuels (substitutes for fossil fuels), and wood products (carbon stored in wood products and substitutes for energy intensive materials such as metals and plastics), only a few forestry offsets can be traded on current carbon markets. The Kyoto Protocol recognizes the GHG offset potential of afforestation, reforestation, forest management, and avoided deforestation. Yet, currently only afforestation and reforestation offsets are credible under CDM. And CDM projects must be validated and verified by an independent third party and approved by the CDM Executive Board. The CDM afforestation/reforestation projects currently registered are rare. EU ETS currently does not allow forestry offset.

Forestry offsets are allowed on CCX. Eligible forestry offsets on CCX include afforestation/reforestation, avoided deforestation and degradation, increased stand- and landscape-level carbon density from improved forest management, long-lived wood products, and fuel substitution. The CCX Committee on Forestry oversees forestry offset projects. Verification by an approved, independent third party is required before offset credits can be registered and traded. Small GHG offset suppliers (landowners) can participate with assistance of Offset Aggregators.

To be eligible for afforestation offsets, an afforestation project must occur on or after January 1, 1990, involve no biomass removal (harvesting or thinning) during the CCX market period, and maintain the forest on the land for at least 15 years. The offset can be quantified using carbon accumulation tables or through measurement and growth and yield models approved by CCX. Wood products can store carbon and displace fossil-fuel intensive construction materials. Forest landowners can earn offset credits from long-lived wood products if all of their forestlands are certified for sustainable management and the carbon rights are retained via a sales contract. The offset credit is based on the amount of carbon retained in wood products in use and landfills at the end of 100 years.

A forest landowner can also earn offset credits for the additional net carbon sequestrated in managed forest projects. To be eligible for this credit, the management of all forest holdings owned by the landowner must be certified as sustainable by a forest certification scheme endorsed by the PEFC Council or approved by the CCX Committee on Forestry. All managed forest projects must be reviewed and approved by the Committee on Forestry on a case-by-case basis. The quantity of offsets from a managed forest project can be determined by either using a growth/yield model or calculating inventory on an annual basis.

2.2. Trading Volume and Price

Though carbon markets are still emerging, the transaction volume and value have quickly gained momentum in recent years (table 1). Almost 3 Gt CO_2 (equivalent) were transacted in 2007, valued at \$66 billion. Most transactions took place in the compliance markets, which accounted for 97.8% of the volume and 99.5% of the value globally. Total transaction volume and value increased by 73% and 65% from 2006 to 2007, respectively. Additionally even though the voluntary market transaction volume and value are much smaller, in recent history they have grown even more rapidly than have the regulated markets.

Table 1. GHG transaction volume and value in 2007

Market	Volume ($MtCO_2e$)	Value ($ million)
Voluntary markets		
Over-the-counter markets	42	258
Chicago Climate Exchange	23	72
Subtotal	65	330
Compliance markets		
EU Emissions Trading Scheme	2,061	50,097
Others	857	15,990
Subtotal	2,918	66,087
Total	2,983	66,417

Source: Hamilton et al. 2008.

Currently there are about $100 million of terrestrial-based carbon transacted annually (Personal communication with Michael Walsh, 2008). Most of the reported transactions take place on voluntary markets, particularly over-the-counter markets. In 2007, $46.5 million of forestland based offsets was traded on OTC markets globally, accounting for 18% of the total transaction value on OTC markets (Hamilton et al. 2008). To date, there are very few CDM afforestation/reforestation projects that have been officially registered and reported.

We examined daily trading prices on the CCX and the European Climate Exchange (ECX). The data show considerable differences in prices and volatility across markets (figure 1). From their inceptions to November 13, 2008, the mean daily CCX closing price was \$2.71/t CO_2 with a standard deviation of \$1.48/t CO_2 while the ECX price[1] was \$29.68/t CO_2 with a standard deviation of \$6.49/t CO_2. The CO_2 trading price on average has been about 10 times higher on ECX whereas the coefficient of variation is about 40% of that on CCX. This suggests that the regulatory environment substantially increases the GHG price while reducing its volatility. In other words, the value of GHG abatement might not be fully realized in a voluntary market.

Despite differences in price and volatility, the two markets seem highly correlated. The correlation coefficient between the CO_2 prices on CCX and ECX was 0.33 and statistically significant at the <1% level. A unit root, I(1), was found for both prices series on CCX and ECX using the Dickey-Fuller unit root test (Dickey and Fuller 1979, 1981). The cointegration test (Engle and Granger 1987) also revealed statistical evidence of comovement of these two price series on CCX and ECX as the null hypothesis of CI(1, 1) could not be rejected at the 5% significance level. Hence, price movements on the voluntary and regulated markets are interrelated. This could be of value for designing future GHG mitigation policies and mechanisms as we are encountering the challenge in reaching a global agreement on means to reduce GHG emissions. Though voluntary and regulated markets differ in terms of functioning and outcomes, they are interrelated and can work together to achieve the common goal.

[1] The European price was measured in €/t CO_2, which was converted to \$/t CO_2 based on the daily exchange rate.

Figure 1. Daily trading prices of CO_2 (equivalent) on CCX and ECX (Data source: http://www.chicagoclimatex.com/).

3. Forestry Carbon Measurement

3.1. Forestry Carbon Pools

Forest carbon pools include above-ground biomass, below-ground biomass, deadwood, litter, soils, and harvested wood products (Nabuurs et al. 2007). These five pools are interrelated through biomass/carbon transfers. The above-ground biomass can be transferred into harvested wood products via harvesting and deadwood due to mortality plus logging residuals. Dead wood, branches, foliage, and roots will be further transferred into litter, which will decompose into soils. Additionally, carbon transfers between the system (five pools) and the outside environment will also take place. Sequestration into the system increases via biomass growth; while decomposition and plant respiration release carbon back to the atmosphere.

3.2. Measurement Tools

IPCC (Eggleston et al. 2006) has developed guidelines for accounting GHG in these pools. Carbon stock changes within a stratum can be estimated by considering the carbon cycle process between the five carbon pools. Annual carbon stock change in a given pool can be calculated using two methods: the gain-loss method and the stock-difference method. The gain-loss method, also called the process method, measures annual **carbon stock change** (ΔC) in a pool using the difference between carbon gains (C_G) and losses (C_L), i.e., $\Delta C = C_G - C_L$. The stock-**difference method estimates annual stock change** (ΔC) **based on the difference of** stock estimates at two time points, i.e., $\Delta C = (C_{t2} - C_{t2})/(t2 - t1)$ where C_{t1} = carbon stock in the pool at time t1, and C_{t2} = carbon stock in the pool at time t2. The gain-loss method can better smooth out inter-annual variability in the estimates than the stock-difference method as the former incorporates modeling approaches using the models fit with empirical data. The stock-difference method assumes the land area in a category does not change over the estimation period; otherwise some adjustment is needed to account for the change in land area. Both methods are valid so long as they are capable of representing actual disturbances as well as continuously varying trends, and can be verified by comparison with actual measurements.

The annual change in the carbon stock of harvested wood products pools can be estimated using $\Delta C(t) = C(t+1) - C(t)$ where $C(t)$ is the carbon stock at time (year) t. The carbon stock can be computed by

$$C(t+1) = e^{-k}\, C(t) + \left(\frac{1-e^{-k}}{k}\right) Inflow(t) \tag{1}$$

where k is the decay constant (units/year) of first-order decay, i.e., $k = \ln(2)/HL$ with HL being the half-life (years) of the wood product pool; and Inflow(i) is the inflow (Gg C/year) to the harvested wood products pool during year t. When using the above formula, the carbon stock in the year 1900 (base year) is assumed to be zero (Eggleston et al. 2006).

Four forest carbon measuring tools/approaches have been developed and applied in the U.S. They are General Technical Report NE-343 (GTRNE-343), the Carbon Calculation Tool (CCT), the Carbon On-Line Estimator (COLE), and the Forest Vegetation Simulator (FVS). The major features including advantages and disadvantages of these tools are summarized by Ingerson and Loya (2008). They all rely on Forest Inventory and Analysis data. GTRNE-343 (Smith et al. 2006) estimates carbon density (t/ac) in forest pools by stand age and forest type as well as carbon in wood products for major regions. CCT (Smith et al. 2007) estimates annual carbon stocks and fluxes in major forest carbon pools for each state and the nation since 1990. COLE (Spinney et al. 2005) uses a similar approach in GTRNE-343 to estimate forest carbon density by stand age and forest type, but it can be applied to county-level estimation. FVS (USDA Forest Service 2008) originally developed as a forest stand simulation model can be used to measure carbon stocks and fluxes at the stand level and simulate the impact of forest management activities on carbon stocks and fluxes. Though these tools provide useful information about forest carbon pools, they lack the calibration and resolution required for registries on carbon markets.

Pearson et al. (2007) develop the guidelines for measuring forest carbon sequestration to support reporting by public and private entities to GHG registries. These guidelines include those for defining boundaries; measuring, monitoring, and estimating changes in carbon stocks; planning for carbon measuring and monitoring; and implementing quality assurance and quality control. Their proposed methods for measuring carbon stock and change, including both direct (sampling) and indirect (look-up tables and modeling) measurements, are summarized in table 2.

Table 2. Methods for measuring forest carbon stock for carbon registries

Carbon pool	Suggested method
Living above-ground biomass	
Trees	Direct Method: Link above-ground biomass (AGB) to growing stock volume (GSV) based on regression equations. Namely, AGB (t/ha) = F x (G + (1-exp(-GSV (m^3/ha) /H)) where F, G, H = regression coefficients. Biomass Expansion Factor (BEF) Method: AGB (t/ha) = GSV (m^3/ha) × BEF (t/m^3)
Nontree vegetation	Sampling and regression equations
Below-ground biomass	Regression equations: *Boreal:* BGB = exp(-1.0587 + 0.8836 ln *ABD* + 0.1874) *Temperate:* BGB = exp(-1.0587 + 0.8836 ln *AGB* + 0.2840) *Tropical:* BGB = exp(-1.0587 + 0.8836 ln *AGB*), where BGB = below-ground biomass density (t/ha) and AGB = above-ground biomass density (t/ha).
Dead organic matter	
Forest floor	Sampling
Dead wood	Line-intersect method: Estimate density in each of the three classes (sound, intermediate, and rotten) Estimate biomass volume in each density class Volume (m^3/ha) = π^2 [($d_1^2 + d_2^2$…….d_n^2)/8L] where d_1, d_2, d_n = diameter (cm) of each of the n pieces intersecting the line, and L = the length of the line (100 m recommended) Determine biomass stock Sum the products of density (from (a)) and volume (from (b)) in all classes Determine carbon stock Convert biomass stock to carbon using the carbon content of the biomass. Note: Standing deadwood should be measured by the same criteria used for live trees except for possible biomass difference in branches and leaves.
Soil carbon	Sampling

Source: Pearson et al. 2007.

CO2FIX is probably the most comprehensive stand-level carbon estimation tool (Scehelhaas et al. 2004). The CO2FIX V 3.1 can simulate carbon stocks and fluxes in a forest using the full carbon accounting approach (i.e., calculating changes in carbon stocks in all carbon pools over time). It consists of six modules: biomass, soil, wood products, bioenergy, financial, and carbon accounting modules. The biomass, soil, and wood products modules estimate carbon stored in living biomass (both above- and under-ground), soil organic matter, and wood products, respectively. The interrelationships among these three carbon pools are incorporated. The bioenergy module calculates the avoided GHG emissions resulting from substituting fossil fuels with forest bioenergy. The carbon accounting module uses different carbon accounting methods to record all fluxes to and from the atmosphere and assess the effects of the different scenarios. The financial module evaluates the profitability of various scenarios including timber values and carbon credits under different GHG accounting systems. This tool has been used for assessing carbon stocks and fluxes of different forestry projects in many parts of the world.

3.3. Additionality, Leakage, Permanence and Uncertainty

As in measuring many other nonforestry carbon sinks, forestry carbon accounting is also confronted with issues related to additionality, leakage, permanence and uncertainty. Additionality means new or additional sequestration beyond what would be under "business as usual". The additionality requirement entails reliable estimates of baselines. There are two methods for determining baselines: base-year or moving baseline approaches. In the base-year approach, the sequestration or GHG stock level at a given year (or an average over a span of several years) is used as the benchmark. A moving baseline is what the level of carbon stock would have been if the offset activity were never implemented. This baseline setting approach could be problematic for some forest carbon sinks such as avoided deforestation. The base-year baseline approach would be easier to use for quantifying additionality of many forestry carbon pools. The moving baseline also raises the issue of partial additionality. In particular a practice may be currently being adopted but a carbon price may accelerate adoption and then one may be able to claim only some part of the activity carbon benefits (see discussion in Smith et al)

Leakage is the offsetting of emission reductions developed under an offset project by increased emissions stimulated outside of the project's boundaries, undermining the net benefits of the offset activity. Leakage could be caused by market and nonmarket forces. Leakage of forestry offsets could be substantial. IPCC (Nabuurs et al. 2007) reports that leakage associated with afforestation ranges from 0 to 42% and that leakage related to avoided deforestation range from 0 to 92% based on surveys and partial equilibrium analyses conducted in several tropic and temperate countries. Using general equilibrium modeling, Gan and McCarl (2007) estimate that transnational leakage associated with forest production also is significant (42-95%) and varies across countries. Using landscape-level (instead of stand-level) carbon inventory could help alleviate some (especially local) leakage concern particularly at the national and global levels, yet reliable landscape-level carbon measurement tools need to be developed.

Carbon may not be permanently stored in some pools, and uncertainty is also evident. For instance, forest carbon could release due to timber harvest (impermanence) and disturbances such as fire, insect/disease infestations, storms, etc. (uncertainty). The time path of carbon offsets from afforestation and reforestation projects often varies and could be hard to predict. These concerns about the uncertainty and impermanence of certain forest carbon sinks add to the complications of carbon accounting. Carbon discounting can be used to account for impermanence and uncertainty of GHG offsets (Kim et al. (2008).

3.4. Forestry Carbon Registries

A carbon registry is a legal way to document and track carbon sequestration from an offset project over time. Carbon registries produce quantifiable and verifiable carbon credits, a prerequisite for carbon trade. There are several carbon registries for forestry offset activities. Besides those described earlier as related to carbon markets, other existing registries include

- Environmental Protection Agency Climate Leaders Program

- California Regional Greenhouse Gas Initiative
- Regional Greenhouse Gas Initiative
- The Climate Trust
- World Resources Institute and World Business Council for Sustainable Development Project Protocol
- The Green Exchange

These registries vary considerably in terms of eligible carbon pools, carbon accounting methods, and verification requirements, among others. Call and Hayes (2007) provide detailed descriptions and comparisons of these registries and their protocols.

4. FOREST MANAGEMENT

4.1. Costs of Forestry Offsets

Cost competitiveness will define the role of forestry offsets in mitigating global GHG emissions. Costs of forestry offset activities vary tremendously with the type and location of forestry projects. Extensive work has been done in estimating the costs of forest carbon sequestration via afforestation, reforestation, and improved forest management (Richards and Stokes 2004, Nabuurs et al. 2007). Through a mega analysis, van Kooten and Sohgen (2007) summarize these costs from published estimates global-wide. These costs range from \$0.1/t CO_2 to \$485/t CO_2 with mean of \$24/t CO_2 and standard deviation of \$61/t CO_2. Incorporating the opportunity cost of land would significantly increase carbon sequestration costs whereas considering fossil fuel substitution would reduce the costs. Costs in tropical regions are lowest where those in Europe are highest (\$48/t-\$280/t CO_2). Costs in the U.S. range from about \$2/t to \$77/t CO_2. The boreal zone has a little higher cost (<\$100/t CO_2) than the Great Plains and Cornbelt in the U.S. Tree planting is more cost effective in sequestrating carbon than improving forest management or forest conservation. At the global level, accounting for the opportunity cost of land and fuel substitution, the marginal cost is \$0/t-\$49/t CO_2 for tree planting, \$48/t-\$77/t CO_2 for forest management, and \$47/t-\$195/t CO_2 for forest conservation (van Kooten 2008).

There are relatively few studies that attempt to estimate the cost of GHG offsets through forest biofuels and wood products. Gan (2007) estimates GHG offset costs from substituting coal with logging residues in electricity generation. The marginal cost ranges from \$8/t to \$56/t CO_2 depending upon the quantity of production and whether full or partial shares of logging costs are counted in residue procurement costs. The median marginal cost is \$12/t CO_2 when only additional costs beyond regular logging operation are counted toward residue procurement, and \$16/t CO_2 when full costs are included. Substituting energy intensive building materials such as metals, plastics, and concrete with wood products show great potential for mitigation GHG emissions (Lippke et al. 2004). Yet, work on estimating the costs of this GHG displacement is limited.

4.2. Forest Management Responses

Forest carbon has not really been valued in the marketplace though it is a joint product/service provided by all kinds of forests. Market emergence creates an opportunity for forestry carbon along with other carbon (GHG) sinks to add value to forests. Consequently, it will have a wide spectrum of implications for forest management.

Many studies have examined the implications of carbon values for forest management, including afforestation, deforestation, forest management, and wood products. Carbon credits will encourage afforestation on marginal land and/or help retain forestland for forest use (avoided deforestation) because forests generally store more carbon than other land use alternatives (Adams et al. 1993, Adams et al.1999). Forest rotation length will be extended if carbon value is added to timber benefits (van Kooten et al. 1995).

Most of the existing studies on the impact of carbon credits on forest management have focused on stand-level analysis of even-aged management. It is necessary to go one step further to address a generic forest management scenario that encompasses both even- and uneven-aged forest management at either stand or landscape level. Results from such an analysis will have broader implications for forest management. Consider the forest management problem that deals with production of both timber and carbon offset. For simplicity, ignore the carbon associated with harvested wood products momentarily. The landowner intends to maximize the net discounted revenue from production of timber and carbon offset on one unit of forestland over an infinite time horizon subject to forest stock constraints. Namely,

$$Max \int_0^{\infty} [p\boldsymbol{h}(t) + \rho\beta x(t) - \omega E(t) - r] e^{-\delta t} dt \tag{2}$$

$$\text{s.t. } \dot{x} = g(x, E) - \boldsymbol{h} \tag{3}$$

$$x(0) = x_o \tag{4}$$

where x(t) is the forest stock at time t; h is the amount of timber harvested at time t; E(t) is the forest management effort made at time t; p is the timber (stumpage) price; ρ is the carbon rent (i.e., the rental value of carbon, which is positively related to the carbon price); β is the ratio of the total biomass carbon to the forest stock[2]; ω is the unit cost of forest management effort; r is the land rent per unit of forestland; g(x, E) is the growth rate of the forest stock; and x_o is the initial forest stock.

The current-value Hamiltonian for the above optimization problem is

$$\mathcal{H}(\cdot) = p\boldsymbol{h}(t) + \rho\beta x(t) - \omega E(t) - r + \mu(t)[g(x, E) - \boldsymbol{h}] \tag{5}$$

---

[2] The meaning of β can be expanded to include carbon sequestrated in wood products and displaced by fuel substitution. Yet in this case the effect of h(t) on the carbon pool also needs to be accounted for. As a result, equation (2) will need to be modified.

where is the current value of the shadow price of the forest stock. The first-order conditions for an interior optimal solution are

$$\frac{\partial \tilde{\mathcal{H}}}{\partial h} = p - \mu(t) = 0 \tag{6}$$

$$\frac{\partial \tilde{\mathcal{H}}}{\partial E} = -\omega + \mu(t) g_E = 0 \tag{7}$$

$$\dot{\mu} - \delta\mu(t) = -\frac{\partial \tilde{\mathcal{H}}}{\partial x} = -\rho\beta - \mu(t) g_x \tag{8}$$

Equation (6) implies $p = \mu(t)$. Substituting μ(t) with p in (8) and after some manipulations, we derive

$$g_x = \delta - \frac{\rho\beta}{p} \tag{9}$$

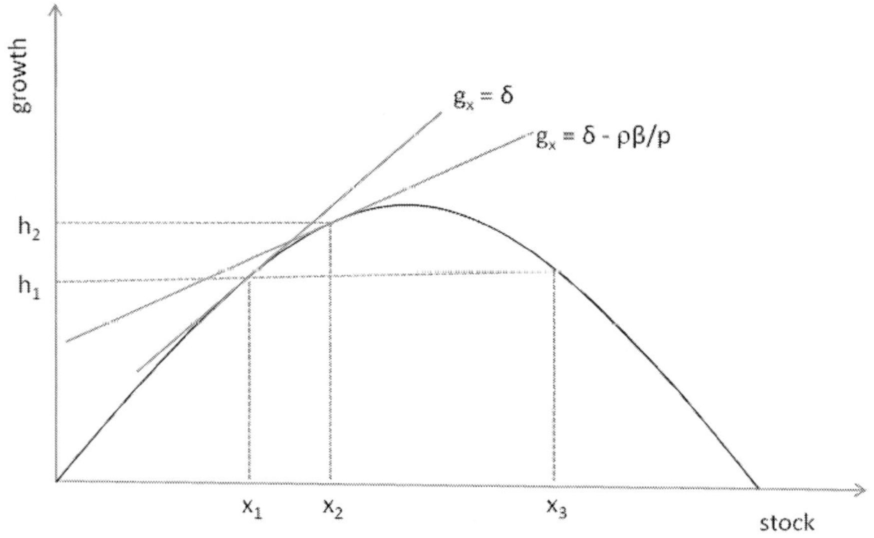

Figure 2. Forest stock and harvest with and without valuing carbon benefits.

The above relationships illustrate how consideration of carbon credits will affect forest management decisions, specifically on timber harvest and forest stock to be remained. Compared with a timber production only case (ρ=0), g_x at the optimal forest stock decreases with carbon value added to timber value. For a logistic forest growth function[3], this suggests that forest stock (not necessarily the steady state stock) will increase when carbon benefits are

[3] Most forests follow a logistic or cubic growth functions (Gan et al. 2001). For a cubic growth function, one can derive similar results.

valued along with timber value. And adding carbon credits can increase the steady state timber harvest [$h = g(E, x)$] (as long as $\rho\beta/p < 2\delta$ for the case of the logistic growth of forest stock) or decrease it when the carbon rent is become sufficiently large such that forest stock exceeds x_3 (figure 2). The effects of β on the optimal forest stock and the stead state timber harvest are similar to those of the carbon rent. Increases in β can be achieved by sequestrating more carbon in forest soils, non-growing stock, harvested forest products, and biofuels (displacing fossil fuels).

5. Summary and Discussion

We review and analyze markets for greenhouse gases (GHG), measurements of forestry carbon offsets, and implications of carbon credits for forest management. Major findings can be summarized as follows:

- Forestry carbon offsets are recognized on both voluntary and regulated carbon markets, yet only few types of forestry offsets are currently traded. The major forestry carbon offsets transacted so far are afforestation/reforestation projects, largely traded on voluntary markets and to a less extent in forms of CDM and JI.
- The carbon price on the regulated market (ECX) tends to be much higher and less volatile than that on the voluntary market (CCX). Yet the prices on the two markets are highly correlated.
- There are several forestry carbon measurement tools/methods, yet many of these tools lack calibration and cannot meet the registry requirements on carbon markets. There are also a few carbon registries that allow forestry offsets, but the standards and scope of eligible forestry offset activities vary considerably across different registries.
- The costs of supplying forestry carbon offsets vary tremendously across projects and locations.
- Accounting for carbon value along with timber value will enhance forest stock and the long-run timber production, suggesting that managing forests for carbon can complement timber production in the long run and other forest services that are positively related to stock.

These results have several implications for improving forestry carbon measurement and forest management.

- First, standardized and valid forestry carbon measuring tools/methods must be developed. In addition to the tools for measuring forest carbon, methods for accounting GHG offsets stored in wood products and displaced by wood products and biofuels are especially needed. It is important to make these tools/methods as simple as possible for measurement and verification while ensuring that GHG estimates are accurate.

- Second, compliance with GHG emissions regulations are likely to boost carbon prices and reduce their volatility, which would help encourage forestry offset activities.
- Finally, given the wide range of forestry offset costs and (joint) outputs/services provided by forests, forestry offset potential and forest management strategies must be developed in a systematic manner. Forestry offset potential should be realistically determined with simultaneous considerations of ecological, economic, and social consequences.

References

Adams, R.M., Adams, D.M., Callaway, J.M, Chang, C.C., & McCarl, B.A. (1993). Sequestering carbon on agricultural land: Social cost and impacts on timber markets. *Contemporary Policy Issues, 11,* 76-87.

Adams, D.M., Alig, R.J., McCarl, B.A., Callaway, J.M., & Winnett, S.M. (1999). Minimum cost strategies for sequestering carbon in forests. *Land Economics*, 75, 360-374.

Call, J., & Hayes, J. (2007). A description and comparison of selected forest carbon registries: A guide for states considering the development of a forest carbon registry. Washington, DC: USDA Forest Service.

Dickey, D., & Fuller, W. (1979). Distribution of the estimates for autoregressive time series with a unit root. *Journal of American Statistics Association, 74,* 427-431.

Dickey, D., & Fuller, W. (1981). Likelihood ratio statistics for autoregressive time series with a unit root. *Econometrica, 49,* 1057-1072.

Easterling, W., Aggarwal, P., Batima, P., Brander, K., Erda, L., Howden, S., Kirilenko, A., Morton, J., Soussana, J., Schmidhuber, J., & Tubiello, F. (2007). Food, fibre and forest products. In M.L. Parry, O.F. Canziani, J.P. Palutikof, P.J. van der Linden, & C.E. Hanson (Eds.), *Climate change 2007: Impacts, adaptation and vulnerability, contribution of Working Group II to the Fourth Assessment Report of the Intergovernmental Panel on Climate Change (pp. 273-313)*. Cambridge, UK: Cambridge University Press.

Eggleston, H.S., Buendia, L., Miwa, K., Ngara, T., & Tanabe, K. (2006). *2006 IPCC guidelines for national greenhouse gas inventories.* http://www.ipcc-nggip.iges.or.jp/public/2006gl/index.html.

Engle, R., & Granger, C. (1987). Co-integration and error correction: Representation, estimation and testing. *Econometrica, 55,* 251-276.

Gan, J. (2007). Supply of biomass, bioenergy, and carbon mitigation: Method and application. *Energy Policy, 35,* 6003-6009.

Gan, J., Kolison, S.H., Jr., & Colletti, J.P. (2001). Optimal forest stock and harvest with valuing non-timber benefits: a case of US coniferous forests. *Forest Policy and Economics, 2,* 167-178.

Gan, J., & McCarl, B.A. (2007). Measuring transnational leakage of forest conservation. *Ecological Economics, 64,* 423-432.

Hamilton, K., Sjardin, M., Marcello, T., & Xu, G. (2008). *Forging a frontier: State of the voluntary carbon markets 2008.* NY: New Carbon Finance and Washington, DC: Ecosystem Marketplace.

Ingerson, A., & Loya, W. (2008). Measuring forest carbon: Strengths and weaknesses of available tools. Science and Policy Brief. Washington, DC: The Wilderness Society.

IPCC (2000). *Land use, land-use change and forestry.* Special report of the Intergovernmental Panel on Climate Change. Cambridge, UK: Cambridge University Press.

Kauppi, P., Sedjo, R., Apps, M., Cerri, C., Fujimori, T., Janzen, H., Krankina, O., Makundi, W., Marland, G., Masera, O., Nabuurs, G., Razali, W., & Ravindranath, N. (2001). Technical and economic potential of options to enhance, maintain and manage biological carbon reservoirs and geo-engineering. In B. Metz *et al.* (Eds.), *Mitigation 2001: The IPCC Third Assessment Report.* Cambridge, UK: Cambridge University Press.

Kim, M-K., McCarl, B.A., & Murray, B.C. (2008). Permanence discounting for land-based carbon sequestration. *Ecological Economics, 64,* 763-769.

Lippke, B., Wilson, J., Perez-Garcia, J., Brown, J., & Meil, J. (2004). CORRIM: Life-cycle environmental performance of renewable building materials. *Forest Products Journal, 54(6),* 8-19.

Malmsheimer, R., Heffernan, P., Brink, S., Crandall, D., Deneke, F., Galik, C., Gee, E., Helms, J., McClure, N., Mortimer, M., Ruddell, S., Smith, M., & Stewart, J. (2008). Forest management solutions for mitigating climate change in the United States. *Journal of Forestry, 106,* 115-173.

Nabuurs, G.J., Masera, O., Andrasko, K., Benitez-Ponce, P., Boer, R., Dutschke, M., Elsiddig, E., Ford-Robertson, J., Frumhoff, P., Karjalainen, T., Krankina, O, Kurz, W.A., Matsumoto, M., Oyhantcabal, W., Ravindranath, N.H., Sanz Sanchez, M.J., & Zhang, X. (2007). Forestry. In B. Metz, O.R. Davidson, P.R. Bosch, R. Dave, and L.A. Meyer (Eds.), *Climate change 2007: Mitigation, contribution of Working Group III to the Fourth Assessment Report of the Intergovernmental Panel on Climate Change.* Cambridge, UK and NY, USA: Cambridge University Press.

Pearson, T., Brown, S., & Birdsey, R. (2007). *Measurement guidelines for the sequestration of forest carbon.* Gen. Tech. Rep. NRS-18. Newtown Square, PA: USDA Forest Service, Northern Research Station.

Richards, K.R., & Stokes, C. (2004). A review of forest carbon sequestration cost studies: A dozen years of research, *Climatic Change, 63,* 1-48.

Scehelhaas, M.J., Esch, P.W., Groen, T.A., Kanninen, M., Liski, J., Masera, O., Mohren, G.M.J., Nabuurs, G.J., Pedroni, L., Pussinen, A., Vallejo, A., Palosuo, T., & Vilén, T. (2004). CO2FIX V 3.1 – A modeling framework for quantifying carbon sequestration in forest ecosystems. ALTERRA Report 1068. Wageningen, The Netherlands.

Sedjo, R. (1993). The carbon cycle and global forest ecosystem. *Water, Air, and Soil Pollution, 70,* 295-307.

Smith, J.E., Heath, L.S., & Nichols, M.C. (2007). *U.S. forest carbon calculation tool: forest-land carbon stocks and net annual stock change*. Gen. Tech. Rep. NRS-13. Newtown Square, PA: USDA Forest Service, Northern Research Station.

Smith, J.E., Heath, L.S., Skog, K.E., & Birdsey, R.A. (2006). *Methods for calculating forest ecosystem and harvest carbon with standard estimates for forest types of the United States*. Gen. Tech. Rep. NE-343. Newtown Square, PA: USDA, Forest Service Northeastern Research Station.

Spinney, M.P., Van Deusen, P.C., Heath, L.S., & Smith, J.E. (2005). COLE: Carbon On-Line Estimator, version 2. In L.S. Kenefic, & M.J. Twery (Eds.), *Changing forests - challenging times: Proceedings of the New England Society of American Foresters 85th Winter Meeting*. Gen. Tech. Rep. NE-325. Newtown Square, PA: USDA Forest Service, Northeastern Research Station.

UNFCCC (1997). *Kyoto Protocol to the United Nations Framework Convention on Climate Change*. http://unfccc.int/resource/docs/convkp/kpeng.html.

USDA Forest Service (2008). Forest Vegetation Simulator. http://www.fs.fed.us/fmsc/fvs/index.shtml.

van Kooten, G.C. (2008). A perspective on carbon sequestration as a strategy for mitigating climate change. *Choices, 23(1),* 24-27.

van Kooten, G.C., Binkley, C.S., & Delcourt, G. (1995). Effect of carbon taxes and subsidies on optimal forest rotation age and supply of carbon services. *American Journal of Agricultural Economics, 77,* 365-374.

van Kooten, G.C, & Sohngen, B. (2007). Economics of forest carbon sinks: a review. *International Review of Environmental & Resource Economics*, *1(3),* 237-269.

In: Global Change and Forestry
Editors: J. Gan, S. Grado and I. A. Munn
ISBN: 978-1-60876-262-0

Chapter 2

ECONOMIC EFFECTS OF PROJECTED CLIMATE CHANGE ON OUTDOOR RECREATION IN TENNESSEE

Donald G. Hodges*[*,1], Jonah Fogel[2], Virginia H. Dale[3], Karen O. Lannom[4] and M. Lynn Tharp[5]

[1] Department of Forestry, Wildlife and Fisheries, The University of Tennessee, TN USA
[2] Virginia Cooperative Extension, Virginia Tech University, VA, USA
[3] Sciences Division, Oak Ridge National Laboratory, TN, USA
[4] Department of Forestry, Wildlife and Fisheries, The University of Tennessee, TN, USA
[5] CompSci Consulting, LLC, McRae, GA, USA

ABSTRACT

Climate change projections from three General Circulation Models were used to adjust the temperature and precipitation in 2030 and 2080 in each of five ecological provinces in Tennessee to estimate the direct economic effects of the projected changes on recreation using the Tourism Climatic Index. The indirect effects on recreation were evaluated qualitatively, based on current demand for the unique values associated with current conditions. The results of the direct impact evaluation reveal that climate change will have variable effects on recreational activities in Tennessee. The magnitude and direction of the effects vary by the recreational activity involved, patterns of precipitation and temperature regimes, and specific location in Tennessee. Recreational activities such as rock climbing, winter activities independent of snow, and whitewater boating are likely to benefit from projected climate changes due to increased temperatures in the winter months. Summer-based activities such as lake recreation and camping are likely to decline with increasing seasonal temperatures. The indirect effects of climate change on recreation are likely to have a larger effect than the direct impacts of climatic variables.

Keywords: Climate change; Outdoor recreation; Tourism climatic index.

* Email: dhodges2@ utk.edu, phone: (865) 974-2706, Natural Resource Policy Center, Department of Forestry, Wildlife and Fisheries, 247 Ellington Plant Science Building, The University of Tennessee, Knoxville, TN.

1. INTRODUCTION

Global climatic change has been a significant scientific and natural resource topic for several decades. During the past ten years, the issue has received increasing attention due to growing evidence of changing temperature and precipitation regimes, and concern about their effects. As evidence for climate change increases, attention has shifted toward projecting the magnitude of potential consequences of climate change and identifying adaptation strategies to address these consequences. This paper presents information on how climate change would affect the forests of Tennessee and identifies the potential economic impacts associated with changes to outdoor recreational activities as a result of global climatic change.

Tennessee was selected for this analysis because its forests are important from recreational, economic, and ecological perspectives. The state offers a wide range of outdoor recreational opportunities and contains a number of key recreational areas such as the Great Smoky Mountains National Park and Big South Fork National River and Recreation Area. The high diversity of tree species in Tennessee's forests supports a large number and variety of plants and animals. Mixed mesophytic forests in Tennessee are among the most diverse in North America, and Tennessee is one of the most diverse inland states in the United States in terms of the number of species found. This high diversity is largely a result of the variation in topography, climate regimes, and by consequence vegetation types across the state.

Tennessee's diverse and unique forests have been affected by land use, land management, non-native species, outbreaks of native insects, and natural disturbances, and climate change is likely to further alter them. There is more forest cover in Tennessee today than there was 50 years ago, as land cut over in the 1940s and 1950s regenerated and is now mature (Wheeler, 1952; Oswalt et al., 2009). The relatively undeveloped, forested regions of the state have become a prime target of retirement and second home developments in the past 10 years, and this trend is expected to exacerbate land-use change in future decades. As a result, Tennessee's forests are being viewed as not only a source of wood products, biodiversity, and other environmental services, but now are also in great demand for development, placing increasing pressure on the lands.

Climate change can have a wide range of potential effects, both localized and global. Responses to these effects vary widely as well, compounding efforts to assess the consequences of change. Local recreational use, for example, can be affected by a changing climate, but forecasting the net effects of climate change is difficult due to the availability of alternative sites or recreational activities which may result in simple shifts in recreational activity. This paper presents the results of an assessment of potential changes in forest cover, recreational use, and the economic implications for Tennessee for two future periods: 2030 and 2080. Based on a range of future climate scenarios, changes in climatic conditions for recreational activities were estimated, as was the effect of these changes on future demand for forest-based recreational use. The same climate scenarios were used to project forest change with the LINKAGES forest ecosystem model (Post and Pastor, 1996). LINKAGES simulates changes in forest composition and structure based on changing biogeochemistry for a 1/12 ha plot. Finally, some of the macroeconomic effects for recreation are reviewed.

2. METHODS

2.1. Climate Change and Tourism Climatic Index

This study was focused on the five ecological provinces in Bailey's ecoregions analysis (Bailey, 1995) that occur in Tennessee (figure 1). Climate change projections from three General Circulation Models (GCMs) were used to adjust the temperature and precipitation in 2030 and 2080 in each ecological province. The temperature and precipitation change projections for 2030 and 2080 for Tennessee were provided by the National Center for Atmospheric Research (NCAR) from three GCMs (selected from 18 possible GCMs) that provide a range of potential climate conditions for Tennessee (personal communication, August 1, 2006, J.B. Smith and C. Wagner, Stratus Consulting Inc., Boulder, CO). The "2030" projections are model simulations of the years 2020-2039, and the "2080" projections are model simulations of the years 2070-2089. The selected GCM outputs represent three conditions:

1. Wet [National Center for Atmospheric Research's Community Climate System Model (ccsm3) (Collins et al., 2006)]
2. Middle [National Center for Atmospheric Research's Parallel Climate Model (PPM) (Washington et al., 2000)]
3. Dry [Center Climate System Research (The University of Tokyo), National Institute for Environmental Studies, and Frontier Research Center for Global Change (JAMSTEC) Model for Interdisciplinary Research on Climate, medium resolution (miroc.medres)].

The GCMs and global climate modeling approaches are described in IPCC (2007). Additional information on the climate projections and forest simulations for Tennessee are described in Dale et al. (2009).

Several recent studies have highlighted the potential impact of climate change on recreational use (Loomis and Crespi, 1999; Richardson and Loomis, 2003; Jones and Scott, 2006; Morris and Walls, 2009). Loomis and Crespi (1999) note that climatic change can have direct effects, including changes in temperature or precipitation regimes, or indirect effects, such as impacts due to changes in the quality or quantity of the natural resources supporting recreation, on recreational use. The direct economic effects of the projected changes on recreation were evaluated using the Tourism Climatic Index (TCI) developed by Mieczkowski (1985). The TCI assesses how changes in several climate variables, including total precipitation, maximum and minimum temperatures, relative humidity, and wind, affect the suitability of a given location for tourism and recreation. The sub-indices used to estimate the TCI are described in table 1.

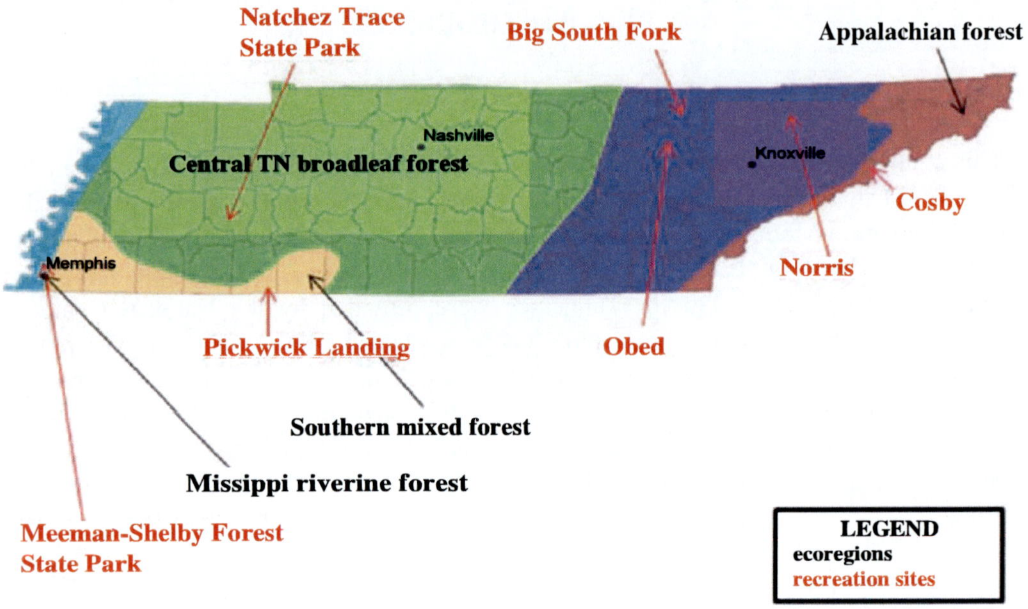

Figure 1. Ecoregions and recreation sites of Tennessee used in analysis.

Table 1. Sub-indices within Tourism Climatic Index[a]

Sub-Index	Monthly Climate Variables	Influence on TCI	Weighting in TCI
Daytime Comfort Index (CID)	Maximum daily temperature & minimum daily relative humidity	Represents thermal comfort when maximum tourist activity occurs minimum daily	40%
Daily Comfort Index(CIA)	Mean daily temperature & mean daily relative humidity	Represents thermal comfort over the full 24 hour period, including activities and holiday enjoyment	10%
Precipitation (P)	Total precipitation	Reflects the negative impact that this element has on outdoor activities and holiday enjoyment	20%
Sunshine (S)	Total hours of sunshine	Rated as positive for tourism, but acknowledged can be negative because of the risk of sunburn and added discomfort on hot days	20%
Wind (W)	Average wind speed	Variable effect depending on temperature (evaporative cooling effect in hot climates rated positively, while 'wind chill' in cold climate rated negatively	10%

[a] Adapted from Scott and McBoyle (1965).

The TCI scores were estimated with the results for the 2030 and 2080 climate projections and applied to visitation numbers for several recreation sites across Tennessee (see figure 1). The sites were selected to assess the potential impacts to recreation in the five ecological regions of Tennessee. In addition, the sites are representative of recreational activities most common in their respective regions. Future recreational use was estimated with a "no" climate change scenario and with the "dry"," middle", and" wet" scenarios. The "no" climate change use scenario was determined by multiplying current use estimates by the projected growth multipliers developed by Bowker et al. (1999). Sole reliance on TCI scores could result in summer-based activities being degraded artificially. Therefore, average monthly visitation numbers were utilized to normalize the impact of the TCI scores, while accurately depicting the lower comfort levels in the summer. The "dry", "middle", and "wet" use scenarios were developed by applying the Bowker et al. (1999) multipliers to the future use estimates, adjusted by the future climate scenarios. Bowker at al. (1999) projected demand only to 2050. The trends, however, were projected forward to 2080 as a simple linear extension.

2.2. Potential Economic Impacts

The direct economic impact of climate change on future recreational use was estimated by comparing the total consumer surplus associated with the "no" climate change scenario to those of the three change forecasts. The consumer surplus values for all recreation sites except the Obed Wild and Scenic River were calculated by multiplying the projected visits for each site by the average consumer surplus values for recreational activities in the southeastern U.S., as reported by Loomis (2005). The values for the Obed area were calculated similarly but used the consumer surplus estimates reported for the area in Sims et al. (2004). All values reported by Loomis (2005) and Sims et al. (2004) were estimated via stated preference (willingness to pay) surveys. Loomis summarizes net average values for 30 different recreational activities for regions within the U.S., while Sims et al. (2004) reports values for the Obed only. Estimates of economic value were applied to the changes in potential recreational demand for various activities.

The indirect effects of climate change on recreation are likely to be substantial, for they are due primarily to changes in forest composition associated with the potential loss of key species or forest types. The impacts of these changes on recreation were evaluated qualitatively, based on current demand for the unique values associated with current conditions.

3. Results

3.1. Projected Forest Composition

In all cases, forest species composition shifted. The hickories (*Carya* species) and hackberry (*Celtis laevigata*) became more dominant, basswood (*Tilia heterophylla*) attained greater biomass in some scenarios and declined in others, and both chestnut oak (*Quercus prinus*) and black oak (*Q. velutina*) declined in their biomass contribution to the forest. The contribution of tree species to total biomass changed for all forest provinces in all climate change scenarios as compared to the 1989 equilibrium. The Southern Mixed Forest experienced the greatest alteration in forest composition among the provinces considered. Under the "dry" scenario, the Southern Mixed Forest becomes dominated by four species [loblolly pine (*Pinus taeda*) and three oaks]. The "middle" and "wet" scenarios both resulted in a greater dominance of hackberry and less biomass of basswood. The East TN Broadleaf Forest maintained high diversity under all cases, but with significant shifts in species composition. The "dry" scenario resulted in dramatic changes in forest composition in both 2030 and 2080. Under both the "middle" and the "wet" scenario, biomass increased for basswood and decreased for chestnut oak, black oak, and yellow buckeye. The Appalachian Forest increased in diversity and dominant species shifted. Under all scenarios, chestnut oak increased initially and then declined in terms of its contribution to stand biomass, basswood increased in biomass, and hickory diversity and biomass also increased. For the "middle" and "wet" scenarios, the Central TN Broadleaf Forest stands were dominated by six hickory species, hackberry, and basswood. For the "dry" case, the forest stands consisted mostly of four hickory species, white oak, southern red oak, and American beech. Projected biomass for the MS Riverine Forest declined in chestnut oak, black oak, basswood, and shumard oak (*Q. shumardii*). For the "dry" scenario, red maple (*Acer rubrum*), hickory, southern red oak, loblolly pine, and beech assumed dominance. For the "middle" and "wet" scenarios, hickory species represent about 40% of the stand biomass and hackberry about 30%.

3.2. Recreational Impacts of Climate Changes

The TCI scores for the region were estimated for key recreational sites within each region and by climate scenario (figures 2-6). Generally, the scores for current conditions, 2030, and 2080 were higher than the lowest 'acceptable' score of 50. The exception is the "dry" scenario for the 2080 Southeastern Mixed Forest for the summer months, which slightly dropped below 50. The scores, however, provide a measure of how the changing climate may affect future comfort levels relative to current conditions. As would be expected with increasing summer temperatures, the 2030 and 2080 scores for all regions fell below the current scores for the summer months. Conversely, the 2030 and 2080 scores exceeded current comfort levels from late fall to early spring.

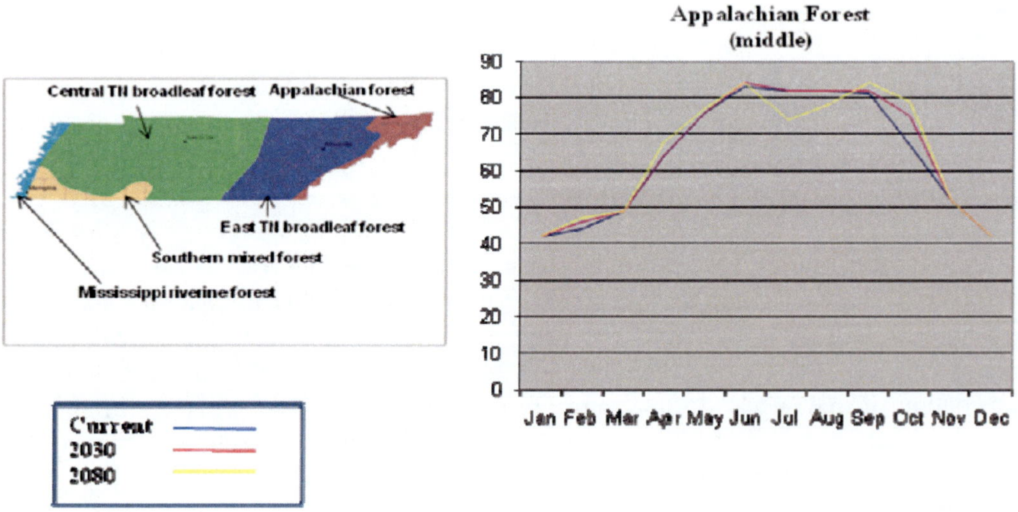

Figure 2. Continued on next page.

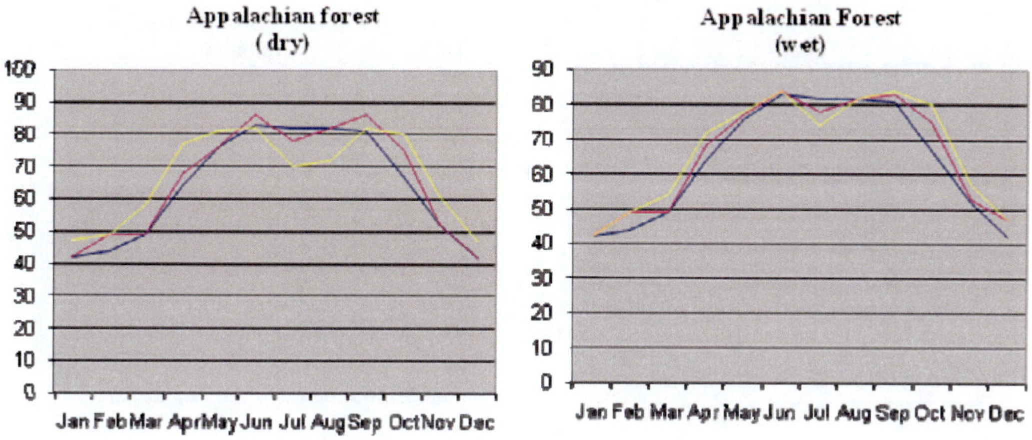

Figure 2. TCI scores for the Appalachian forest.

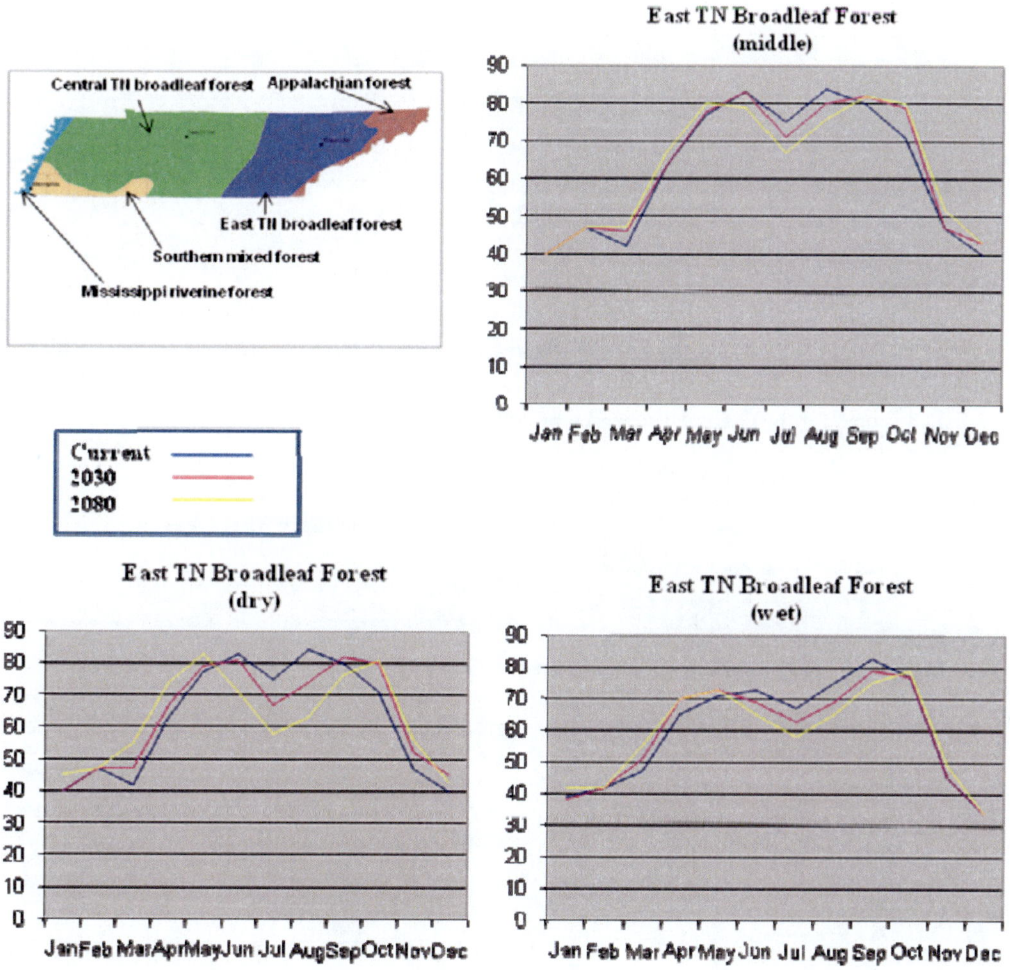

Figure 3. TCI scores for the East Tennessee broadleaf forest.

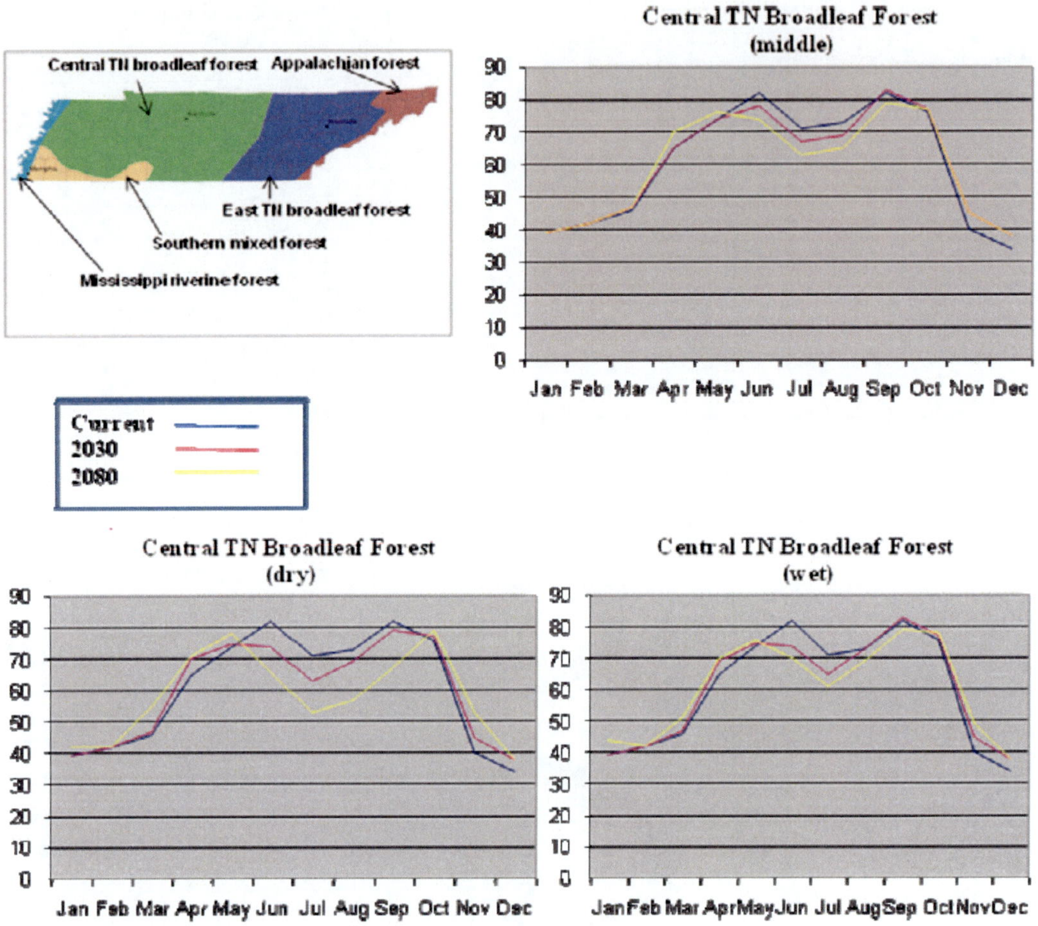

Figure 4. TCI scores for the Central Tennessee broadleaf forest.

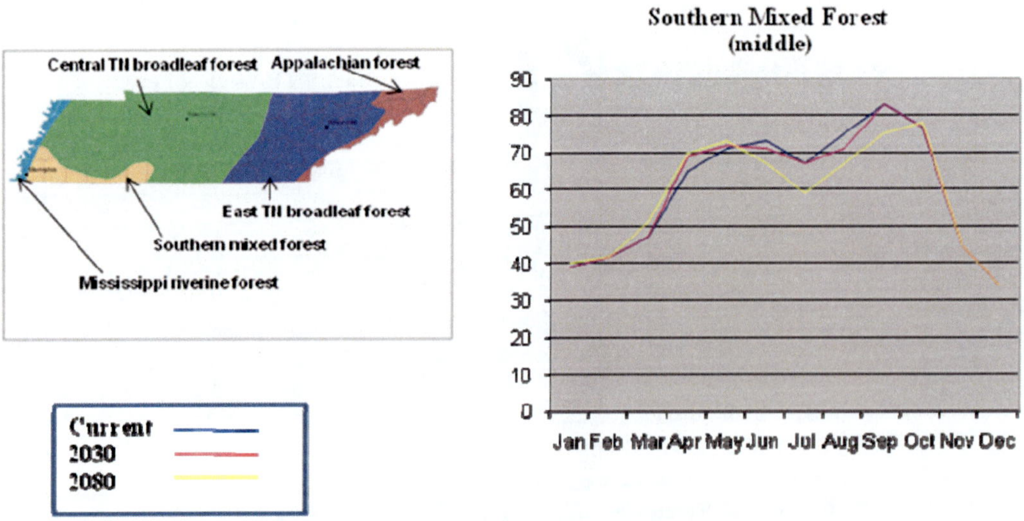

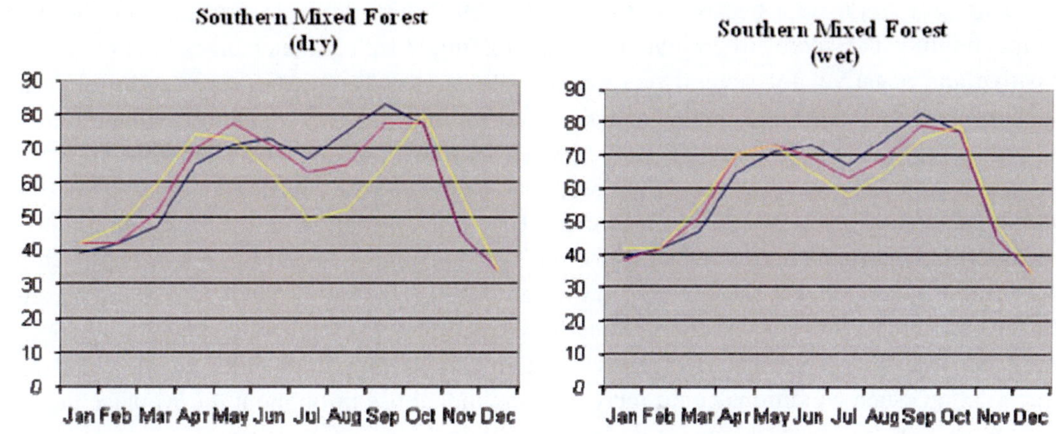

Figure 5. TCI scores for the Southern mixed forest.

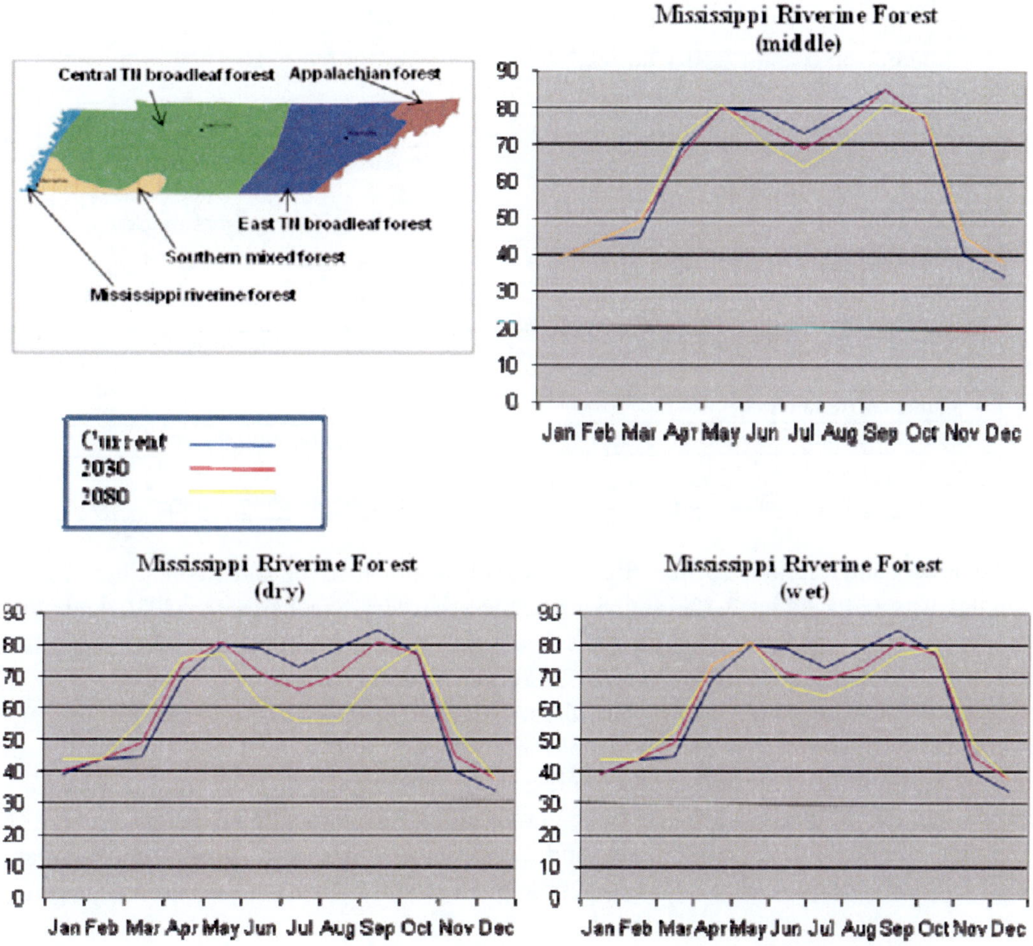

Figure 6. TCI scores for the Mississippi riverine forest.

Using the trends reported by Bowker et al. (1999), future use projections were adjusted by the monthly TCI scores to project how visitation might reflect the changes in climate (both positive and negative) for several recreational sites across Tennessee. As stated previously, the sites were selected based on the availability of average monthly visitation estimates and to provide a representative sample of use across all regions (table 2). The visitation projections for all regions reflect changes in recreational use and associated economic values (as measured by consumer surplus) of less than 10 % (which represents the maximum average deviation in historical annual visitation for the sites examined) for all but one scenario (table 3). The one exception was the "dry" scenario for Cosby. This can be attributed to the fact that the TCI scores for the "dry" scenario equal or exceed the "No Climate Change" scenario for most months, due largely to increased temperatures during the winter months and less precipitation. Such conditions could increase recreational use throughout the Appalachians of Tennessee, which is an extremely important recreational region for the state.

Table 2. Recreational Sites for TCI Assessment

Forest Types	Primary Activities
Appalachian Forest	
Great Smoky Mountains National Park (Cosby)	camping, hiking
East Tennessee Broadleaf Forest	
Norris State Park	boating, general recreational use
Big South Fork National River and Recreation Area	camping, hiking, horseback riding
Obed Wild and Scenic River	rock climbing
Central Tennessee Broadleaf Forest	
Natchez Trace State Park	camping, fishing, horseback riding
Southern Mixed Forest	
Meeman-Shelby Forest State Park	camping, boating, fishing
Mississippi Riverine Forest	
Pickwick Landing State Park	camping, boating, fishing

Total consumer surplus for the Big South Fork National River and Recreation Area, for example, was reduced by 2.1% and 6.1% under the "dry" scenario for 2030 and 2080, respectively. The "wet" scenario for the same area resulted in a slight (< 1%) increase in value for 2030 and a similar decrease by 2080. Two locations, Cosby and Obed, exhibited increased values for both periods and all scenarios. The Cosby site increased primarily because of warmer fall and winter conditions, which would extend the comfortable use season. On the Obed, fall and spring represent the busiest seasons for rock climbing – a period when the TCI scores are greater in future climate scenarios than current scores. The remaining sites exhibit both increases and decreases in consumer surplus, most notably decreases for the "dry" scenario with less precipitation and higher temperatures than current conditions or other scenarios evaluated.

Beyond the direct effects of climate change on recreation, several indirect effects, which are more difficult to assess quantitatively, could impact recreational use. Most notably, the potential decline of species, such as trout and high elevation spruce-fir forests, could have substantial economic impacts on forest-based recreation. More specifically, the trout fishery in the southern Appalachians is a significant component of the regional economy. Increases in stream temperature and the loss of hemlocks as a major component of riparian forests are projected to decrease trout populations throughout the southern Appalachian region (Evans, 2002), which in turn would result in a decline in the fishing economy of the region. Ahn et al. (2000) estimated that the welfare loss per angler is between \$5.63 and \$53.18 per visit, under future climate scenarios. The economic impact of the loss of the high-elevation spruce-fir forests is more difficult to assess because most visitors visit these areas as part of multi-purpose trips. Visitors to the Great Smoky Mountain National Park, for example, are likely to visit other areas of the park or surrounding natural and human-made attractions. Moreover, no data are available identifying the relative importance of these forests in trip decisions or the values visitors place on such areas. Visitation patterns within the Great Smoky Mountains National Park, however, demonstrate the popularity of the spruce-fir forests, and any changes to this resource could affect recreational uses such as bird watching and backpacking significantly.

4. Conclusion

The results of the direct impact evaluation reveal that climate change will have variable effects on recreational activities in Tennessee. The magnitude and direction of the effects vary by the recreational activity involved, patterns of precipitation and temperature regimes, and specific location in Tennessee. Recreational activities such as rock climbing, winter activities independent of snow, and whitewater boating are likely to benefit from increased temperatures in the winter months. Summer-based activities such as lake recreation and camping are likely to decline with increasing seasonal temperatures. These conclusions are based largely on the TCI scores for all regions and climate scenarios, which depict a bimodal distribution of scores, with the lowest scores occurring in the summer. Geographically, recreation in the eastern portion of the state will not experience as much decline in use levels or economic value as the western portion of the state. Some locations may even experience an increase in certain types of recreational pursuits. This difference can be attributed largely to the fact that the higher elevations and recreational pursuits of the east are less affected by heat than recreational activities in the west.

The total consumer surplus for the 7 sites evaluated across Tennessee totaled more than \$10.5 million for 2030 and almost \$15.0 million for 2080, in the "no climate change" scenario (table 3). The total change in consumer surplus for 2030 was estimated to be 0.01 percent for the "dry" scenario, 0.51 percent for the "middle" scenario, and 0.90 percent for the "wet" scenario. The changes in consumer surplus were considerably different for 2080, where the projected changes were -0.83 percent for the "dry" scenario, -0.81 percent for the "middle" scenario, and 1.26 percent for the "wet" scenario. These differences reflect the effect of the ongoing changes in precipitation and temperature patterns for the 50-year period. By using a recent estimate of the total annual economic impact of outdoor-based tourism in Tennessee as \$9.6 billion (Cho et al., 2007), the scenarios could translate into a change in economic activity in the state that ranges from a loss of \$80 million for the 2080 "dry" scenario to a net gain of \$121 million for the 2080 "wet" scenario. It is important to note that the sites were evaluated by examining the consumer surplus, or net benefit, of the recreation activities. Conversely, the total economic activity estimates from Cho et al. (2007) reflect the direct and indirect impacts of tourism in Tennessee that is based on outdoor activities.

Table 3. Percent Change in Consumer Surplus by Site and Climate Scenario

Forest Type	Dry	Middle	Wet
Appalachian Forest			
Great Smoky Mountains National Park (Cosby)	%Δ	%Δ	%Δ
2030 *($1,271,550)a*	3.22	1.65	4.73
2080 *($1,900,200)*	11.09	2.57	7.59
East Tennessee Broadleaf Forest			
Norris State Park			
2030 *($183,375)*	-1.04	0.18	-0.46
2080 *($242,050)*	-5.70	-1.16	-2.33
Big South Fork National River and Rec. Area			
2030 *($959,275)*	-2.13	-0.98	0.64
2080 *($1,710,600)*	-6.06	-2.67	-0.29
Obed Wild and Scenic River			
2030 *($823,650)*	1.60	1.23	4.12
2080 *($1,294,500)*	4.45	1.67	5.27
Central Tennessee Broadleaf Forest			
Natchez Trace State Park			
2030 *($2,073,660)*	-0.24	0.87	1.25
2080 *($2,879,900)*	-0.35	-0.08	2.14
Mississippi Riverine Forest			
Meeman-Shelby Forest State Park			
2030 *($2,243,750)*	0.06	0.73	0.50
2080 *($3,067,450)*	-1.50	-0.50	0.95
Southern Mixed Forest			
Pickwick Landing State Park			
2030 *($2,911,500)*	-0.94	-0.11	-1.47
2080 *($3,850,250)*	-5.69	-3.26	-2.69
TOTAL CONSUMER SURPLUS			
2030 *($10,466,760)*	0.01	0.51	0.90
2080 ($14,944,950)	-0.83	-0.81	1.26

[a] Figures represent estimated consumer surplus by year with "no" climate change scenario.

The indirect effects of climate change on recreation are likely to have a larger effect than the direct impacts of climatic variables. Changes in the environments where many recreational pursuits occur may limit such activities. The loss of hemlocks in riparian forests due to the hemlock wooly adelgid, coupled with projected increases in water temperature, could have a substantial detrimental effect on streams in the southern Appalachians (Evans, 2002). The result could be a loss of one of the more popular recreational pursuits in the region, trout fishing, with significant economic implications as described by Ahn et al. (2002). In addition to trout populations, climate change will change forest species composition and potentially alter the attractiveness of some areas for recreation. As mentioned previously, the potential loss or reduction of the high-elevation spruce-fir forests in the eastern border of the state could eliminate a unique environment that attracts a significant portion of the visitors to the Great Smoky Mountains and other public lands for a number of recreational activities dependent on or enhanced by forests.

Not all factors were considered in the analysis of recreational impacts. The analysis was confined to considering changes only within the state boundaries. Recreationists may substitute alternative sites for their activities to locations in other states in search of more favorable climatic conditions or environments, in most cases to more northerly environs. While this substitution can occur for some activities, other recreational pursuits are often site-specific and substitutes may not exist. Tennessee offers a wide range of outdoor recreational experiences unique to the state or small regions of bordering states. Rock climbing, whitewater boating, and backpacking opportunities, for example, are available throughout the eastern one-third of the state and in the states and regions bordering Tennessee. Therefore, changes to these sites and unique areas such as high-elevation forests could eliminate recreational opportunities from the region. Moreover, individuals who historically have recreated in areas south of Tennessee are likely to seek opportunities in the state, negating a portion of the lost activities and the accompanying economic value. Another consideration in evaluating climate change impacts that was not addressed is the influence of changes in climate-dependent insects and diseases threatening human health. A number of threats such as Lyme disease, Rocky Mountain spotted fever, encephalitis, and hantavirus are likely to be affected by climate change (NRC, 2001; WHO, 2003) and could discourage recreational activities.

Acknowledgments

This research was supported by a grant from the National Commission on Energy Policy. Bob Abt, Paul Hanson, Rameen Nasrollahi, Mac Post, Joel Smith, and Stan Wullschleger provided useful suggestions for this study and on early drafts of the report. The General Circulation Model data and PDFs were supplied by Seth McGinnis of the National Center on Atmospheric Research to Stratus Consulting Inc. and then to this project.

References

Ahn, S., de Steiguer, J.E., Palmquist, R.B., & Holmes, T.P. (2000). Economic analysis of the potential impact of climate change on recreational trout fishing in the southern Appalachian mountains: an application of a nested multinomial logit model. *Climate Change. 45*, 493-509.

Bailey, R.G. (1995). *Description of the ecoregions of the United States* (2nd ed.). Misc. Pub. No. 1391. Fort Collins, CO: USDA Forest Service Rocky Mountain Research Station.

Bowker, J.M., English, D. B.K., & Cordell, H.K. (1999). Projections of outdoor recreation participation to 2050. In H.K. Cordell, C. Betz, J.M. Bowker, et al (Eds.), *Outdoor recreation in American life: a national assessment of demand and supply trends.* (pp. 323-351). Champaign, IL: Sagamore Publishing.

Cho, S-H., Hodges, D.G., & Roberts, R. (2007). *Valuing Tennessee's natural resources*. Final report submitted to The Nature Conservancy and Trust For Public Lands. Knoxville, TN: The University of Tennessee.

Collins, W. D., Bitz, C.M., Blackmon, M.L., Bonan, G.B., Bretherton, C.S., Carton, J.A., Chang, P., Doney, S.C., Hack, J.J., Henderson, T.B., Kiehl, J.T., Large, W.G., McKenna, D.S., Santer, B.D., & Smith, R.D. (2006). The Community climate system model version 3 (CCSM3), *Journal of Climate. 19*, 2122-2143.

Dale, V.H., Lannom, K.O., Tharp, M.L., Hodges, D.G., & Fogel, J. (2009). Effects of climate change, land-use change, and invasive species on the ecology and tourism of the Cumberland Forests. *Canadian Journal of Forest Research. 39*, 467-480.

Evans, R. A. (2002). An ecosystem unraveling? In B. Onken, R. Reardon, & J. Lashomb (Eds.), *Proceedings of Symposium on the Hemlock Woolly Adelgid in Eastern North America.* (pp. 23-33). NJ: Rutgers University.

IPCC (Intergovernmental Panel on Climate Change). (2007). *Climate Change 2007: The physical science basis*. Geneva, Switzerland: IPCC.

Jones, B., & Scott, D. (2006). Climate change, seasonality, and visitation to Canada's national parks. *Journal of Park and Recreation Administration. 24(2),* 42-62.

Loomis, J. (2005). *Updated outdoor recreation use values on national forests and other public lands*. Gen. Tech. Rep. PNW-GTR-658. Portland, OR: USDA Forest Service, Pacific Northwest Research Station.

Loomis, J.B., & Crespi, J. (1999). Estimated effects of climate change on selected outdoor recreation activities in the United States. In R. Mendelsohn, & R.E. Neumann (Eds.), *The impact of climate change on the United States economy* (pp. 289-314). Cambridge, UK: Cambridge University Press.

Mieczkowski, Z. (1985). The tourism climatic index: a method of evaluating world climates for tourism. *The Canadian Geographer. 29*, 220-33.

Morris, D., & Walls, M. (2009). *Climate change and outdoor recreation resources.* Washington, DC: Resources For The Future Backgrounder.

National Research Council (NRC). (2001). *Climate change science: an analysis of some key questions*. Washington, DC: National Academy Press.

Oswalt, C.M., Oswalt, S.N., Johnson, T.G., Chamberlain, J. L., Randolph, K.C., & Coulston, J.W. (2009). *Tennessee's forests, 2004*. Resour. Bull. SRS–144. Asheville, NC: USDA Forest Service, Southern Research Station.

Post, W.M., & Pastor, J. (1996). Linkages — an individual-based forest ecosystem model. *Climatic Change. 34*, 253-261.

Richardson, R.B., & Loomis, J.B. (2005). Climate change and recreation benefits in an alpine national park. *Journal of Leisure Research. 37(3)*, 307-320.

Sims, C.B., Hodges, D.G., & Scruggs, D. (2004). Linking outdoor recreation and economic development: a feasibility assessment of the Obed Wild and Scenic River, Tennessee. In P.T. Tierney & D.J. Chavez (Eds.), *Proceedings of the Fourth Social Aspects and Recreation Research Symposium.* (pp. 72-78). San Francisco, CA: San Francisco State University.

Skinner, M., Parker, B.L., Gouli,S., & Ashikaga, T. (2003). Regional responses of hemlock woolly adelgid (Homoptera: Adelgidae) to low temperatures. *Environmental Entomology. 32*, 523-528.

Washington, W.M. , Weatherly, J.W., Meehl, G.A., Semtner, A.J., Bettge, T.W., Craig, A.P., Strand, W.G., Arblaster, J., Wayland, V.B., James, R., & Zhang, Y. (2000). Parallel climate model (PCM) control and transient simulations. *Climate Dynamics. 16*, 755-774.

Wheeler, P.R. (1952). *Forest statistics for Tennessee, 1950.* Forest Survey Release 70. New Orleans, LA: USDA Forest Service, Southern Forest Experiment Station.

World Health Organization (WHO). (2003). *Climate change and human health - risks and responses (summary).* Geneva, Switzerland: WHO.

In: Global Change and Forestry
Editors: J. Gan, S. Grado and I. A. Munn

ISBN: 978-1-60876-262-0

Chapter 3

LAND USE DYNAMICS ALONG URBAN-RURAL GRADIENT: A COMPARISON OF MODELING APPROACHES[1]

Maksym Polyakov[1] and Daowei Zhang[2*]
[1] School of Agricultural and Resource Economics, University of Western Australia, Australia
[2] School of Forestry and Wildlife Sciences, Auburn University, AL, USA

ABSTRACT

Modeling land use dynamics is an important component of the landscape level analysis of socio-ecological drivers at the urban-rural interface. We constructed a dataset consisting of 5313 sample points characterizing land cover in 1992 and 2001 across three contiguous counties in West Georgia. We examined two specifications of conditional logit model based the land rent theory: a model of land use allocation, which is similar to models of land use share that utilize aggregated data, and a model of land use change. The land use change model showed better goodness of fit as it controls for cross-sectional effects and allows parameters of explanatory variables to capture temporal effects. The results have implications on land use modeling at watershed or sub-watershed levels.

Keywords: Land use change; Urban-rural interface; Logit model; Georgia.

1. INTRODUCTION

Land use changes, while driven by landowners seeking maximization of economic benefits, sometimes produce negative externalities such as air and water pollution, loss of

[1] This study is supported by the National Research Initiative of the Cooperative State Research, Education and Extension Service, USDA, Grant #USDA-2005-3540015262 and the Auburn University Center for Forest Sustainability.

* Email: zhangd1@auburn.edu, phone: (334)844-1067, School of Forestry and Wildlife Sciences, Auburn University, Auburn, AL 36830.

biodiversity, wildlife habitat fragmentation, and increased flooding (Stavins & Jaffe, 1990). In places where the majority of the land base is privately owned, like the U.S. South, it is important to understand how socioeconomic and natural factors affect private landowners' decisions concerning land use.

Most existing studies of land use in the U.S. are based on the classic land use theory developed by David Ricardo and Johann von Thünen (Ricardo, 1817; von Thünen, 1826). This theory explains land use patterns in terms of relative rent to alternative land uses, which depends on land quality, location, and socio-economic factors. Due to data limitations, most econometric studies of land use utilized data describing areas or proportions of certain land use categories within a well defined geographic area, such as a county, as a function of socioeconomic variables and land characteristics aggregated at the level of geographic unit of observation (Alig & Healy, 1987; Plantinga, Buongiorno, & Alig, 1990; Stavins & Jaffe, 1990). Some studies modeled shares of an exhaustive set of land uses within a specified land base using a binomial or multinomial logit model, which allows restricting shares to unity (Parks & Murray, 1994; Hardie & Parks 1997; Ahn, Plantinga, & Alig, 2000; Nagubadi & Zhang, 2005; Zhang & Nagubadi, 2005).

While being widely used, the model of aggregated land use shares can have some limitations related to its ability to reveal the temporal effect of economic variables on land use and, thus, its usability for the inferences about land use dynamics. Comparing pooled, fixed effects, and random effects specifications of the cross-sectional time series land use share model at county level, Ahn, Plantinga, & Alig (2000) concluded that the pooled specification does not adequately control for cross-sectional variations in dependent variables. As a result, the models' parameters measure a combination of spatial and temporal effects and cannot be used for inferences regarding land use change or land use change predictions. They suggested that a specification with cross-sectional fixed effects provides a better measure of temporal relationship. However, the use of cross-sectional fixed effects requires relatively long time series, limits number of cross-sectional elements, and prevents the use of explanatory variables such as land quality and slope that do not have temporal variation.

A number of recent studies utilized disaggregate approach to modeling land use by using parcel level (Irwin & Bokstael, 2002; Carrión–Flores & Irwin, 2006), sample plots (Kline, Moses & Alig, 2001; Lubowski, Plantinga & Stavins, 2006), or remotely sensed (Chomitz & Gray, 1996; Turner, Wear and Flamm, 1996) data. This approach allows better using of physical attributes affecting land use decision by connecting them to specific units of observation. Furthermore, this approach allows modeling land use on a small scale, such as within one or several counties (Carrión–Flores & Irwin, 2006), which was not possible with aggregate data models.

Some of the studies utilizing disaggregate approach modeled the probability of a parcel or sample plot to be allocated to a particular land use as a function of physical and economic variables (Chomitz & Gray, 1996; Nelson et al., 2005). This model is similar to pooled models of aggregate land use shares. Another group of studies used plot- or parcel-based observation of land characteristics over at least two points in time in order to directly measure land use transitions, or, in other words, probability of a parcel to be allocated to a particular land use was modeled as a function of physical and economic variables, as well as previous land use (Bockstael, 1996; Lubowski, Plantinga & Stavins, 2006; Polyakov & Zhang, 2008). This allowed isolating temporal effects of the factors driving land use dynamics, similar to the fixed effect model of aggregate land use shares.

In this chapter, we apply a discrete choice model to disaggregate data and model land uses at a sub-watershed level in a three-county region. We compare performance of two approaches used to model land use. Our results show that a model that takes into account previous land use of sample plots (land use change model) performs better than a land use allocation model. In the next section we lay out a discrete choice model of land use allocation and a discrete choice model of land use change as well as their corresponding econometric models. The third section describes the study area and data. The fourth section provides estimation results of conditional logit models of land use allocation and land use change. The final section concludes.

2. THE MODELS

We assume that spatial patterns of land use and their changes are results of decisions of the owners of individual land parcels or cells in the landscape. A landowner chooses to allocate a parcel of land of uniform quality to one of several possible alternative uses. We further assume that the landowner's decision is based on the maximization of net present value of future returns generated by the land. The landowner's expectation concerning future returns generated by different land uses are drawn from the characteristics of the parcel and historical returns.

Let W_{nj} be the net present value of parcel *n* in use *j*, which is not directly observable for individual parcels. However, there are observable attributes of a plot, $\mathbf{x}_n$, that are related to, and depend on, characteristics of a parcel such as land quality and location, as well as economic conditions. The landowner's utility of keeping parcel *n* in use *j* is $U_{nj} = W_{nj}(\mathbf{x}_n) + \varepsilon_{nj}$, where ε_{nj} captures factors that affect utility and assumed to be random.

The parcel will be allocated to land use with greatest utility to the landowner. Depending on assumptions about the density distribution of random components of utility, several different discrete choice models could be derived from this specification (Train 2003). Assuming random components are independent and identically distributed (IID) with a type I extreme value distribution, we obtain a conditional logit model (McFadden 1973):

$$P_{njt} = \frac{\exp(W_{nj})}{\sum_{k=1}^{J} \exp(W_{nk})} = \frac{\exp(\alpha_j + \boldsymbol{\beta}_j{}'\mathbf{x}_{n,t-1})}{\sum_{k=1}^{J} \exp(\alpha_k + \boldsymbol{\beta}_k{}'\mathbf{x}_{n,t-1})} \tag{1}$$

where P_{njt} is the probability of allocating parcel *n* to land use *j* at time *t*, $\boldsymbol{\beta}_j$ is a vector of land use specific parameters, and α_j are land use specific constants ($\boldsymbol{\beta}_{Jn} - 0$, $\alpha_{Jn} - 0$ to remove an indeterminacy in the model). This is a model of land use allocation similar to one used by Nelson et al. (2005).

However, the situation when the owner allocates to use *j* parcel of land, which is not already allocated to another productive use, is very uncommon. More frequent is the situation when the land is changed from one productive use to another, for example, converting forestland to development or agricultural land to forestry use. Let us consider the utility of converting parcel n from land use i to land use j. This involves return to old and new land uses W_{ni} and W_{nj}, as well as one time conversion cost C_{nij}, all of which depend on the particular land uses that the parcel is being converted from and to, as well as the characteristics of the parcel:

$$U_{nj|i} = W_j(\mathbf{x}_n) + \varsigma_{nj} - W_i(\mathbf{x}_n) - \upsilon_{ni} - C_{ij}(\mathbf{x}_n) - \sigma_{nij}$$

where ς_{nj}, υ_{ni}, and σ_{nij} capture factors that affect utility, and $\varepsilon_{nj|i} = \varsigma_{nj} + \upsilon_{ni} - \sigma_{nij}$ is assumed to be random. The parcel could be converted to land use j if $U_{nj|i}$ is positive. Furthermore, a parcel will be converted to the land use for which utility of conversion is greater. The parcel will remain in its current land use ($C_{nii} = 0$; $U_{ni|i} = 0$) if $U_{nj|i} < 0 \; \forall \; j \neq i$. The conditional logit model of the probability of converting parcel n from land use i to land use j is (see Polyakov & Zhang (2008) for details):

$$P_{nj,t|i,t-1} = \frac{\exp\left(\alpha_{ij} + \boldsymbol{\beta}_j{}'\mathbf{x}_{n,t-1} - \boldsymbol{\beta}_i{}'\mathbf{x}_{n,t-1}\right)}{\sum_{k=1}^{J}\exp\left(\alpha_{ik} + \boldsymbol{\beta}_k{}'\mathbf{x}_{n,t-1} - \boldsymbol{\beta}_i{}'\mathbf{x}_{n,t-1}\right)} = \frac{\exp\left(\alpha_{ij} + \boldsymbol{\beta}_j{}'\mathbf{x}_{n,t-1}\right)}{\sum_{k=1}^{J}\exp\left(\alpha_{ik} + \boldsymbol{\beta}_k{}'\mathbf{x}_{n,t-1}\right)} \qquad (2)$$

where α_{ij} is a transition specific parameter.

Both models (1) and (2) predict probability of parcel n being allocated to land use j in period t. However, in model (1) this probability is conditional on the characteristics of parcel n in period t–1, while in model (2) it is conditional on the characteristics of parcel n in period t–1 and the particular land use of parcel n in period t–1. Formally, the difference between these models is the categorical variable indicating previous land use. This leads to two conceptual differences in these models. First, model (2) takes into account transition costs. As we mentioned, all the changes (or non-changes) of land use are influenced by the conversion costs, which are ignored by the model (1). Second, in both models, there are unobserved characteristics of plots, which influence returns to alternative land uses, and assumed to be random. However, in model (2), we compare differences of returns to two alternative land uses, and effects of unobserved characteristics of plots on two land uses are partially cancelled. This is similar to the fixed effect model. Thus, model (2) has a potential to perform better than model (1) in predicting future land use.

Although both these models predict probability of future land use, because model (2) takes into account previous land use, in this chapter it will be referred to as the "model of land use change", while model (1) will be referred to as the "model of land use allocation".

Due to the size of the study area, the prices and costs do not vary and therefore do not affect relative rents and land use choice behavior (Turner, Wear, & Flamm, 1996; Bockstael, 1996). The factors that are variable within the study area and influence relative rents are location of sample point relative to employment, market and population centers, transportation networks, and physical site characteristics. We hypothesize that by affecting relative rents to alternative land uses, location (accessibility to jobs, markets, and population) influences changes both between rural and developed land uses, and between agricultural and forestry uses.

Model evaluation of discrete choice models is challenging because estimated models yield probabilities of land uses, while the actual outcome is represented by a set of categorical values (Bockstael, 1996). To account for these issues, we evaluated the forecasting performance of the conditional logit model using information indices and statistics developed for evaluating performance of discrete choice models by Hauser (1978). The information index, $I(\mathbf{A};\mathbf{X})$, quantifies the additional information provided by the explanatory variables through the estimated model in comparison to a null model:

$$I(\mathbf{A};\mathbf{X}) = \frac{1}{N}\sum_{n=1}^{N}\sum_{j=1}^{J}\delta_{nh}\ln\left(\frac{p(a_j \mid \mathbf{x}_n)}{p(a_j)}\right),$$

where $p(a_j)$ is the prior likelihood of the land use j (based on the null model), $p(a_j \mid \mathbf{x}_j)$ is the land use h predicted by the model, and δ_{nj} is the binary variable indicating land use j observed at sample plot n. The information index can be compared to the expected information provided by the model:

$$EI(\mathbf{A};\mathbf{X}) = \frac{1}{N}\sum_{n=1}^{N}\sum_{j=1}^{J}p(a_j \mid \mathbf{x}_n)\ln\left(\frac{p(a_j \mid \mathbf{x}_n)}{p(a_j)}\right).$$

The information index $I(\mathbf{A};\mathbf{X})$ is normally distributed with a mean of $EI(\mathbf{A};\mathbf{X})$ and a variance of $V(\mathbf{A};\mathbf{X})$:

$$V(\mathbf{A};\mathbf{X}) = \frac{1}{N}\sum_{n=1}^{N}\left\{\sum_{j=1}^{J}p(a_j \mid \mathbf{x}_n)\left[\ln\left(\frac{p(a_j \mid \mathbf{x}_n)}{p(a_j)}\right)\right]^2 - \left[\sum_{j=1}^{J}p(a_j \mid \mathbf{x}_n)\ln\left(\frac{p(a_j \mid \mathbf{x}_n)}{p(a_j)}\right)\right]^2\right\},$$

which allows testing the accuracy of the model.

The index of the prior entropy,

$$H(\mathbf{A};\mathbf{X}) = -\sum_{j=1}^{J} p(a_j)\ln(p(a_j))$$

defines the uncertainty inherent in the null model and allows measuring proportion of uncertainty explained by the estimated model:

$$U^2 = \frac{I(\mathbf{A};\mathbf{X})}{H(\mathbf{A};\mathbf{X})}.$$

Furthermore, the log–likelihood ratio $LLR = 2n \times I(\mathbf{A};\mathbf{X})$ is χ^2 distributed with degrees of freedom equal to the number of coefficients in the model and allows testing the significance of the empirical model.

3. Study Area and Data

Our study area is in the Georgia Piedmont, a region that is developing rapidly and ranks among the highest regions in terms of percentage increase in developed land area during the 1990s. Within this region we study land use change in three contiguous counties: Muscogee, Harris, and Meriwether. Despite being contiguous, these counties exhibit broad ranges of population pressure and patterns of land uses and land use change from urban and moderately developing (Muscogee county) to suburban and fast developing (Harris county), and to rural (Meriwether county). Columbus, located in Muscogee County, is the third largest city in Georgia. Muscogee County accounts for 80% of the population of the three county region. However, it only had a moderate population growth during the 1990s. The population of Harris County, which is located north of Muscogee County and is becoming its bedroom community, increased by one-third during the same period, while the population of Meriwether County increased only slightly (Table 1).

Table 1. Population and land use statistics for Harris, Meriwether and Muscogee counties 1990-2000 (U.S. Census Bureau, 2002; Natural Resources Conservation Service, 2000)

Characteristics		County			Total
		Harris	Meriwether	Muscogee	
Population:	Person, 2000	23,695	22,534	186,291	232,520
	Person/km2, 2000	19	17	325	75
	Annual % change, 1990-2000	3.3%	0.1%	0.4%	0.6%
Agricultural lands:	% of land base, 1997	6.3%	10.2%	5.5%	7.8%
	Annual % change, 1992-1997	-0.3%	-3.1%	-4.7%	-2.5%
Forest lands:	% of land base, 1997	78.3%	80.5%	24.8%	69.3%
	Annual % change, 1992-1997	-0.4%	0.8%	-2.1%	0.0%
Developed lands:	% of land base, 1997	6.9%	5.9%	29.8%	10.7%
	Annual % change, 1992-1997	4.6%	4.1%	3.8%	4.1%

To develop a model of land use transitions, we need information about land uses for a set of sample points for at least two points in time. We used USGS National Land Cover Datasets (NLCD) 1992, generated from leaf-on Landsat TM data centered on a nominal collection date 1992, and NLCD 2001, generated from multi-season Landsat 5 and Landsat 7 imagery centered on a nominal collection year of 2001 (USGS, 1999, 2003; also see Homer et al. (2004) for details). However, these two datasets cannot be used directly to model land use transition on a point (pixel) basis for several reasons.[2] First, these datasets use slightly different classification schemes; many land cover types of 1992 cannot be matched with land cover types of 2001. Second, the accuracy is not good enough to model land use/cover transition on a pixel basis. Finally, NLCD land cover classification does not identify clearcuts and young plantations among other (non-forest) barren/grasses/shrubs land cover types. Clearcuts and young plantations have different land cover change patterns than non-forestry barren land, grasses, or shrubs.

In order to overcome these problems, we used subset of the data, which had been visually validated using high resolution aerial photographs. The aerial photographs allow much better precision of identification of land cover types than Landsat images. We performed a systematic sampling of the original data by placing a 750–meters rectangular grid over the study area, yielding 5313 30×30 meter sample points. These sample points were assigned land cover values from the NLCD 1992 and NLCD 2001 datasets. A GIS layer with 30×30 meter polygons representing sample points was overlaid with black and white aerial orthophotos with 1 m resolution dated 1992, and with color aerial orthophotos with 0.8 m resolution dated 2003. The land cover values of sample points were then visually validated, if necessary corrected or reclassified according to the NLCD 2001 classification scheme of 21 types. For this study, we aggregated these 21 land use/cover types to Developed, Forest, Agriculture, Wetlands, Water Bodies, and Others.

There are several ways to quantify the effect of location when multiple populated places of different sizes influence each parcel of land simultaneously. Regional scientists traditionally evaluate and compare the influence of multiple populated places using gravity potential, which is proportional to the population of the populated place and inversely proportional to the squared distance between the populated place and the parcel of interest. Because the influences of multiple populated places on a given parcel are additive, Hoover (1971) suggests aggregating gravity potentials into a single index. This approach has been used by a number of land use change studies (Kline et al., 2001, 2003; Polyakov & Zhang, 2008). There are various specifications of gravity index (Song, 1996). We calculated Population Gravity Index (PGI) using the traditional specification suggested by Hoover (1971):

$$G_i = \sum_k \frac{P_k}{D_{ki}^2} \quad \forall k : D_{ki} \leq 50,$$

where G_i is the population gravity index for sample point i, P_k is the population of census block k, and D_{ki} is the distance between sample point i and census block k in miles. The

[2] Direct comparison of NLCD 1992 and NLCD 2001 is not recommended by USGS, see

1990 and 2000 Censuses of Population census block data were taken from ESRI Data and Maps (ESRI, 1999, 2005).

Other explanatory variables used in our models are per capita income, slope and proximity to transportation network. Per capita income on the census tract level was obtained from US Census 2000 (U.S. Census Bureau, 2002). The slope affects suitability of land for development and agriculture. The value for the slope attribute is derived from the Digital Elevation Model (DEM) obtained from the Georgia Spatial Data Clearinghouse (GSDI, 2006). The factor that affects accessibility of a parcel is its proximity to a transportation network. We hypothesized that proximity to roads and highways might have different effects on relative rents to different land uses. In particular, proximity to highway may be irrelevant for rural land uses, but for the developed (residential) use could have both positive effect (need for access to transportation) and negative externality effect (noise nuisance and bad air quality). Distances from each sample plot to the nearest road and to the nearest highway are calculated using TIGER/Line spatial data from the US Census Bureau (2006).

4. Results

We modeled land use in these three counties in 2001 using the land use allocation and land use change models. Because there is virtually no transition to and from such land uses as wetlands, and water bodies, we excluded them from the consideration in both models. Transition to developed land is practically irreversible; therefore developed lands were excluded from the list of initial land use types in the second model. As a result, in our land use allocation model we consider four (j) land use types (developed, forest, agricultural, other), while in our land use change model we consider three initial (i) and four final (j) land use types.

The models were estimated using SAS 9.1 (SAS Institute, Inc., 2004). The results are presented in Table 2. "Other" land use type is a reference type ($\boldsymbol{\beta}_{Jn} = 0$ to remove indeterminacy in the model). The regression coefficients for the rest of the land use types indicate effects of particular variables on allocation/conversion to respective land use types relative to the reference type ("Other lands"). As expected, the land use change model has a substantially better fit than the land use allocation model, as indicated by McFadden pseudo-R^2.

To evaluate forecasting performance of our model, we conducted within-sample predictions of land uses for 2001 using both land use allocation and land use change models. We utilized two *a priori* null models of the land uses: $p_0(a_j)$ is equal probability for all land uses, and $p_1(a_j)$ is a probability proportional to the occurrence of land uses in the sample, which is equivalent to the probability predicted by the choice model using only an intercept. Information indices and statistics are presented in Table 3. The U^2 values suggest that the proportion of uncertainty explained by the empirical model is relatively high and in tests of full models versus equal probability null these indices are comparable with McFadden pseudo-R^2 (Table 2). Proportion of uncertainty explained by the land use change model vs.

http://www.mrlc.gov/multizone.php.

intercept-only null is much higher than proportions of uncertainty explained by the land use allocation model vs. intercept-only null. The analysis of information index, expected information index and its variance suggest that the empirical models are accurate, while the log-likelihood ratios (LLR) indicate that models are statistically significant.

It is difficult to interpret the coefficients in a conditional logit model because the effect of the variable on a particular transition probability is jointly determined by all the coefficients for this variable. In Table 4 we presented marginal effects of the explanatory variables on the transition probabilities.[3]

Table 2. Estimation Results of Conditional Logit Models of Land Use in West Georgia

Variables	Model of land use change				Model of land use allocation			
	Final land use (*j*):				Final land use (*j*):			
	Developed	Forestry	Agricult	Other	Developed	Forestry	Agricult	Other
Intercept					5.4216	11.0925†	25.2582‡	
Initial Forestry	-24.0534‡		-1.6864	-20.9026†				
Initial Agriculture	-27.6295‡	-6.0344		-23.1974†				
Initial Other		22.9644‡		9.9049				
Log (PGI)	0.4561	-0.5868†	-0.1611		0.4867‡	-0.6855‡	-0.7654‡	
Log (PGI rate of change)	4.0989†	-1.1492	-1.0554		0.0826	-0.8115	-0.6939	
Log Per Capita Income	0.7651	-0.9774	-1.8630*		-0.1089	-0.4960	-2.0320‡	
Log Distance to Road	-0.6473‡	-0.2786	0.0461		-0.4410‡	0.0346	0.1883	
Log Distance to Highway	-0.1613	-0.0039	-0.0380		-0.6501‡	0.3071‡	-0.0327	
Slope	-0.1780*	0.0281	-0.1922*		-0.0668	0.0905*	-0.1941‡	
McFadden Pseudo R^2	0.8706				0.5961			

* significant at 10%; † significant at 5%; ‡ significant at 1%.

Table 3. Information Indices and Statistics

Model	Test	$H(\mathbf{A};\mathbf{X})$	U^2	$I(\mathbf{A};\mathbf{X})$	$EI(\mathbf{A};\mathbf{X})$	$V(\mathbf{A};\mathbf{X})$	*LLR*
Land use allocation	Full model vs. equal prob.	1.39	0.60	0.82632	0.82632	0.00016	7761
	Full model vs. intercept only	0.73	0.23	0.16867	0.16864	0.00005	1584
Land use change	Full model vs. equal prob.	1.39	0.88	1.21975	1.22022	0.00010	11456
	Full model vs. intercept only	0.73	0.77	0.56210	0.56202	0.00006	5279

[3] The marginal effect of attribute m of a sample plot on the probability of transition to land use type j is $\partial P_j / \partial x_m = P_j \left(\beta_{jm} - \sum_{k=1}^{J} \beta_{km} P_k \right)$.

Table 4. Marginal Effects of Explanatory Variables in Conditional Logit Models of Land Use in West Georgia

Variables	Model of land use change				Model of land use allocation			
	Final land use:				Final land use:			
	Developed	Forestry	Agricult	Other	Developed	Forestry	Agricult	Other
Intercept					-0.2091†	-0.7723‡	1.2026‡	-0.2212†
Initial Forestry	-0.3287‡	0.4903‡	-0.0105	-0.1511†				
Initial Agriculture	-0.2962‡	0.3634‡	0.0571	-0.1243*				
Initial Other	-0.3118‡	0.6030‡	-0.1991‡	-0.0921				
Log (PGI)	0.0142‡	-0.0220‡	0.0036	0.0042*	0.0359‡	-0.0367‡	-0.0111	0.0119‡
Log (PGI rate of change)	0.0721‡	-0.0801‡	0.0001	0.0079	0.0267*	-0.0465	0.0058	0.0140
Log Per Capita Income	0.0240‡	-0.0230*	-0.0081	0.0071	0.0161	0.1012‡	-0.1286‡	0.0113
Log Distance to Road	-0.0051‡	0.0002	0.0029†	0.0021	-0.0151‡	0.0016	0.0141‡	-0.0006
Log Distance to Highway	-0.0022	0.0024	-0.0003	0.0000	-0.0283‡	0.0575‡	-0.0247‡	-0.0044†
Slope	-0.0028‡	0.0049‡	-0.0019‡	-0.0002	-0.0040‡	0.0279‡	-0.0229‡	-0.0011

* significant at 10%; † significant at 5%; ‡ significant at 1%.

Population pressure (as reflected by the marginal effects of PGI) and population growth (as indicated by the marginal effects of PGI rate of change) increased the probability of allocation and conversion to developed use, and decreased the probability of allocation and conversion to forestry use. The "elasticity" of developed lands with respect to population change in these three counties is close to 7 (4.1%/0.6%; See table 1).

Steeper slope negatively affected both allocation and conversion to agricultural land use. It also negatively affected the probability of conversion to developed use. Slope positively affected the probability of conversion to forest. Development was found to be closer to highways and roads. Proximity to roads decreases the probability of forestry use and increased the probability of agricultural use, as expected. Finally, higher per capita income was related to higher probability of development and lower probability of agricultural use/conversion to agricultural use.

The directions of marginal effects of all explanatory variables were similar in both models. However, there was a substantial difference in their magnitudes. Consider variables that represent level and change of urbanization. The marginal effects for log of population gravity index have greater absolute values in the land use allocation model than in the land use change model. The coefficients for log of PGI rate of change are greater in the land use change model, and in the land use allocation model, only the coefficient for developed use is significant. This is likely due to the fact that coefficients (and marginal effects) in allocation model reflect both spatial and temporal influences of the factors driving land use change, while coefficients (and marginal effects) in the change model reflect only temporal effects. Similar phenomenon can be seen in Ahn et al. (2000), when in model with fixed effects the magnitude of variables driving land use change (rents to alternative land uses) had lower magnitudes than in pooled model, which does not control for cross-sectional effects. Thus, model of land use allocation could potentially yield biased results when used to predict future land use.

To highlight the differences in performance of analyzed models, we mapped the forecast errors at a sub-watershed level. For this, we aggregated our in-sample projections to 12–digits hydrologic units,[4] calculated predicted proportion of land uses (based on predicted probabilities of land uses for each sample plot within hydrologic unit), and calculated actual proportions of land uses for these units. Figure 1 presents deviation of proportions of developed land use predicted by each of the models from the actual proportions in 2000 for each hydrologic unit. The map of population density is shown as a reference. Comparison of two maps shows that our model of land use allocations under-predicts developed land use close to population centers (for example, Columbus/Bibb city and Warm Springs/Woodburry/Greenville). The land use allocation model and, to a lesser extent, the land use change model over-predict developed land use on the outskirts of population concentrations (east and southeast of Columbus). Overall, the land use change model shows much better results, which is consistent with the goodness of fit statistics of conditional logit models and our expectations.

5. Conclusion

The use of disaggregate data in land use modeling becomes more common with the availability of quality remote sensing data because it brings a possibility to incorporate spatially referenced socioeconomic and biophysical information and allows analysis on a small scale. Unfortunately, time series of disaggregate data are often unavailable, and researchers sometimes make inferences about allocation of land to alternative land uses in the future or effects of various policies on future land use based on models that utilize cross-sectional or pooled cross-sectional data. However, extreme caution should be exercised when interpreting results of the land use modeling where cross-sectional variation is not adequately controlled for. Due to a large number of cross-sectional elements, the use of fixed effect for the purpose of controlling for cross sectional effects is not feasible; the alternative way is to incorporate information about previous land use.

In this study we compared application of conditional logit model of land use with and without using information about previous land use at each sample plot ("land use allocation" and "land use change" models). The results have shown that the model that incorporates previous land use thus controlling for cross-sectional variation, does a much better job predicting land use. The information theory statistics reveal that the land use allocation model explained about one quarter of uncertainty, while the model of land use change explained about of three-quarters of uncertainty in in-sample prediction of 2001 land use in the study region. Furthermore, the model of land allocation that does not control for cross-sectional effects could produce biased prediction of future land use because of changing compositions of land uses.

[4] A hydrologic unit is a topographically defined area of land, the boundaries of which are ridge tops. A 12-digit hydrologic unit is a level 6 sub-watershed. The sizes of the 12-digit hydrologic units within the study area range between 3 to 12 thousand hectares.

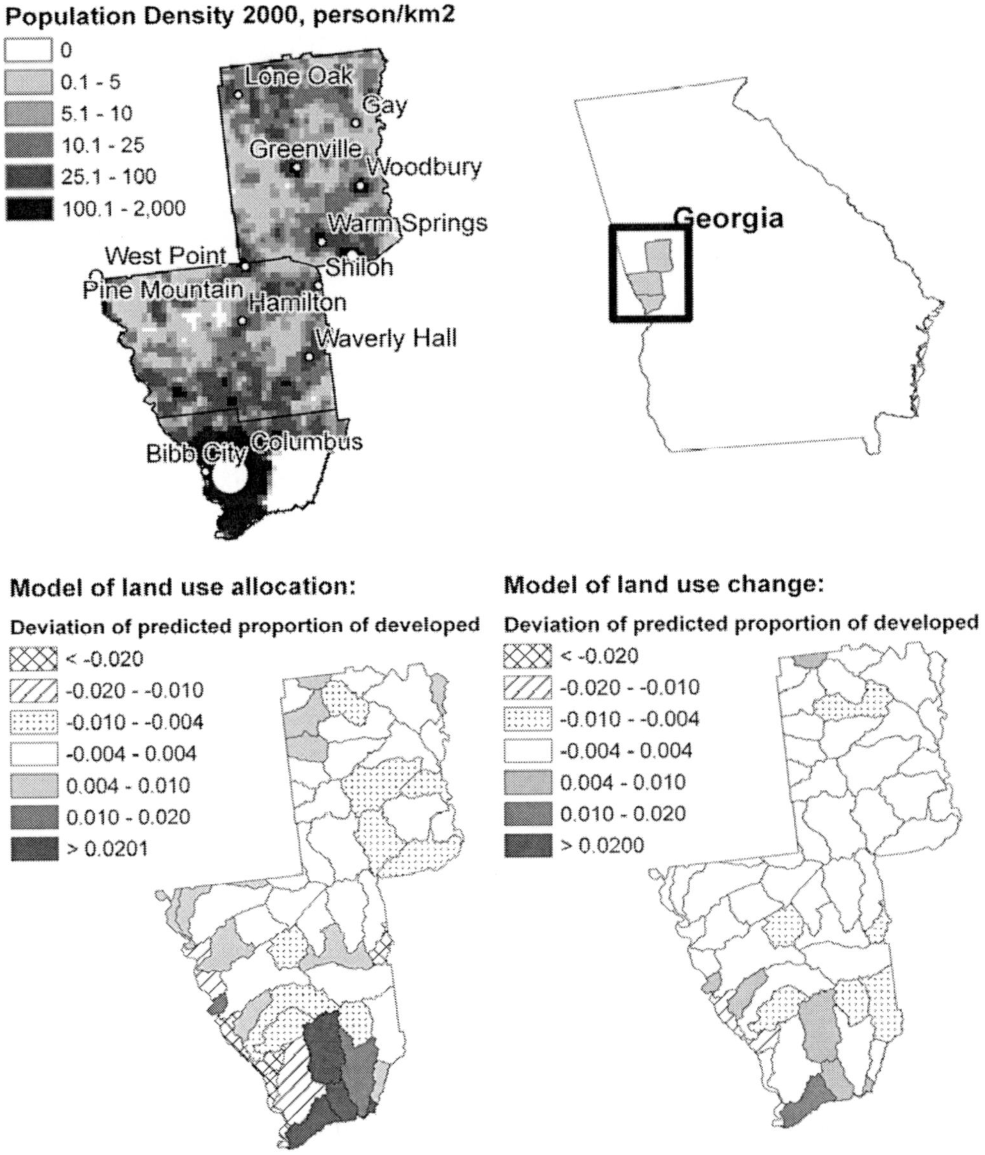

Figure 1. Spatial patterns of population density and deviation of proportions of developed lands predicted by two models from actual values at a watershed level.

There are several limitations of this study primarily due to our desire to make the models as simple as possible for the purpose of comparison. The conditional logit model assumes that the independence of irrelevant alternatives (IIA) property holds. This is a strong assumption. It can be relaxed by applying nested or random parameter logit models. Furthermore, we do not take into account zoning and other regulations, which determines, among other things, possibility and maximum density of development.

References

Ahn, S., Plantinga, A., & Alig, R. J. (2000). Predicting future forestland area: A comparison of econometric approaches. *Forest Science, 46*(3), 363-376.

Alig, R. J., & Healy, R. G. (1987). Urban and built-up land area changes in the United States: An empirical investigation of determinants. *Land Economics, 63*(3), 215-226.

Barlowe, R. (1978). *Land Resource Economics, third ed.* Englewood Cliffs, NJ: Prentice-Hall.

Bockstael, N. E. (1996). Modeling economics and ecology: The importance of a spatial perspective. *American Journal of Agricultural Economics, 78*(5), 1168–1180.

Carrión–Flores, C. & Irwin, E. G. (2004). Determinants of Residential Land Use Conversion and Sprawl at the Rural–Urban Fringe. *American Journal of Agricultural Economics, 86*(4), 889–904.

Chomitz, K. M. & Gray, D. A. (1996). Roads, Land Use, and Deforestation: A Spatial Model Applied to Belize. *World Bank Economic Review, 10*(3), 487–512.

GSDI (Georgia Spatial Data Clearinghouse). (2006). Internet site: https://gis1.state.ga.us/. (Accessed October 14, 2006).

Hardie, I. W., & Parks, P. J. (1997). Land use with heterogeneous land quality: An application of an area-based model. *American Journal of Agricultural Economics*, *79*(2), 299–310.

Hauser, J. R. (1978). Testing the accuracy, usefulness, and significance of probabilistic choice models: An information–theoretic approach. *Operations Research*, *26*(3), 406–421.

Hoover, E. M. (1971). *An Introduction to Regional Economics*. New York, NY: Knopf.

Homer, C., Huang, C., Yang, L., Wylie, B., & Coan, M. (2004). Development of a 2001 national Land-cover database for the United States. *Photogrammetric Engineering and Remote Sensing*, 70, 829-840.

Irwin, E. G., & Bockstael, N. E. (2002). Interacting Agents, Spatial Externalities, and the Endogenous Evolution of Residential Land Use Patterns. *Journal of Economic Geography*, *2*(1), 31–54.

Kline, J. D., Moses, A., & Alig, R. J. (2001). Integrating urbanization into landscape-level ecological assessments. *Ecosystems, 4*(1), 3-18.

Lubowski, R. N., Plantinga, A. J., & Stavins, R.N. (2006). Land-use change and carbon sinks: Econometric estimation of the carbon sequestration supply function. *Journal of Environmental Economics and Management, 51*(2), 135–152.

McFadden, D. (1973). Conditional logit analysis of quantitative choice models. In P. Zarembka (Ed.) *Frontiers of Econometrics*. New York, NY: Academic Press.

Nagubadi, V. R., & Zhang, D. (2005). Determinants of timberland use by ownership and forest type in Alabama and Georgia. *Journal of Agricultural and Applied Economics, 37*(1), 173–186.

Natural Resources Conservation Service. (2000). *1997 National Resources Inventory.* Revised December 2000. Washington, D.C.: Natural Resources Conservation Service, U.S. Department of Agriculture.

Nelson, G., De Pinto, A., Harris, V., & Stone, S. (2005). Land use and road improvements: A spatial perspective. *International Regional Science Review, 27*(3), 297–325.

Parks, P. J., & Murray, B. C. (1994). Land attributes and land allocation: Nonindustrial forest use in the Pacific Northwest. *Forest Science, 40*(3), 558-575.

Plantinga, A. J., Buongiorno, J., & Alig, R. J. (1990). Determinants of changes in non-industrial private timberland ownership in the United States. *Journal of World Forest Resource Management, 5,* 29-46.

Polyakov, M., & Zhang, D. (2008). Property tax policy and land use change. *Land Economics, 8*(3), 396–408.

Ricardo, D. (1817). *The Principles of Political Economy and Taxation.* London: John Murray.

SAS Institute, Inc. *SAS/ETS User's Guide, Version 9.1.* Cary, NC, 2004.

Stavins, R. N., & Jaffe, A. B. (1990). Unintended impacts of public investments on private decisions: The depletion of forested wetlands. *American Economic Review, 80,* 337-352.

Train, K. E. (2003). *Discrete Choice Methods with Simulation.* Cambridge, UK: Cambridge University Press.

Turner, M. G., Wear, D. N. & Flamm, R. O. 1996. Land ownership and land–cover change in the Southern Appalachian Highlands and the Olympic Peninsula. *Ecological Applications, 6*, 1150–1172.

US Census Bureau. (2006). *TIGER/Line.* Internet site: http://www.census.gov/geo/www/tiger/. (Accessed October 20, 2006).

U.S. Census Bureau. (2002). *Census of Population and Housing, 2000* [United States]: Summary File 3, Georgia [Computer file]. Washington, DC: U.S. Census Bureau [producer]. Ann Arbor, MI: Inter-university Consortium for Political and Social Research [distributor].

U.S. Geological Survey (USGS). (1999). *Georgia Land Cover Data Set. Edition 1.* Sioux Falls, SD: U.S. Geological Survey ftp://edcftp.cr.usgs.gov/pub/data/ landcover/states/georgia_FGDC.txt

U.S. Geological Survey (USGS). (2003). *National Land Cover Database Zone 54 Land Cover Layer. Edition 1.0.* Sioux Falls, SD: Geological Survey. http://www.mrlc.gov

von Thünen, J. H. (1826). *Der Isolierte Staat in Beziehung auf Landwirtschaft und Nationalökonomie*, Hamburg, Germany: Perthes. English translation: *The Isolated State*, Oxford, UK: Pergammon Press.

Zhang, D. & Nagubadi, R. V. (2005). Timberland Use in the Southern United States. *Forest Policy and Economics*, *7*(3), 721–731.

In: Global Change and Forestry
Editors: J. Gan, S. Grado and I. A. Munn

ISBN: 978-1-60876-262-0

Chapter 4

A NESTED LOGIT APPROACH TO MODELING FOREST LAND USE CHANGE

Indrajit Majumdar*[*,1], Maksym Polyakov[2], Larry Teeter[3] and Brett J. Butler[4]

[1] Ontario Ministry of Natural Resources, Sault Ste. Marie, Ontario, Canada
[2] School of Agricultural and Resource Economics, University of Western Australia, Australia
[3] Forest Policy Center, Auburn University, AL, USA
[4] USDA Forest Service, Northern Research Station, PA, USA

ABSTRACT

The chapter focuses on exploring the impacts of urbanization on changes in forestland use/land cover in Alabama for the period between 1972 and 2000. Nested logit analysis of the discrete land use choices made by the private landowners show that initial forest type and population gravity index significantly explain the variation in forest type transition. Anthropogenic factors influence the decision in favor of forestland conversion to non-forest use. Softwood stands were more preferred for harvests relative to hardwood while hardwood was the more preferred choice for maintaining land in forest cover near the population centers relative to softwood. Per capita income turned out to be a better predictor of land use change than population density or population gravity index suggesting that efforts in acquiring accurate income information may be more useful than expending resources in the acquisition of other demographic measures such as population density in land use models.

Keywords: Nested logit; Urbanization; Land use; Population gravity index.

---

[*] Email: Indrajit.majumdar@ontario.ca, Ontario Ministry of Natural Resources, Sault Ste. Marie, Canada.

1. INTRODUCTION

Changing land use patterns and the dynamics of land allocation to various uses by landowners viz. forestry, agriculture and urban uses have been in the forefront of research for the past few decades. Prompted by the re-allocation of scarce resources to satisfy growing human needs, land use change impacts numerous ecosystem services and functions such as watershed protection, biogeochemical cycling, soil degradation, and habitats for other species. Land use change and respective change of land cover is attributed to human activities and is a common phenomenon associated with population growth, market development, technical and institutional innovation and policy action. Veldkamp and Fresco (1996) note, land use "is determined by the interaction in space and time of biophysical factors (constraints) such as soils, climate, topography, etc., and human factors like population, technology, economic conditions, etc." Vitousek (1994) identified land cover changes by humans as the primary effect of humans on natural systems. Few forested areas on our planet remain that have not been influenced by human actions, yet our understanding of the long term human influences on forestland use remains superficial.

Currently Alabama with more than 70% of its land area in forests (Hartsell & Brown 2002) is the national leader in total roundwood pulpwood production (10.2 million cords) (Johnson et al. 2007). The objective of this chapter is to explore the effect of increasing population pressures on choices made by the private forest landowners of Alabama. Alabama ranks second (after Georgia) in the southern U.S. in acres of forestland, and the forests of the state account for 13% of the total timber removals in the South (Smith et al 2002). The impacts on forestry land use and its conversions into agriculture and urban lands will have significant effects on the ability of Alabama's forests to provide both timber and non-timber amenities in the future. While most of the developed land comes from previous forestland, there are simultaneous conversions of forestland to and from agricultural land. Forestland conversions due to urbanization and increases in population not only impacts the forestland base, but also forest management (Munn et al. 2002), environmental changes (Bockstael 1996) and may also produce negative externalities such as habitat fragmentation, increased flooding and loss of tree diversity (Polyakov et al. 2008), decreased timber production and supply of recreation and amenities.

In the last couple of decades, research on modeling land use changes has mainly focused on an area based approach where the proportion of land area in a particular use to the total land area is explained by the impacts of economic, socio demographic and biological factors. In this chapter we analyze the determinants of forestland changes in the state of Alabama using Forest Inventory and Analysis (FIA) plot level data as opposed to regional aggregated studies. Spatially referenced socioeconomic variables, land and topography characteristics, market forces and ownership characteristics are used to explain the probability of forestland change to primarily urban and agricultural use. Results show that human factors play a significant role in affecting land use changes. Changes in land use, particularly forest use, have important consequences for the future availability of timber, wildlife habitat, and other benefits provided by forests.

2. Literature Review

Empirical land use models have been derived mostly based on the classic Thünen-Ricardo land rent theory that describes land-use patterns as a result of variability in geophysical land attributes and differences in location and transport costs. Two approaches have been predominantly used in modeling land use patterns depending on data availability, scope and scale of research and the extent of detail sought by the researcher in understanding how landowner decisions affect land use outcomes. The first is the aggregated approach that models areas or proportions of land in different use categories such as forestry, agriculture and urban (Alig 1986, Alig & Healy 1987, Parks & Murray 1994, Hardie & Parks 1997) or different forest types such as softwood, mixed hardwood, hardwood, agriculture and urban land (Zhang & Nagubadi 2005) within a well defined geographic region such as a county as a function of socioeconomic variables and land characteristics aggregated at the level of the geographic unit of observation. The second is the spatially explicit approach that explicitly models land use change on the basis of pixels, parcels, or sample points (Bockstael 1996, Chomitz & Gray 1996, Turner et al. 1996, Munn & Evans 1998, Wear & Bolstad 1998, Kline et al. 2001, Irwin & Bockstael 2002, Carrion-Flores & Irwin 2004, Lubowski et al 2006). While the aggregated approach has the disadvantage of averaging the physical land characteristics for the unit of study, the spatially explicit approach has often found it difficult to obtain spatial socio-demographic data at scales finer than the census tract level. Also in the former approach, the coefficients of the model capture simultaneously both the spatial and temporal effects of the explanatory variables and does a poor job in projecting land use shares through time (Ahn et al. 2000). In contrast, the spatially explicit approach models change directly by taking into account the dynamic nature of the land use change decision.

3. Empirical Land Use Model

Researchers have extensively used multinomial logit models (Chomitz and Gray 1996, Turner et al. 1996, Hardie and Parks 1997) for explaining landowners' choice of land use without taking into consideration the possibility of correlation between alternative choices. A feature of this chapter is the use of the less restrictive nested logit econometric framework which relaxes the assumption of Independent and Irrelevant Alternatives (IIA)[1] to account for the possible substitution patterns amongst alternative choices.

It is assumed that a landowner starting with an initial forest type chooses from among the five possible discrete alternatives the one that maximizes his utility. The alternative choice set includes either converting forest into non-forest use, regenerating into one of the three forest types (hardwood, softwood or mixed) following harvest and a no harvest[2] decision to maintain the initial forest type. A landowner's utility gained from choosing a particular alternative depends on the attributes associated with each forest plot. For models of land use

[1] McFadden (1974) suggested that IIA implies that conditional and multinomial logit models should only be used in cases where the outcome categories can plausibly be assumed to be distinct and weighed independently in the eyes of each decision maker.

[2] This study does not assume type transition if there is no harvest and considers the forest type as fixed until harvest occurs.

change, the vector of plot characteristics, $\mathbf{x}$, typically consists of data on land quality, socio-demographic and socio-economic decision environment attributes ,and rent (return) to alternative land use choices. In this discrete choice framework, a risk neutral landowner is assumed to choose for parcel i an alternative j from a set of J alternatives that maximizes his utility at time t.[3] Assume that the landowner's utility function for choice j is given by:

$$V(\boldsymbol{\beta}_j, \mathbf{x}) = v(\boldsymbol{\beta}_j, \mathbf{x}) + \varepsilon_j \tag{1}$$

where $\mathbf{x}$ is the vector of attributes of plot characteristics and $\boldsymbol{\beta}_j$ is a vector of preference parameters on the observable portion of the landowner's utility function for the alternative j, $v(\boldsymbol{\beta}_j, \mathbf{x})$. Finally, ε_j is the unobservable portion of the landowner's utility function and is assumed to be a function of certain forest plot characteristics and the characteristics of the decision maker. The landowner then compares all potential choices in his choice set J and chooses the best land use alternative j such that:

$$V(\boldsymbol{\beta}_j, \mathbf{x}) > V(\boldsymbol{\beta}_k, \mathbf{x}) \quad \forall \quad j \in J, \quad k \in J, \quad k \neq j \tag{2}$$

The challenge is to take the model given by (1) and (2) and develop a statistical model that will enable the recovery of the parameters $\boldsymbol{\beta}$. The structure of the model will depend heavily on the assumptions about the form of the distribution of error terms. Assuming error terms ε_j are independent and identically distributed (i.i.d.) with a Type I Generalized Extreme Value distribution (GEV)[4], (1) and (2) are expressed as a multinomial logit model:

$$\text{Prob}(k) = \frac{\exp(\boldsymbol{\beta}_k' \mathbf{x})}{\sum_{j \in J} \exp(\boldsymbol{\beta}_j' \mathbf{x})} \tag{3}$$

This denotes that the ratio of probabilities of choices k and j would remain unchanged with a change in the parameters of choices other than k and j (IIA). In reality, that might not be the case. For example, a change in the stumpage price of hardwood might influence the ratio of the probability of a transition to pine plantation vs. probability of a transition to agricultural land. A study by Lubowski (2002) on the economic and policy determinants of land use change using a nested logit model supports the need for exploring alternative nesting structures in land use studies. We use a three level nested logit model, which assumes that decisions are made at three hierarchical levels (figure 1).

---

[3] For notational simplicity the subscripts i and t will be dropped from the equations.

[4] Type I GEV also known as the Gumbel distribution is based on simplifying assumptions such as independent and identical distribution (iid) of random components and the absence of heteroscedasticity and autocorrelation in the model (see Mcfadden (1974) for details).

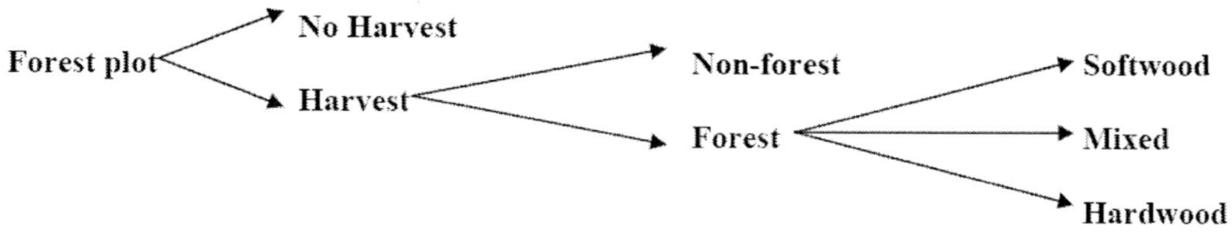

Figure 1. Three level-nested representation of the landowner decision set.

The decision at each of these three levels is modeled as an outcome of separate utility maximizing decisions. The decision to harvest or not to harvest at the uppermost level of the nested tree can be modeled as a binary logit model. Assuming the landowner makes the decision to harvest, he has to make another decision at the medium level of the nested model, which is whether to keep the land in forest or convert it to non-forest use. This can also be modeled as a binary logit model. Finally, assuming the landowner decides to keep the land in forest use, he decides whether to regenerate it to softwood, mixed or hardwood type of forest. Each of these decisions is taken with a view of maximizing utility. The three level nested model decomposes the choice probability into three components, the marginal probability of choosing a particular subgroup (nest) s at the uppermost level, S=1,2 for harvest or no harvest, the marginal probability of choosing a particular sub-nest l within the nest s, where L=1,2 for non-forest or forest, and the conditional probability of choosing a particular alternative j at the lowest level within the alternative set $J = 1 \ldots J_{ls}$ in the sub-nest l and nest s conditional on the choice of that sub-nest and nest. Given this, the probability that a landowner i is observed choosing alternative j at time t in the nested logit formulation requires the decomposition of the choice probability in (3) into three components: the marginal probability P_{is} of choosing a particular nest s (s=1,2) and conditional probabilities $P_{il|s}$ and $P_{ij|l,s}$ of choosing a particular sub-nest l (l=1,2) conditional on the choice of that nest s and choosing a particular alternative j from within the alternatives (j=1,2,3,4,5) conditional on the choice of that nest and sub-nest. The probability defined in (3) thus becomes:

$$P_{islj} = P_{is} \times P_{il|s} \times P_{ij|l,s} = \frac{\exp(\boldsymbol{\delta}_s^{'} \mathbf{y}_i + \tau_s I_{is})}{\sum_{k \in S} \exp(\boldsymbol{\delta}_k^{'} \mathbf{y}_i + \tau_k I_{ik})} \times \frac{\exp(\boldsymbol{\gamma}_l^{'} \mathbf{z}_i + \sigma_{l|s} I_{il})}{\sum_{m \in L} \exp(\boldsymbol{\gamma}_m^{'} \mathbf{z}_i + \sigma_{m|s} I_{im})} \times \frac{\exp(\boldsymbol{\beta}_j^{'} \mathbf{x}_i)}{\sum_{n \in J} \exp(\boldsymbol{\beta}_n^{'} \mathbf{x}_i)} \tag{4}$$

where τ_s and $\sigma_{l|s}$ are the parameters associated with the Inclusive Value (IV) for nest s and sub-nest l defined as

$$I_{is} = \ln \sum_{m \in L} \exp(\boldsymbol{\gamma}_m^{'} \mathbf{z}_i + \sigma_{m|s} I_{im}) \tag{5}$$

and

$$I_{il} = \ln \sum_{n \in J} \exp(\boldsymbol{\beta}_n^{'} \mathbf{x}_i) \tag{6}$$

where, $\mathbf{y}_i$ are the observed plot attributes influencing the choice of the nest, $\mathbf{z}_i$ are the observed plot attributes influencing the choice of the sub-nest and $\mathbf{x}_i$ being the observed plot attributes influencing the decision to keep land in an alternative forest type conditional on the choice of the nest and sub-nest. The inclusive value for nest s and sub-nest l defined in (5) and (6) is the log of the denominator of the conditional probabilities in (4) and measures the average utilities of the alternatives within that subset of alternatives for the choice of a particular nest s and sub-nest l. If the parameters $\boldsymbol{\delta}_k$ and $\boldsymbol{\gamma}_m$ are zero and the inclusive value parameters τ_k, $\sigma_{m|s}$ are jointly equal to one, then the model will collapse into a multinomial logit model shown in (3).

4. Data and Variables

A close look at past land use studies reveals that biophysical factors such as land quality and topography; economic factors such as population, market conditions, proximity to population centers and income; and institutional factors such as government policy are the major determinants of land use change. The main sources of data used come from Forest Inventory and Analysis (FIA)[5] program of the U. S. Department of Agriculture (USDA) Forest Service, USDA Economic Research Service (ERS), Bureau of Census and the Regional Economic Information System (REIS) of the Bureau of Economic Analysis (BEA). We used Alabama FIA data for the census years 1972, 1982, 1990 and 2000 and the Census Bureau data on population demographics for the same periods.[6] REIS provided us with the per capita personal income by county for the corresponding years. All the plots considered were restricted to be in forest use at the beginning of the period and privately owned. The total number of observations for the period (1972-2000) that included three transition periods (1972-1982, 1982-1990 and 1990-2000) was 10383.

All the explanatory variables in the model, associated with the FIA plots were lagged values based on the previous period 't-1' to incorporate the general trends in the variable's effect on the landowners' discrete choices as observed at the current period t. For example a FIA plot observed in a particular land use for the FIA survey year 1982 had all the corresponding explanatory variables from the FIA survey 1972 and the population census for the year 1970 and so on. From among the array of variables used the key variable that represents the influence of humans on forestland use change is the Population Gravity Index (PGI). Past studies have reported a negative correlation between harvest intensity and increasing development (Munn & Evans 2002, Barlow et al. 1998). PGI was constructed by utilizing information on the location of the FIA plots in relation to the location of Census

---

[5] Historically FIA provides detailed data on forest inventory for all the states on approximately 10-year periodic cycle with each plot roughly representing a 3×3 mile grid pattern.

[6] Census collects decennial data and so for the FIA counterpart of 1972 and 1982 we used its closest census counterpart which was 1970 and 1980.

populated places within 100 km. The geographic location of census places[7] was taken from ESRI Data and Maps, 2005 (http://www.esri.com/data/about/data_maps_media.html). Other variables in the model include the initial forest type dummy for the three classes of forest type denoted by the variable names SW (softwood), MX (mixed) and HW (hardwood) for each FIA plot. Volume in cubic feet of all trees within a FIA plot divided by the plot acres is denoted by VOL and was included as a potential measure of the propensity to harvest for the plot. We also included growing stock removals in cubic feet (from FIA county data) per unit of county land area in acres (from ERS) as a proxy for forestland use return (RET) hypothesized as one of the chief economic drivers of land use change in almost all of the previous land use models. SLOPE in percent for the FIA plots was included to examine the potential influence of topography on landowner choice. Finally, to explore the full potential of the urbanization pressures acting on forestland use change, county level estimates of per capita personal income (INC) from REIS of the BEA deflated by the Consumer Price Index (Urban South, 1982=100), and county level estimates of population density (PD) were also included in the model. A list of the variables used in the analysis with their sources and standard statistical summary is given in table 1.

Table 1. Mean and standard deviations of the explanatory variables along with their description and source

Variable	Definition	Source	Mean	Std.dev.
PGI	Number of persons/Km^2 around each FIA plot within 100 Km radius of each FIA plot	FIA plot and Census Bureau	136.03	99.60
VOL	Average volume in cubic feet per acre for the FIA plots	FIA plot data	1027.19	965.92
SW	Initial forest type dummy for Softwood forest	FIA plot data	0.35	0.47
MX	Initial forest type dummy for Mixed forest	FIA plot data	0.44	0.50
HW	Initial forest type dummy for Hardwood forest	FIA plot data	0.20	0.40
INC	Real (1982=100) per capita personal income by county in $	BEA	111.95	23.48
RET	Growing stock removals in cubic feet per acre of county land area	FIA county data and ERS	37.53	18.40
SLOPE	Slope in percent for FIA plot	FIA plot data	9.93	10.75
PD	Number of persons per unit of land area by county	Census Bureau	73.28	97.75

[7] Bureau of Census definition for a place is "concentration of population either legally bounded as an incorporated place, or identified as a Census Designated Place (CDP) including comunidades and zonas urbanas in Puerto Rico. Incorporated places have legal descriptions of borough (except in Alaska and New York), city, town (except in New England, New York, and Wisconsin), or village".

4.1. Population Gravity Index

A 100 km[8] buffer around Alabama incorporating the influence of census places from the four contiguous states of Georgia (GA), Tennessee (TN), Mississippi (MS) and Florida (FL) in addition to all the designated census places within the state of Alabama was created. Population Gravity Index (PGI)[9] for a plot k was specified as

$$PGI_k = \sum_p \frac{P_{pt}}{D_{kp}^{\ 2}} \quad \forall\ p : D_{kp} \leq 100\ km \tag{7}$$

where P_{pt} is the population of populated place p at time t, and D_{kp} is the distance between FIA plot k and populated place p. PGI was previously found to be positively correlated with conversion to non-forest use from forest use (Majumdar et al. 2005).

5. RESULTS

The three utility functions representing the variables likely to influence landowners' decisions at the three decision nodes of the nested tree and the attribute vectors in $\mathbf{y}_i$, $\mathbf{z}_i$ and $\mathbf{x}_i$ are:

$$\text{Pr(no harvest relative to harvest)} \equiv f(VOL, SLOPE, SW) \tag{8}$$

$$\text{Pr(Non-forest relative to Forest)} \equiv f(PGI, INC, RET, PD) \tag{9}$$

$$\text{Pr(Softwood or Mixed or Hardwood)} \equiv f(SW, MX, HW, PGI) \tag{10}$$

We estimated a three level nested logit model in which the landowner decides to either harvest or not to harvest at the top level, then makes the decision to convert the harvested plots into non-forest use or keep them in forest use at the next level, and finally decides on whether the forested plot will be of softwood, mixed hardwood, or hardwood forest type at the lowest level (see figure 1 for the nested tree depiction).

---

8 100 km (representing an approximate 60-minute commute time) from FIA plots) was assumed as the threshold distance and varying this distance did not substantially affect the sign and magnitude of the estimated coefficients of the gravity index and other variables.

9 Kline et al (2001) used a similar formulation of gravity index but with different exponents on the population and distance components of the index and they used three cities with population greater than 5000 and greatest urban influence based on their gravity index on each FIA plot.

The reference category[10] in (8) was harvest and in (9) forest. In (10) the reference category was hardwood (for PGI) and no change in forest type for the variable initial forest types (SW, HW or MX) respectively. The results are summarized in table 2.

Table 2. Nested logit parameter estimates along with their standard errors and odds ratio for the three-level landowner decision set depicted in figure 1

Variable	Coefficient	Standard Error	Odds Ratio	t-statistic
No Harvest Vs. Harvest				
SW × C_{NH}	–0.5714	0.059*	0.56	–9.61
SLOPE × C_{NH}	0.0217	0.002*	1.02	10.12
VOL × C_{NH}	0.0009	0.00003*	1.00	32.31
Non-Forest Vs. Forest				
PGI × C_{NF}	0.0011	0.0005**	1.00	2.16
RET × C_{NF}	–0.0172	0.004*	0.98	–4.40
PD × C_{NF}	0.0019	0.0005*	1.00	3.99
INC × C_{NF}	–0.4266	0.0397*	0.65	–10.75
Forest Type				
MX × C_{SW}	–1.9815	0.1225*	0.14	–16.17
MX × C_{HW}	-2.3806	0.1032*	0.09	–23.06
HW × C_{SW}	–0.2388	0.1184**	0.79	–2.02
HW × C_{MX}	0.4010	0.0909*	1.49	4.41
SW × C_{MX}	–0.9931	0.0857*	0.37	–11.59
SW × C_{HW}	–1.2661	0.0874*	0.28	–14.48
PGI × C_{SW}	–0.0045	0.0005*	0.99	–8.72
PGI × C_{MX}	–0.002	0.0003*	0.99	–6.88
IV[e] (Forest)	0.87	0.1360*		6.42
IV[e] (Harvest)	0.85	0.1311*		6.48
Log likelihood	10096			
McFadden's LRI	0.39			
Observations	10383			

[a] C_{SW}, C_{MX}, C_{HW}, C_{NF}, C_{NH} represent the dummies for the choice alternatives softwood, mixed, hardwood, non-forest and no harvest respectively.

[e] IV were constrained to be the same at each nest level for model identification, moreover for degenerate branches the such as Non-forest and No harvest IVs cannot be identified (for detail see Hunt 2000), * $p < 0.10$, ** $p < 0.05$.

The nesting structure in figure 1, together with equations (4)-(6) and (8)-(10) can be used to formulate appropriate log likelihood function to estimate the parameters of the model. The nested logit model was estimated using full information maximum likelihood estimation in SAS 9.1. Maddala (1983, page 73) states that if the IV parameters lie outside the range of zero to one then this should be considered as evidence for a specification error and warrants re-examination of the model. Further McFadden (1981) states that if the dissimilarity coefficients (IV coefficients) are larger than 0 and not statistically larger than 1, it can be concluded that the nested model is consistent with stochastic utility maximization. The results support the choice of a nested logit model, over a more restrictive multinomial logit model that does not allow for correlation within nests. The estimated maximum likelihood nested

[10] With all the explanatory variables being characteristics of the FIA plot and not the alternative land use choices we used interactions of each variable with the dummy of choice alternatives and hence had to remove a particular choice and make it as a reference base for model identification.

logit model had a reasonable fit with McFadden likelihood ratio index statistic (pseudo-R2) being 0.39.

5.1. No Harvest relative to Harvest

SW, representing the initial forest type as pine, had the expected negative sign and indicates less likelihood of no harvest of a pine plot. In other words there is a greater likelihood that a pine plot will be harvested relative to hardwood or mixed plot.

SLOPE had the expected positive sign with a statistically significant parameter estimate which indicates that with an increase in slope there is a greater likelihood of no harvest due to a possible hindrance to accessibility of logging equipment and associated increase in harvesting cost. Moreover steep slope also constrains excessive harvests to prevent erosion. The result is consistent with previous studies (Wear & Flamm 1993).

VOL, denoting the average volume per acre in cubic feet for the FIA plot had a positive sign. This implies that higher the volume the less likely it will be harvested. This is contrary to our expectation that greater average volume would lead to a greater probability of harvest. A close examination[11] reveals that most of the harvests took place in the softwood plantation type, which typically has lower average volume in comparison to the hardwood plots.

5.2. Non-Forest Relative to Forest

The population gravity index (PGI), representing the developmental pressure on forestland, had a statistically significant positive coefficient indicating that with an increase in PGI there is a greater likelihood of forestland conversion to non-forest use. This is an expected result since in general demand for developed land near the population centers with higher PGI is high in comparison to the demand for forests. This result is consistent with findings by Polyakov and Zhang (2008). Researchers have found other measures of urbanization like increase in population density (Nagubadi & Zhang 2005) and decrease in distance from the center of the county to the nearest city (Ahn et al 2002) to favor an increasing non-forest share of land.

INC had a statistically significant negative coefficient suggesting that counties with higher real per capita income are more likely to maintain their forests, with less inclination for conversion to non-forest use, ceteris paribus. This is contrary to the expectation of a casual observer and inferences drawn from previous research (Zhang & Nagubadi 2005). The intuitive explanation could be that with an increase in income the landowner may perceive returns from consumptive uses (aesthetics, amenities) of his forestland as higher in comparison to the returns from conversion to a developed use (intuitively somewhat like an environmental Kuznets curve).

[11] Separate models had to be estimated which could include interaction terms of VOL and the initial forest type keeping the VOL main effect for each forest type due to collinearity problem and results showed the coefficient of the interaction term of pine with the average volume as negative while that of the hardwood and mixed as positive.

RET, denoting the total amount of removals of growing stock from all FIA plots within a county adjusted for the difference in county land area in cubic feet per acre has a negative coefficient and is statistically different from 0 at the1% level of significance. This result is consistent with our expectation, since counties with higher timber removals represent the timber basket of the state and have less likelihood of forestland conversion to other uses. This result is consistent with the Ricardo-Thünnen land rent theory of land use change that proposes that land is put to the land use alternative that provides the highest land rent. Positive forest use returns (denoted by higher RET) are expected to decrease the likelihood of forest conversion to non-forest use.

PD had a statistically significant positive coefficient reflecting the increased likelihood that a plot will be converted to non-forest use when there is an increase in demand for land for residential purposes, a result consistent with past studies on land use change (Wear et. al 1998, Nagubadi & Zhang 2005).

5.3. Forest Types

The negative parameter estimate for five out of the six (except HW × CMX)[12] initial forest type variables indicates less likelihood of a forest type transition from one type to another relative to its likelihood of remaining as the same type reflects the costs of conversion constraints (Alig & Butler 2004). However the positive estimate for hardwood plots to be regenerated into a mixed type (HW × CMX) following harvest indicates a contrary result. This may be possible since depending on the FIA classification[13] of a forest type a stand classified as hardwood could be retyped as a mixed type in the subsequent census. Zhou et al. (2003) found a significant percentage of FIA plots in the South (upland hardwood), which were not harvested, transitioned to a mixed type in the subsequent survey.

The negative significant parameter estimate for PGI × CSW and PGI × CMX reveals that landowners are not willing to manage forest intensively, and, in particular, invest in (softwood) plantations in a proximity to locations with growing population, because of the higher chance of development in the near future. Furthermore, forests located near populated places are more likely to be managed for amenity values, therefore regeneration of hardwoods is preferable in such places. This corresponds with findings of Wear et al. (1999); Munn et al. (2002); Polyakov, Majumdar and Teeter (2008) and Polyakov and Zhang (2008).

6. CONCLUSION

The nested logit model seems to be an appropriate choice for investigating the discrete choice behavior of the private forest landowner. It is superior to multinomial logit, an econometric technique widely used to model land use, and allows for correlation of the error terms within a nest of similar choices. To our knowledge, application of the nested logit

---

[12] C_{SW}, C_{MX}, C_{HW}, C_{NF}, C_{NH} refer to the choice alternatives: softwood, mixed, hardwood, non-forest and no harvest respectively.

[13] A classification of forest land is commonly based upon, and named for, the tree species that forms the plurality of live-tree stocking.

technique to analyze the forest harvesting decision by the landowner has not been considered previously. Our results show that the initial forest type and population gravity index are significant variables in explaining the variation in type transition. Consistent with previous research findings, population gravity index, a proxy for the anthropogenic influence, favored forestland conversion to non-forest use.

The probability that a forest plot will be converted to non-forest at the mean of all the explanatory variables in the model is 0.02. In the softwood, mixed and hardwood forest types those probabilities increased to 0.05, 0.17 and 0.06 following harvest. The probability of no harvest at the mean of the variables was 0.7. In summary, given the 21.7 million acres of private timberland (Hartsell & Brown 2002) our model projects 434,000 acres to be converted from forest to non-forest use over a period of the next 10 years. For the same period, the acreage of non-harvested forest plots is projected to be 15.19 million acres with 1,085,000 acres, 3,689,000 acres and 1,302,000 acres of harvested timberland projected to be regenerated as softwood, mixed and hardwood forest types respectively. These results are consistent and can be used for short-term predictions.

Our findings suggest that income levels are more important in projecting future conversions of forestland than the population parameters such as population density or population gravity index. We find that while increasing population pressures have a positive correlation with forestland conversion into alternative uses, communities with increase in real per capita income level are more likely to retain their forest. Policy makers and modelers interested in forestland use change should expend greater effort in obtaining accurate income projections than simple population projections. Forests are primarily privately owned in the southern U.S. and provide a multitude of environmental services and amenities along with timber and other non-timber forest product consumables. Alterations of management practices by private landowners have a profound effect on the ability of these forests to continue providing these products and services in the future on a sustainable basis. Increasing population and income levels have opposite effects on conversion of forestland to other developed land or conversion of agricultural land back to forestland. Results from our model suggest that careful examinations of these parameters are important prerequisites for land use models.

From a policy standpoint identifying goals and objectives of forestland owners as guiding factors in the preparation of forest management plans could ensure that policy implementation are made more efficient. For example smaller landowners who have non-timber amenity consumptive objectives might be more inclined to extract biomass for bio-energy and other bio-based applications for environmental reasons rather than supplying to the traditional forest industry for economic reasons. Biomass utilization thus in a previously unutilized forests can help reduce overall costs of silvicultural operations and reduce other risks, such as from fire.

In conclusion we wish to acknowledge a couple of caveats and plausible improvements for future research. First of all, in our model we assume that the utility functions for all types of private landowners' viz., TIMOs, REITs, Forest Industry and NIPFs, have similar correlations amongst the functions' random components for all the choice alternatives. In reality these correlations might vary widely among decision makers and future studies can explore these differences by analyzing the data separately for private ownership categories like non-industrial private owners and industrial owners. Second, we do not make adjustments for the potential spatial correlation of the model error terms and assume that the systematic sampling of plots (roughly on a 3-mile grid with each plot approximately representing 6000 acres) minimizes the likelihood that plots fall under the same ownership. It would be interesting to test this hypothesis and perform regression analysis by taking into consideration the spatial distribution of forest plots to see if plots that are closer are similar than distant ones.

REFERENCES

Ahn, S., Plantinga, A.J., & Alig, R.J. (2000).Future forest land area: A comparison of econometric approaches. *Forest Science, 46(3),* 363-376.

Ahn, S., Plantinga, A.J., & Alig, R.J. (2002). Determinants and Projections of Land Use in the South Central United States. *Southern Journal of Applied Forestry, 26(2),* 78-84.

Alig, R.J., & Butler, B. (2004). Projecting large-scale area changes in land use and land cover for terrestrial carbon analyses. *Environmental Management, 33(4),* 443–456.

Alig, R.J. (1986). Econometric analysis of the factors influencing forest acreage trends in the Southeast. *Forest Science, 32(1),* 119-134.

Alig, R.J., & Healy, R.G. (1987). Urban and built-up land area changes in the United States: An empirical investigation of determinants. *Land Economics, 63(3),* 215-226.

Barlow, S.A., Munn, I.A., Cleaves, D.A., & Evans, D.L. (1998). The effect of urban sprawl on timber harvesting: A look at two southern states. *Journal of Forestry, 96(12),* 10-14.

Bockstael, N.E. (1996). Modeling economics and ecology: the importance of a spatial perspective. *American Journal of Agricultural Economics, 78,* 1168-1180.

Carrión-Flores, C., & Irwin, E.G. (2004). Determinants of Residential Land Use Conversion and Sprawl at the Rural-Urban Fringe. *American Journal of Agricultural Economics, 86(4),* 889–904.

Chomitz, K.M., & Gray, D.A. (1996). Roads, land use, and deforestation: a spatial model applied to Belize. *The World Bank Economic Review, 10(3),* 487-512.

ESRI Data and Maps 2005. (http://www.esri.com/data/about/data_maps_media.htm).

Hardie, I.W., & Parks, P.J. (1997). Land use with heterogeneous land quality: an application of the area base model. *American Journal of Agricultural Economics, 79,* 299-310.

Hartsell, A.J., & Brown, M.J. (2002). Forest statistics for Alabama, 2000. USDA For. Serv., Southern Research Station Resource Bulletin SRS-67, Asheville, NC 28802.

Hunt, G.L. (2000). Alternative nested logit model structures and the special case of partial degeneracy. *Journal of Regional Science, 40(1),* 89-113.

Irwin, E.G., & Bockstael, N.E. (2002).Interacting agents, spatial externalities, and the endogenous evolution of residential land use patterns. *Journal of Economic Geography, 2(1),* 31–54.

Johnson, T.G., & Steppleton, C.D. (2007). Southern pulpwood production, 2005. Resource Bulletin SRS–116. Asheville, NC: U.S. Department of Agriculture Forest Service, Southern Research Station. 42 p.

Kline, J.D., Moses, A., & Alig, R.J. (2001). Integrating urbanization into landscape-level ecological assessments. *Ecosystems, 4(1),* 3-18.

Lubowski, R.N. (2002). Determinants of land-use transitions in the United States: econometric analysis of changes among the major land-use categories. Ph.D dissertation, Harvard University, Cambridge, Massachusetts. 266 p.

Lubowski, R.N., Plantinga, A.J., & Stavins, R.N. (2006). Land-use change and carbon sinks: Econometric estimation of the carbon sequestration supply function. *Journal of Environmental Economics and Management, 51(2),* 135–152.

Maddala, G.S. (1983). Limited-dependent and qualitative variables in econometrics. Cambridge University Press, Cambridge, United Kingdom. 401p.

Majumdar, I., Polyakov, M., & Teeter, L.D. (2005). Landuse change determinants in Alabama. P. 67-70 *in* Proc. of conf. on Emerging issues along urban/rural interfaces: Linking science and society, Laband, D.N. (eds.) Auburn, AL, August 2005.

McFadden, D. (1974). Conditional logit analysis of qualitative choice behavior. P. 105-142 in Frontiers in Econometrics, (P. Zarembka eds.) Academic Press.

McFadden, D. (1981). Econometric models of probabilistic choice. in C.F. Manski and D. McFadden (eds.), Structural Analysis of Discrete Data with Econometric Application, MIT Press: Cambridge, MA.

Munn, I.A. & Evans, D.L. (1998). The southern commercial timberland base: changes and projections. Presented at the first international conference on geospatial information in agriculture and forestry, June, Florida.

Munn, I.A., Barlow, S.A., Evans, D.L., & Cleaves, D. (2002). Urbanization's impact on timber harvesting in the south central United States. *Journal of Environmental Management 64(1),* 65-76.

Nagubadi, R.V. & Zhang, D. (2005). Determinants of timberland use by ownership and forest type in Alabama and Georgia. *Journal of Agricultural and Applied Economics, 37(1),* 173-186.

Parks, P.J., & Murray, B.C. (1994). Land attributes and land allocation: Nonindustrial forest use in the Pacific Northwest. *Forest Science, 40(3),* 558-575.

Polyakov, M., Majumdar, I., & Teeter, L. (2008). Spatial and temporal analysis of the anthropogenic effects on local diversity of forest trees. *Forest Ecology and Management, 255,* 1379-1387.

Polyakov, M., & Zhang, D. (2008). Population growth and land use dynamics along urban-rural gradient. *Journal of Agricultural and Applied Economics, 40(2),* 649-666.

Smith, W.B., Miles, P.D., Vissage, J.S., & Pugh, S.A. (2002). Forest resources of the United States. USDA For. Serv. Gen. Tech. Rep. NC-241. St. Paul, MN.

Turner, M.G., Wear, D.N., & Flamm, R.O. (1996). Land ownership and land-cover change in the Southern Appalachian Highlands and the Olympic Peninsula. *Ecological Applications, 6(4),* 1150-1172.

Veldkamp, A., & Fresco, L.O. (1996). CLUE: a conceptual model to study the conversion of land use and its effects. *Ecological Modeling, 85,* 253–270

Vitousek, P.M. (1994). Beyond global warming: ecology and global change. *Ecology, 75,* 1861-1876.

Wear, D.N., & Bolstad, P. (1998). Land-use changes in southern Appalachian landscapes: spatial analysis and forecast evaluation. *Ecosystems, 1,* 575-594.

Wear, D.N., & Flamm, R.O. (1993). Public and private forest disturbance regimes in the southern Appalachians. *Natural Resource Modeling, 7(4),* 379-397.

Wear, D.N., Liu, R., Foreman, J.M., & Michael, R.M. (1999). The effects of population growth on timber management and inventories in Virginia. *Forest Ecology and Management, 118,* 107–115.

Zhang, D., & Nagubadi, R.V. (2005). The influence of urbanization on timberland use by forest type in the Southern United States. *Forest Policy and Economics, 7,* 721-731.

Zhou, X., Mills, J.R., & Teeter, L.D. (2003). Modeling forest type transitions in the South Central region: Results from three methods. *Southern Journal of Applied Forestry, 25(3),* 190-197.

In: Global Change and Forestry
Editors: J. Gan, S. Grado and I. A. Munn
ISBN: 978-1-60876-262-0

Chapter 5

AN ANALYSIS OF FORESTLAND CHANGE ON THE NORTHERN CUMBERLAND PLATEAU: BRIDGING THE BROAD UNITS AND FINE UNITS DATASETS IN A LANDUSE MODEL[1]

Neelam C. Poudyal[1], Seong-Hoon Cho[2], Jeffery D. Strickland[3] and Donald G. Hodges[4*]

[1] Department of Forestry, Wildlife and Fisheries, The University of Tennessee, TN, USA, and Daniel B. Warnell School of Forestry and Natural Resources, University of Georgia, GA, USA
[2] Department of Agriculture Economics, The University of Tennessee, TN, USA
[3] Molpus Woodlands Group, Hattiesburgh, MS, USA
[4] Department of Forestry, Wildlife and Fisheries, The University of Tennessee, TN, USA

ABSTRACT

This paper analyzes the forest landuse change in the Cumberland Plateau of Tennessee, which is a rapidly developing rural landscape due to influx of retirees and second home owners. The model presented in this study investigates the factors of forestland conversion by bridging the broad socio-demographic and market force information measured at the census-block group level in conjunction with site-specific information of forest land use change measured at the satellite image pixel level. The results indicate that the surface terrains, resident's education level, market forces from urban centers in proximity, and distances to streams as well as major transportation corridors are likely to affect the probability of conversion of forestland into both agricultural and urban uses. Findings also suggest that the role of forest is changing from timber, fodder, and fiber toward amenity benefits on the Plateau.

---

[1] A version of this chapter first appeared in the International Journal of Ecological Economics and Statistics 10 (W08): 53-65.

[*] Email: dhodges2@ utk.edu, phone: (865) 974-2706, Natural Resource Policy Center, Department of Forestry, Wildlife and Fisheries, 247 Ellington Plant Science Building, The University of Tennessee, Knoxville, TN.

Keywords: Forest conversion; Market forces; Satellite imagery; Census data; Cumberland Plateau; Tennessee.

1. INTRODUCTION

Forests in the Southeastern United States are the major supply source of timber and fiber in the nation. Economic growth during the recent decades has accelerated urbanization around many metropolitan areas in this region. The growing competition among the major land use practices is indicated by a significant amount of land use change in recent years (Reynolds 2001, Alig et al. 2003). Projected increases in population, income, and other factors in the South have been expected to decrease the forest area by 6 % by the middle of this century (Alig et al. 2003).

Studies have shown that most of the urban development in the area has occurred at the expense of private forests. As almost 90% of the forests in the southern states are under private ownership, socio-demographic and economic factors are likely to affect forestland change. The loss of forests also translates into economic and ecological consequences. The ecological impacts include the loss of biodiversity and endangered species due to habitat loss or deterioration, while the economic impacts include the decrease in timber and outdoor recreation supply. The challenge for land use policy-makers and planners is balancing the land use in such a way that it adequately meets the increasing and competitive needs for the long-term supply of farm products, timber and non-timber benefits, and housings and industrial facilities. Moreover, it is also important for the landuse managers and environmental planners to understand the factors that cause forestland loss.

Earlier literature on land use analysis focused on identifying drivers of land use change. von Thunuen (1826) originally developed a basic model that observed the relationship among markets, production, spatial distance and the land allocation for various uses. On the other hand, a Bid-rent function model proposed by William Alonso, extends the von Thunen model and explains the land use and land rent as a function of distance to the central business district (Alonso 1964). Other more recent studies were based on the locational rent model of von Thunen for urban use and land rent model of David Ricardo for rural use, (i.e., Wear and Bolstad 1998, Miller and Platinga 1999, Hardie et al. 2000). These models emphasized the role of land quality, and the spatial location of land in relation to market centers in determining the land rent which arguably influences the land use decision of an individual landowner.

There have been a number of studies to explore the drivers of forest landuse change in the region. Earlier, Alig (1986) used the forest survey unit data to examine the determinants of forest acreage change in the Southeast. Using a system of equations for different land uses, he estimated the effects of change in demographic and economic factors on different land use types such as agriculture, forestry, and urban uses. Reynolds (2001) found that the population growth in the South is fueling the conversion of forestland into urban expansion. Another study by Ahn et al. (2002) projected the future forestland use based on the predicted value of stumpage price and population growth. A recent study by Zhang and Naguvadi (2005) observed the influence of urbanization on timberland in the South. They calculated the elasticity of change in different types of forests (hardwood, softwood, etc.) in response to demographic, economic, and urban development-related variables. Nagubadi and Zhang (2005) analyzed the factors of forestland change by species and ownership type (private, public, etc.) in Alabama and Georgia.

Morita et al. (1997), Alig et al. (1988), Miller and Platinga (1999), and Ahn et al. (2002), observed the share of various land use types as a single owner's problem to allocate the land into different uses based on expected return. Another model used by Seto and Kaufman (2003) utilized a linear regression on panel data of land use change and socio-economic determinants. Miller and Platinga (1999) applied the maximum entropy model that is derived from the traditional logistic regression to apply on aggregate data of land use share at county level. Bhattarai et al. (2006) used a multinomial logistic framework in watershed level data; in which the land decision in alternative uses was explained as a function of socio-economic characeteristics and spatial factors.

Land use change models in earlier studies have been applied at the spatial scale of both broad and fine units. The models using broad unit of land use examine the patterns of land use change from a macro viewpoint. These models generally use states, counties, or county groupings as units to observe how the socioeconomic and biophysical factors influence the land use allocations. The socioeconomic factors include those reflecting the demographic, social, and economic pattern. The biophysical factors include those capturing the biological and physical characteristics (e.g., vegetation, slope, aspect, and elevation). A group of studies apply broad unit models of analysis in aggregated data (Alig 1986, Hardie and Parks 1997, Miller and Platinga 1999, Hardie et al. 2000, Zhang and Nagubadi 2005). The models of fine units on the other hand, provide analysis of spatially explicit land use decisions. These models estimate the direct influence of site-specific factors of land use at a fine resolution. For example, the influence of road construction on land development (e.g., Dale et al. 1993, Chomitz and Gray 1996, Nelson and Hellerstein 1997) and the influence of spatial location, topographic features and ownership (e.g., Spies et al. 1994, Turner et al. 1996) are analyzed in this framework. Although each type of model independently serves as a valuable function, both have limitations as well. The broad-scale analysis cannot capture the information in a spatially explicit framework, whereas micro-scale analysis fails to detect the changes taking place at the broader physical and social scale.

This paper attempts to analyze the forest land use change by bridging the broad socioeconomic information at the census-block group level in conjunction with site-specific information of forest conversion measured at the pixel level. The study focuses on the classical von Thunen model and finds determinants of forestland conversion to other two non-forest use types of urban and agriculture. These two types are the major land use transitions in the study area. A pixel of 30 meter by 30 meter size, which is the smallest homogenous unit mapped in the remotely sensed satellite imagery of land cover, was used in the model. The census-block group was used as a neighborhood boundary reflecting socio-economic characteristics of neighborhood.

2. METHODS

2.1. Empirical Model

This study used a multinomial logistic regression model in which the probability of forestland conversion to urban or agricultural use was estimated in relation to various socio-economic characteristics of the neighborhood, and bio-physical characteristics of the location. The dependent variable was a discrete choice of no conversion, a conversion of forestland to urban use, or a conversion of forestland to agriculture use between 1992 and 2000. Among these three alternatives, the category of no conversion was considered as a base in relation with the probability of conversion into other category (Greene 2003, p. 720). The probability model for conversion of forest into other use can be estimated by using the equation 1

$$\Pr ob(Y_i = j / X_i) = P_{ij} = \frac{\exp(X_i'\beta_j)}{1+\sum_{j=1}^{2}\exp(X_i'\beta_j)} \qquad for\ j=1,2 \tag{1}$$

where Y_i was a vector of discrete choice representing the type of conversion and X_i represented the vectors of explanatory variables for the pixel i. The term P_{ij} was the probability of forest in ith pixel to be converted into jth land use type (j=0,1,2). For example, if a forested pixel in 1992 remained forested in 2000, the dependent variable for that particular pixel took the value of 0, and β represented parameters to be estimated. Similarly, the probability model for this category was estimated using equation 2.

$$\Pr ob(Y_i = 0 / X_i) = P_{i0} = \frac{1}{1+\sum_{j=1}^{2}\exp(X_i'\beta_j)} \tag{2}$$

Since, the dependent variable in this model was a log of odd ratio, direct interpretation of regression coefficient was difficult. The marginal effects of the individual characteristics needed to be estimated. Equation 3 below is the marginal effect obtained by partially differentiating the equation 2 with respect to x_i.

$$\frac{\partial P_j}{\partial x_i} = P_j\left[\beta_j - \sum_{k=0}^{j} P_k\beta_k\right] = P_j(\beta_j - \overline{\beta}) \tag{3}$$

Because of the complexity in its function, the marginal effects are rarely used for interpretation of individual effect in logistic regression (Long 1997). Instead, following Zhang and Nagubadi (2005), the elasticity was calculated using equation 4.

$$e_{ijx} = \frac{\partial y_{ij}}{\partial x_{ij}} \cdot \frac{x_i}{y_i} = (\beta_{jx} - \sum_{k=0}^{2} y_{ij} \beta_{jx}) x_{ij} \qquad (4)$$

where e_{ijx} was the elasticity of forest conversion in pixel i to land use j, with respect to change in value of independent variable x. This elasticity can be used as percentage change in the probability of forest conversion into particular non-forest use type because of a percentage unit change in the respective explanatory factor.

A set of explanatory variables included socio-demographic variables and bio-phyical variables. Socio-demographic variables included the change in population density and education level of residents. The population density was expected to capture the pressures of demand of rural land for urban use. The education level was included to test the role of education in forest loss. Further, the influence of market factors was measured in terms of average household income and change in number of building construction. The household income variable was expected to capture the effect of change in economic condition of neighborhood on forest loss. For instance, the change in level of income can influence the landowner's behavior and priority with the land use decision. Gunter et al. (2002) also included education and income in their land use model. The change in number of building construction at the census-block group level was included to test if new structures induce more land conversion.

Also included among the measures of market forces was the gravity index, which captures the regional urban influence in the area. The gravity index captured the combined economic influence of urban population living in the major metropolitan cities nearby (Reilly 1929). The gravity index was used in this model considers the change in average influence from four nearby urban centers including Knoxville, Oak Ridge, Cookeville, and Crossville. The index can be calculated using equation 5 (Kline et al. 2001).

$$Gravity\ Index = \sum_{m=1}^{4} \frac{(population_m)^{0.5}}{distance_m} \qquad (5)$$

where m represents the metropolitan areas considered in the model to measure their urban influence. The indexes used in this analysis were computed by Strickland (2003).

Similarly, the variables capturing biophysical characteristics of the area including the average slope at the pixel location, and distances of pixel to the nearest road, water stream, and Knoxville downtown were included as explanatory variables in the model. Turner et al. (1996) analyzed the effects of slope, elevation, and distance to roads on land use change. The slope is a degree of surface terrain around the pixel that serves as a proxy measure of the land quality for intensive land use practices such as agriculture farming. The distance to road captures the accessibility of locations. Following the von Thunen model of land rent for urban use, the distance to Knoxville city was included to capture the effect of proximity to the central business district. This can capture the effect of market-related influences as the easier access to markets for forest and farm products may influence the land use pattern in the area. The proximity to streams or river may create amenity values for residential housing market whereas it could induce negative externality for other uses such as agriculture because the conservation policies often restrict land use near water bodies. Definitions and statistical descriptions of explanatory variables are reported in table 1.

Table 1. Variable definition and descriptive statistics

Variables	Definition	Mean	Std. Dev.
POP	Change in population density in census block group (person/sq. km.)	4.6	40.23
ED	Change in percent of residents with college education in census block group (%)	8.97	8.18
INCP	Change in average household income in census block group (%)	47.60	17.78
BLDG	Change in construction in the census block group (%)	88.39	108.90
SLOPE	Average slope at the pixel location (degree)	6.4	6.09
ROAD	Distance to the nearest road from the pixel location (km)	0.32	0.35
WATER	Distance to the nearest stream from the pixel location (km)	0.31	0.22
KNOX	Distance to the Knoxville city center from the pixel location (km)	91.74	84.50
GINDEX	Square root of city population divided by the distance (averaged for 4 nearest cities)	13.10	52.71

2.2. Study Area and Data

Cumberland and Morgan Counties on the Cumberland Plateau of Tennessee were chosen for this research as a case study for three reasons. First, there is a growing concern of forest loss and fragmentation of large tracts of contiguous forest in the area. Second, forests on the plateau watershed serve as the headwaters for many of the downstream rivers such as the Tennessee, Elk, and Duck Rivers. Therefore, the water quality in these rivers heavily depends on the conservation of forests on the Plateau. Third, the land resources in the region are under increasing market influences in recent decades. Those included the demand for timber, fiber and other forest products from the private forests, and also the demand for low-density housing development for retirees and second home owners. US Census records show that the number of retiree households in these two counties increased by 22.8 % between 1990 and 2000. Forestland accounts for 79% of the total land area, whereas about 4% of total forestland has been lost between 1992 and 2000 (Strikland 2003). The trend of change indicates that the forests in this area have been experiencing high development pressure with an accelerating rate of forest conversion to urban land uses.

This study uses two types of dataset. Remotely sensed satellite imageries were used to measure the forestland change at pixel level, and to measure the biophysical information such as slope, distances, etc. US Census datasets were used to measure the socio-economic characteristics of the neighborhood, in which the pixel was located. GIS datasets used included the land cover dataset of 1992 and 2000, layer of roads, stream and city center, digital elevation model (DEM). US Census dataset used included census-block group (CBG) information for 1992 and 2000. Specifically, the forest land conversion between 1992 and

2000 was taken from a forest change map classified and developed by Stickland (2003) using Landsat thematic mapper (TM) data of 1992 and Landsat enhanced thematic mapper plus (ETM+) data of 2000. Altogether, 806 pixel samples were randomly picked with the corresponding location in the forest change map indicating no conversion, conversion from forestland to urban land, and conversion from forest to agriculture land between 1992 and 2000. The GIS data layers of road, stream and city center were used to calculate proximities of those sample pixels to nearest road, city center, and stream. The source of these data layers was the TIGER 2000 data from the U.S. Census Bureau. The DEM was obtained from US Geological Survey. The demographic and economic data for 1990 and 2000 were obtained from the US Census Bureau. It should be noted that census dataset of 1990 were used as proxy for 1992. The census-block group level data for population density, percentage of population with college degree, average household income, and number of building constructed were obtained from the census data and combined with the categorical variable of forestland conversion in the corresponding pixel location. Note that the analysis of forest conversion in this study excludes non-private forests because the land use decision in public forests and protection zones might be influenced by some other factors and decision criteria (Zhang and Naguvadi 2005).

3. Results

The results from the multinomial logit regression are presented in table 2. While we have reported the coefficients, marginal effects, and elasticity for each explanatory variables, our discussion focuses on the interpretation of the elasticity. The result indicates that the different factors are responsible for forestland conversion to urban and agriculture uses. All variables except change in household income were found to influence the conversions. The surface terrain, education level of residents, and proximity to streams and roads were found significantly related to forestland conversion to both agriculture and urban uses at the level of 10%.

Population density was found to have a negative effect on the forest conversion to urban use whereas it was not a significant factor in the conversion to agriculture. The elasticity of -0.024 indicates that a 10% increase in population density decreases the probability of conversion into urban use by 0.2%, and it increases the probability of forest retention by 0.06 %, ceteris paribus. The result of the negative effect of population density on forest conversion seems to be counter-intuitive but it can be explained by the fact that the area is experiencing housing growth for retiree's population who appreciate amenity values of forestland. The retirees may prefer low-density residential development, and probably oppose deforestation in their neighborssurroundings. Although it contradicts the growth theory, Gottfried et al. (2005) also reported increasing low-density housing development in this region.

Table 2. Parameter estimation from the multinomial logistic regression (N = 806)

	Urban (y = 1)			Agriculture (y = 2)			Forest (y =0)	
Variable	Coeff.	ME	Elasticity	Coeff.	ME	Elasticity	ME	Elasticity
Constant	0.618 (0.581)	---	---	-0.222 (0.423)	---	---		
POP	-0.008** (0.003)	-0.001** (0.000)	-0.024** (0.012)	-0.005* (0.003)	0.000 (0.000)	-0.019 (0.013)	0.001** (0.000)	0.006** (0.003)
ED	-0.041** (0.019)	-0.003* (0.002)	-0.247* (0.128)	-0.045** (0.021)	-0.004* (0.002)	-0.278* (0.146)	0.007*** (0.003)	0.067*** (0.024)
INCP	0.000 (0.007)	0.000 (0.001)	0.048 (0.273)	-0.010 (0.007)	-0.001 (0.001)	-0.447 (0.306)	0.001 (0.001)	0.049 (0.052)
BLDG	-0.002 (0.002)	0.000 (0.000)	-0.103 (0.102)	-0.004** (0.002)	0.000** (0.000)	-0.231** (0.118)	0.000** (0.000)	0.042** (0.020)
SLOPE	-0.209*** (0.035)	-0.019*** (0.003)	-1.095*** (0.206)	-0.114*** (0.031)	-0.008*** (0.003)	-0.502*** (0.178)	0.026*** (0.004)	0.207*** (0.028)
ROAD	-0.442 (0.349)	-0.031 (0.033)	-0.097 (0.102)	-1.043** (0.450)	-0.088** (0.038)	-0.296** (0.134)	0.119** (0.048)	0.049** (0.020)
WATER	0.961** (0.452)	0.079* (0.042)	0.233* (0.124)	1.157** (0.494)	0.092** (0.043)	0.295** (0.137)	-0.171*** (0.060)	-0.067*** (0.024)
KNOX	-0.011** (0.005)	-0.001** (0.000)	-0.941** (0.438)	0.000 (0.000)	0.000 (0.000)	0.120 (0.125)	0.001** (0.000)	0.110** (0.050)
GINDEX	0.004** (0.002)	0.000** (0.000)	0.051** (0.023)	0.000 (0.000)	0.000 (0.000)	-0.052 (0.067)	0.000 (0.000)	0.000 (0.008)

Note: ***, ** and * indicate the significance of parameter at 1 %, 5% and 10% respectively.

The education level of residents was found to have negative effects on the forestland conversion to urban and agriculture uses. The elasticity suggests that a 10% increase in population of college graduates in the area is likely to decrease the probability of forest conversion into urban and agriculture use by 2.4 % and 2.7% respectively. This is consistent with the finding that the education lowers the probability of deforestation (Godoy and et al. 1998). This may be related with the fact that the growing retiree's population with higher education is likely to have higher amenity value for green space in the community. The change in construction of new buildings was negatively related with the forest conversion to agriculture use. The elasticity of -0.231 suggests that a 10% increase in building construction in the surroundings decreases the probability of forest conversion to agriculture use by 2.3%. This is not surprising because the forestland in the neighborhood with higher demand for housing may not suit for agricultural use.

The relative surface terrain measured by slope in degree was found negatively related to forest conversion to both agricultural and urban uses. The level of statistical significance and the elasticity suggest that the slope is the single-most important factor in the deforestation. Results suggest that a 10% increase in slope lowers the probability of forest conversion to urban and agriculture uses by 11% and 5% respectively. As the slope is closely associated with the productivity of land, this finding verifies the fact that the land use decision depends on the productivity of land, and is consistent with the classical land rent model of David Ricardo. The land with higher slope is less likely to be suitable for the use of either agriculture or urban development. The elasticity of 0.207 for the base category indicates that a 10% increase in slope of a forested area is likely to increase the retention of forest by 2.07%.

A decrease in distance of the pixel location to the nearest road increased the probability of forest conversion to both urban and agriculture uses. The elasticity of forest conversion to agriculture use with respect to the distance the nearest road was found to be -0.296. The elasticity for the conversion to urban use was not statistically significant, however. The insignificance may be explained by the fact that the real estate builders in the area build the access roads as they develop new housings. The result indicates that a 10 % decrease in the distance from pixel location to the nearest road increases the probability of forest conversion to agriculture by 2.96%. The result is consistent with the von Thunen's theory of land rent in which accessibility of a site is believed to influence land use decision (Miller and Platinga 1999, Hardie et al. 2000). The elasticity of 0.049 for no conversion suggests that the greater the distance of pixel from the road, the higher the chance for remaining the pixel as forest.

Similarly, the distance to stream was a significant factor for the forest conversion to both agricultural and urban land uses. The result suggests that an increase in distance from the stream increases the probability of forest conversion for the both land uses. The elasticity of 0.233 and 0.295 are reported for the conversion to urban and agriculture use respectively. Potential flood hazards and bank erosion may be reasons for the lower probability of the conversion to urban use for the pixels closer to the stream. The lower conversion to agricultural use for the pixels that are closer to the stream makes sense since land use restrictions along streamside for conservation of riparian buffers and erosion control are common in the area.

The estimated elasticity of -0.941 suggests that a 10% decrease in distance of pixel location from the Knoxville city center increases the probability of forest conversion into urban use by 9.4%. Given that the Knoxville is the nearest local retail center, the accessibility to this regional employment and business hub increases the probability of forestland to convert for urban use. The proximity to the Knoxville was not a significant factor on the conversion of forestland to agriculture use. Further, the gravity index, which was used to measure the spatial influence of urbanization from four nearby urban areas, was also found related to forest conversion to urban use significantly at the level of 5%. It was not significant for the conversion into agriculture use. This reveals that the regional urban influence in the area is toward urban development not increase in agriculture demand.

4. CONCLUSION

This study examined the forest land use change in relation to various socio-demographic and market forces on the Cumberland Plateau, Tennessee. In particular, we attempted to bridge the broad and fine scales of land use analysis by examining the socio-demographic and market forces information at the census-block group level in conjunction with site-specific information of land use change measured at the pixel level. The result indicates that the changes in socio-demographic structure and market forces do have a great deal of effect on the land use pattern in a rural landscape. Forest conversion into urban and agriculture are determined by different factors, in addition to some common factors. The common factors affecting both kinds of conversion include surface terrain, resident's education level, and proximity to stream that are likely to affect the probability of conversion of forest into both agricultural and urban uses. Furthermore, this paper confirms the arguments of literatures including Gottfried et al. (2005) that the forested lands of Cumberland plateaus is experiencing the low-density residential development. This may be an issue of concern because the low-density development may result into rural-urban interface areas, where human-wildlife conflict may be an issue of concern in long run.

Two caveats of this study should be noted. First, we could not include some important variables such as property tax rate, land rent (e.g., per land unit return from different landuse types) because such data were not available at the census block group level. Second, the pixel samples were taken from two counties only. A broader geographical coverage in sampling would have given implications that could be potentially generalized for regions other than the Plateau.

Findings suggest that the role of forests may be changing from source of timber, fodder, and fiber towards aesthetic beauty and amenity benefits. Such benefits are more concerns of urban life and of people who migrated from urban to sub-rural life in areas such as the Cumberland Plateau that experienced a significant increase in inter-state immigrant retiree population. The findings from this study have important policy implications in real estate planning, farmland conservation, and integrated regional development. As the estimated model suggests, policy makers could adopt programs, which encourage or discourage urban and agricultural development in their community depending on the characteristics or preferences of the residents. For example, policy makers could discourage urban and agricultural developments in the community where the highly education educated residents value more open spaces in their neighborhood. Further, the elasticity estimated in the forest conversion model could also be useful to predict future forest conversion for both urban and farmland uses, given the projected value of other variables. Our study also proved that theoretically consistent estimates could be achieved by carefully combining the datasets of two different spatial scales. This would prove useful in refining the landuse models, particularly for areas, where datasets at the same spatial scales are not available.

Two caveats of this study should be noted. First, we could not include some important variables such as property tax rate, land rent (e.g., per land unit return from different landuse types) because such data were not available at the census block group level. Second, the pixel samples were taken from two counties only. A broader geographical coverage in sampling would have given implications that could be potentially generalized for regions other than the Plateau.

ACKNOWLEDGMENTS

The authors are thankful to the participants of Land Use Planning: Understanding Land Conversion Session, in 86th annual conference of the Society of American foresters Foresters (SAF), 24-27 October 2006, Pittsburgh, PA.

REFERENCES

Agresti, A. (1996). *An Introduction to Categorical Data Analysis*. New York. Wiley-Interscience.

Ahn S., Platinga, A. J. and Alig, R. (2002). Determinants and projections of land use in the south central United States. *Southern Journal of Applied Forestry, 26 (2),* 78-84.

Alig, R. J. (1986). Econometric analysis of the factors influencing forest acreage trends in the Southeast. *Forest Science 32 (1)*, 119-134.

Alig, R. J., Platinga, A. J., Ahn, S. and Kline, J. D. (2003). *Land Use Changes Involving Forestry in the United States: 1952 to 1997, With Projections to 2050.* General Report (PNW-GTR-587). Portland, OR. USDA Forest Service, Pacific Northwest Research Station.

Alig, R. J., White, F. C., and Murray, B. C., (1988). *Economic Factors influencing Land Use Changes in the South-Central United States.* Research Paper (SE-272). Southeastern Forest Experiment Station.

Alonso, W. (1964). *Location and Land Use*. Cambridge, MA. Harvard University Press.

Bhattarai, G. R., Zhang, D. and Hatch, U. (2006). Measuring the socioeconomic influences on land use distribution at the watershed level: A multinomial logit analysis. *International Journal of Ecology and Development, 4 (Winter),* 23-34.

Chomitz, K. M. and Gray, D. A. (1996). Roads, land use, and deforestation: A spatial model applied to Belize. *The World Bank Economic Review, 10 (3),* 487-512.

Dale, V. H., O'Neill, R. V. Pedlowski, M., and Southworth, F. (1993). Causes and effects of land use change in central Rhondonia, Brazil. *Photogrammetric Engineering and Remote Sensing, 59 (6),* 997-1005.

Godoy, R., Groff, S. and O'Neill, K. 1998. The role of education in neotropical deforestation: household evidence from Amerindians in Honduras, *Human Ecology, 26 (4),* 649-675.

Gottfried, R., William, D., Lane, M., Butler, C., Lemoine, D. and Willis. K. (2005). Modeling forest change and its ecological consequences in the southern Cumberland Plateau. In: D.

Laband (Ed). *Emerging Issues along Urban Rural Interface: Linking Science and Society*. Conference Proceedings. Atlanta, Georgia, March, 2005.

Greene, W. (2003). *Econometric Analysis*. Fifth Edition. New York, Pearson Education Inc.

Gunter, J. T., Hodges, D. G., Swalm, C. M., and Regens, J. L. (2000). Predicting the urbanization of pine and mixed forest in Saint Tammany Parish, Louisiana. *Photogrammetric Engineering and Remote Sensing, 66 (12),* 1469-76.

Hardie, I. A. and Parks, P. J. (1997). Land use with heterogeneous land quality: an application of an area base model. *American Journal of Agriculture Economics, 1997,* 299-310.

Hardie, I., Parks, P., Gottleib, P., and Wear, D. (2000). Responsiveness of rural and urban land uses to land rent determinants in the U.S.. South. *Land Economics, 76 (4),* 659-673.

Kline, J. D., Moses, A. and Alig, R. J. (2001). Integrating urbanization into landscape-level, ecological assessments. *Ecosystems, 1 (4),* 3-18.

Long, S. (1997). *Regression Models for Categorical and Limited Dependent Variables*. Sage Publications, Inc.

Miller, D. J. and Platinga, A. J. (1999). Modeling land use decisions with aggregate data. *American Journal of Agricultural Economics, 81,* 180-194.

Morita, H., Hoshino, S., Kagatsume, M. and Mizuno, K. (1997). *An Application of the Land use Change Model for the Japan Case Study Area.* Interim Report. IR-97-065/November. International Institute for Applied Systems Analysis. Austria.

Nagubadi, R. V., and Zhang, D. (2005). Determinants of timberland use by ownership and forest type in Alabama and Georgia. *Journal of Agricultural and Applied Economics, 37 (1),* 173-186.

Nelson, G. C. and Hellerstein, D. (1997). Do roads cause deforestation? using satellite images in econometric analysis of land use. *American Journal of Agricultural Economics, 78 (10, 80-88.*

Reilly, W. J. (1929). *Methods for the Study of Retail Relationships.* Bulletin Number 2944. Austin, Texas, The University of Texas.

Reynolds, J. E. (2001). Land Use Change and Competition in the South. *Journal of Agriculture and Applied Economics, 33 (2),* 271-281.

Seto, K. C., and Kaufmann, R. K. (2003). Modeling the Drivers of Urban Land use Change in the Pearl River Delta, China: Integrating Remote Sensing with Socioeconomic Data. *Land Economics, 79 (1),* 106-121.

Spies, T. A., Ripple, W. C. and Bradshaw, G. A. (1994). Dynamics and pattern of a managed coniferous forest landscape in Oregon. *Ecological Applications, 4,* 555-568.

Strikland, J. D. (2003). *Current Assessment and Future Prediction of Forest Cover Change in Cumberland Morgan Counties, Tennessee: A Modeling Technique*. Master's Thesis. Knoxville, TN: The University of Tennessee.

Thunen, J. H. von, (1826). *Der Isolierre Staat in Eziehung auf Landwirtschaft and Nationalokonmie*. Gustav Fisher, Stuffgart; translation by C. M. Wartenburg (1966) *The Isolated State*. Oxford University Press, Oxford.

Turner, M. G., Wear, D. N., Flamm, R. O. (1996). Land ownership and land cover change in the southern Appalachian Highlands and Olympic Peninsula. *Ecological Applications, 6,* 1150-1172.

Vesterby, M. and Heimlich, R. E. (1991). Land use and demographic change: results from fast-growth counties. *Land Economics, 67 (3),* 279-291.

Wear, D. N., and Bolstad, P. (1998). Land use changes in southern Appalachian landscapes: spatial analysis and forecast evaluation. *Ecosystems (1),* 575-594.

Wear, D. N., Turner, M. G., and Naiman, R. J. (1998). Land cover along an urban-rural gradient: implications for water quality. *Ecological Applications 8 (3),* 619-630.

Zhang, D. and Nagubadi, R. V. (2005). The influence of urbanization on timberland use by forest type in the southern United States. *Forest Policy and Economics, 7 (5),* 721-731.

In: Global Change and Forestry
Editors: J. Gan, S. Grado and I. A. Munn
ISBN: 978-1-60876-262-0

Chapter 6

POPULATION, POLICY, AND PROFIT IN TROPICAL DEFORESTATION IN XISHUANGBANNA, CHINA

***Yaoqi Zhang**[*1], **Bin Zheng**[1], **Youxin Ma**[2], **Hongmei Li**[2], **Wenjun Liu**[2] **and Min Cao**[2]*

[1] School of Forestry and Wildlife Sciences, Auburn University, AL, USA
[2] Xishuangbanna Tropical Botanical Garden, Chinese Academy of Sciences, Yunnan 666303, China

ABSTRACT

Deforestation was traditionally driven by local factors such as shifting cultivation practiced by the local people who lived in the area. This situation has been changing as public policy, globalization, and trade are emerging as more important factors. This chapter uses Xishuangbanna, China as a case study to investigate changes of the driving factors in the past century. Prior to the 1950s, the tropical forests in Xishuangbanna were still largely intact and deforestation was slow and primarily driven by shifting cultivation by relatively small and slowly growing local ethnic minorities for foods. From the 1950s to the 1970s, the population grew rapidly by immigration driven by the public policy for rubber tree plantations. Since the 1980s, globalization and free markets have become dominant forces with rising prices driven by market and land tenure reforms promoting an increase in for-profit rubber plantations at the expense of natural forests.

Keywords: deforestation, economic development, land use and land cover change, Mekong River, rubber plantations, shifting cultivation.

1. INTRODUCTION

Over time the driving forces behind tropical deforestation have been changing. To combat deforestation we should understand these changes. If we look at the history of tropical

[*] Email: Yaoqi.zhang@auburn.edu, School of Forestry and Wildlife Sciences, Auburn University, Auburn, AL 36830.

deforestation, initially it was extensive subsistence agriculture, especially shifting cultivation for foods, practiced by the local population that had the most significant impacts on forests. If the population is low, the forests can be largely recovered and maintained. When population densities were low, the slash-and-burn method provided a sensible and low cost way to obtain vital produce from otherwise low agricultural yield regions (Moran and Ostrom 2005). Alcorn (1990) called swidden farming as "managed deforestation."

Population growth can be dramatic if large migration occurs from the outside. Road development encourages movement of people to frontier areas. The growing population would demand more food and need to convert more forest land to cultivation. Therefore, it is not surprise that it was found that population growth was the leading cause of tropical deforestation worldwide (Palo 1994, Marcoux 2000, Carr 2004, Uusivuori et al. 2000).

More and more recent evidence has shown that overpopulation did not come from growing numbers of rural famers, but from major industries and economic globalization with large-scale farming and exotic-tree plantations that led to deforestation (Grau et al. 2005, Butler and Laurance 2008). Such a change was often associated with public policies. For example, the concession system which has often incorporated with corruption allows international logging firms to cut rainforests at lower stumpage prices (Repetto and Gillis 1988). Favorable credit policies for cattle ranchers, rather than population growth, explained the process of deforestation in the Brazilian Amazon (Moran 1993). Between 1990 and 2002, an additional 10.7 million ha of tropical forest were converted to oil palm plantations, mainly in Southeast Asia (Casson 2003).

In this study, we used Xishuangbanna as a case study to demonstrate the changing driving forces of tropical deforestation. Following the general pattern in other regions of the world, deforestation in Xishuangbanna was also initiated by government policies, such as population reallocation for rubber plantations by state-owned farms. This was followed by market forces in response to the increase in rubber prices and the flow of capital and market access.

Xishuangbanna is located in Yunnan Province in southwest China. It is approximately 20,000 km^2, bordering Laos to the south and Myanmar to the southwest. The region has mountain valley topography with the Hengduan Mountains running north-south, with 95% of the region covered by mountains and hills (figure 1). The Mekong River flows through the center of Xishuangbanna, and the region contributes to more than 20 important tributaries, resulting in many river valleys and small basins (Cao and Zhang 1997). Altitude varies from 2,430 to 475 m above sea level.

Due to its location on the northern edges of tropical Asia and its landscape, Xishuangbanna is best known as "Kingdom of Plants and Animals in China" for its biodiversity and "Kingdom of Cultures" for its variety in ethnic groups and culture. According to the 2000 census, Xishuangbanna had 1 million people with a population density of 50 per km^2. In addition, this area has more than a dozen ethnic groups, shaping a variety of economic and cultural landscapes.

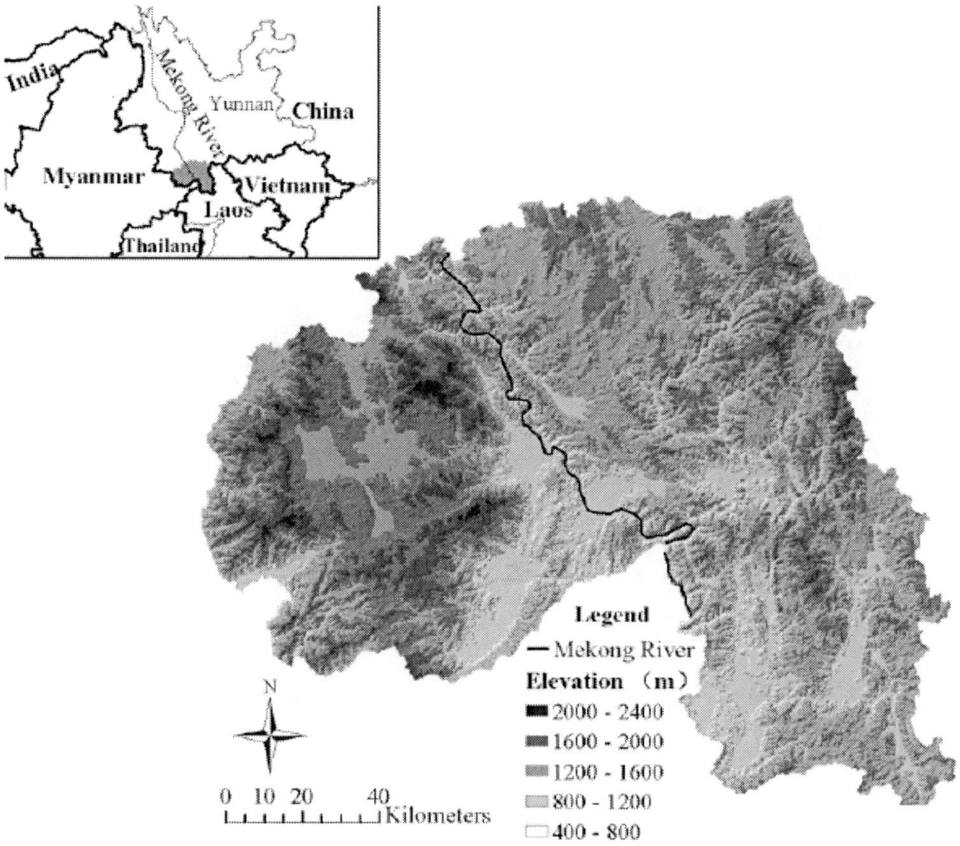

Figure 1. Location of Xishuangbanna, China and its elevation (Li et al. 2007).

2. Ethnic Minorities and Shifting Cultivation: Prior to 1950

Xishuangbanna was covered by tropical and subtropical forests historically and has been the traditional home of upland minority peoples including Dai, Hani, Bulang, and Lahu (figure 2).

In the 1940s, the Dai accounted for 63% of the population. Ancestors of the Dai already lived in what is today Yunnan, Guangxi, Sichuan, parts of Guizhou and Laos, the North of Thailand, Myanmar, and India's Assam area. They later migrated southwest. Dai people were said to be the earliest ethnic group to use ploughs and cultivate rice, and also the earliest to settle in Xishuangbanna. They usually lived on plains or in valleys near the rivers. Therefore, they are often called lowlanders. The name Xishuangbanna comes from the Dai language (it means 12,000 rice fields), describing the dominant land use system.

The Lahu trace their origins to the Qiang tribal group from Northwest China. The Bulang people can be traced back to the ancient "Pu" tribe who were believed to be the earliest known settlers in the Lancang and Nujiang River Valleys. As a consequence of the contact between the Pu tribe and Han as well as their amalgamation into the Han state, these people gradually evolved into the present Bulang ethnic minority. Lahu and Bulang people used to be hunting tribes, whose dwelling places were very remote, often high in the mountains. Their farming systems were subsistence-oriented. They have fewer rice paddies and more fallow land. They usually plant soybeans, peanuts, and corn, all in small plots. Shifting cultivation was widely practiced by those people.

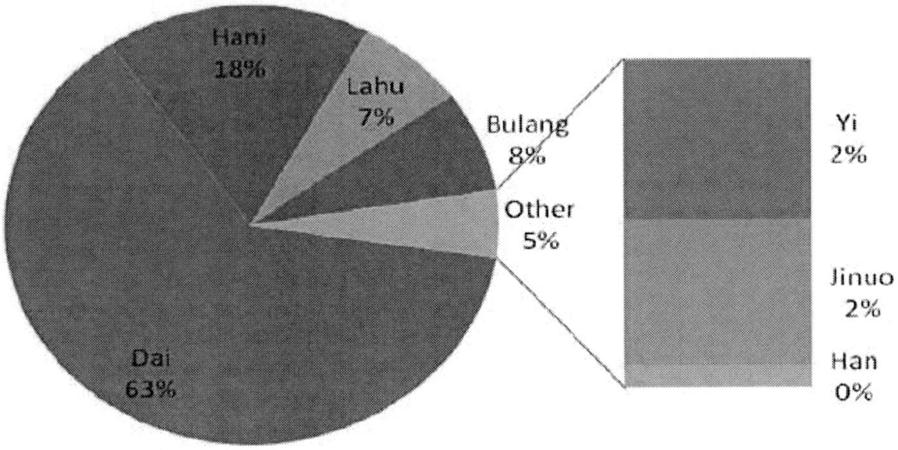

Figure 2. Population composition in Xishuangbanna, China by ethnic groups (1949).

The Hani, like Lahu, originated from the Qiang people who were herdsmen living on the Qinghai-Xizang Plateau and who gradually moved south. One group went from western Sichuan, past Kunming to Southeast Yunnan's Liuzhao Mountain area. The other group went from Northwest Yunnan, passed Dali's Erhai Lake, and pushed further south to the Ailao and Wuliang Mountain regions. They finally reached Xishuangbanna and other areas at the southern tip of Yunnan. Hani land-use patterns were somewhere between that of the Dai and Lahu. They liked to settle mid-way up mountains. They chose neither to live higher up, nor in the valley. In recent years, they have had more access to markets than the Lahu and Bulang. The Hani people practice a swiddening cultivation system that includes traditional tea gardens in the forest, intensively terraced rice paddies in the river bottoms, and livestock grazing and shifting cultivation on the hillsides (Xu et al. 2005).

During the past several centuries in Xishuangbanna, only a few lowland areas were cultivated for rice, with most high land areas still covered by tropical and subtropical forests. These forests included some patches of fallow land after shifting cultivation in a low population density especially the hilly areas. As the majority ethnic group, the Dai have played long and important roles in organizing social institutions in the uplands and establishing economic relationships between valley inhabitants and upland swidden cultivators (Cai 1997, Coward 2002).

Xishuangbanna was a frontier area and almost isolated from other regions because of poor road access. Murrain and malaria and other diseases kept people away and checked population growth. The population grew slowly, from 168,391 in 1924 to 199,300 in 1949. In such low population density the shifting cultivation (or swidden farming) clearing fields allows secondary vegetation to recover for periods of varying length as fallow periods, and maintains a large amount of tropical forests in this region for a long period of time.

3. PUBLIC POLICY AND DEFORESTATION: 1950S-1970S

The situation has changed since 1949 when the People's Republic of China was founded. To increase the domestic supply of rubber production, a large scale Farm Reclamation System (FRS) was set up by the central government to plant rubber in Hainan and Yunnan, the only two regions suitable for rubber tree growth in China (Zhang et al. 2001). Rubber trees were first introduced to China in 1904. In 1948, first rubber plantation with an area of more than 20 hectares in Xishuangbanna was established. By 1950, however, only 91 rubber trees survived due to the Chinese civil war and technical challenges and only12 rubber trees were being tapped in 1955 (Li 2002).

Table 1. Population by ethnic composition, Xishuangbanna, China, 1949-1997

Year	Total	Dai	Han	Hani	Lahu	Bulang	Yi	Jinuo
	(population)	(%)	(%)	(%)	(%)	(%)	(%)	(%)
1949	199,300	57.4	0.3	16.0	6.5	6.8	2.2	2.0
1956	258,645	49.1	8.8	17.8	6.2	7.4	2.3	2.1
1982	646,449	34.9	28.8	19.1	5.1	4.2	2.4	1.7
1985	691,462	34.9	28.9	17.6	5.1	4.3	2.6	2.0
1990	781,710	34.6	25.8	19.7	6.0	4.2	5.0	2.3
1997	838,400	34.7	26.1	18.8	5.7	4.2	4.2	2.1

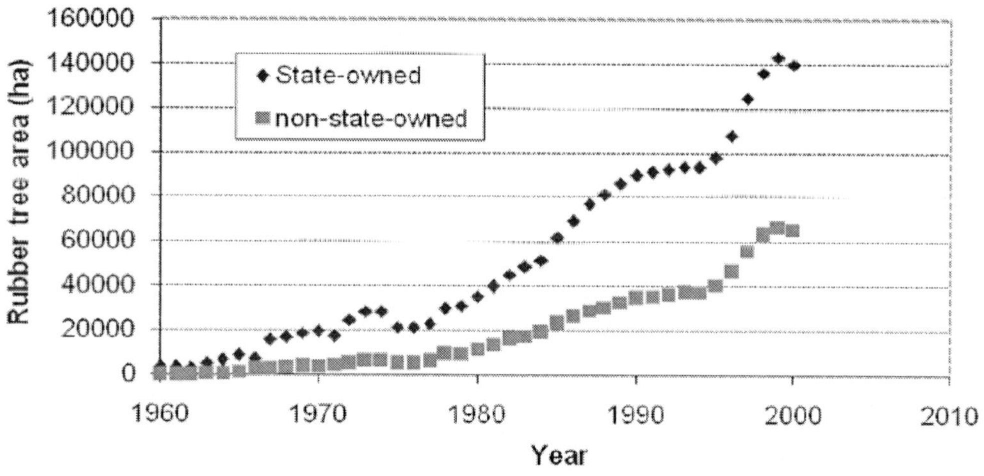

Figure 3. Change in area and proportion between state-owned and non-state-owned rubber plantations in Xishuangbanna, China, 1960-2000.

Rubber tree plantations were accompanied by government investments in roads and population reallocation, and at the expense of tropical forests. The increasing population needed more food, energy, and other land uses. For ethnic minorities, the fallow period for shifting cultivation was too short to allow for the recovery of the forest as the area available to them was shrinking. In the 1950s, Han Chinese farmers from Hunan Province were reallocated to Xishuangbanna to establish state rubber farms. In the late 1960s, another wave of immigration of mostly new high school graduates were sent to Xishuangbanna to reclaim 'wastelands' and defend the borders. In the 1970s, a third but much smaller wave of rubber planting followed the development of rural township enterprises. These immigrations changed the ethnic group composition significantly. In the 1980s, about one-third of the population was Han Chinese (table 1). The influx of this large number of migrants displaced local residents (Xu et al 2005). In 1956, rubber plantation in Xishuangbanna constituted only 3.7 ha, but climbed to 8,000 ha in 1965 (figure 3).

4. Profits and Deforestation Since the 1980s

After 1978, rubber plantations have been primarily driven by the private sector in response to the market and growing demand (Li et al. 2007, Stone 2008). After 1978, farmers were given the right to chose what they wanted to plant under the Household Responsibility System. Agricultural lands, such as paddy fields, were distributed to individual households. Local farmers converted fallow forests (i.e., secondary forests) into rubber plantations as the price of rubber increased dramatically due to domestic protection policy and international markets. New technology also encouraged small-scale farmers to plant rubber as a cash crop, particularly during the 1980s and mid-1990s. The ratio of family-run plantations to the total had increased to about one-third in the late 20th century (figure 3).

Currently, rubber plantations have become one of the most important income sources for the local economy and farmers. Among the 31 townships, 27 townships have rubber plantations, which account for 43% of farmer net income. Total production has reached more than 200,000 tonnes, accounting for 35% of total production in China (http://info.rubber.hc360.com). It was reported that the share of household income from rubber, from 2000 to 2004, had increased from 60 to 80% in Daka and from 21 to 40% in Baka (Fu et al. 2007). Rubber plantation revenues were so significant in this region that the rubber tree was called the cash tree.

The topography in Xishuangbanna, China consists of alternating valleys and hills with elevations ranging from 2,430 m in the north to 475 m in the south. The valleys have a tropical monsoon climate and the hills are subtropical. Summers (June through September) are characterized by high temperatures and humidity, whereas winters (December through March) are characterized by little rain and heavy fog. The region is largely frost free, but cold winds render this region relatively cool and dry in relation to the tropics of mainland Southeast Asia. Spatially, land use and cover, especially rubber plantations, are very much determined by altitude (Liu 2000, Jia 2006, Yu et al. 2007, Li et al. 2008) since altitude largely determines the temperature that is a biological constraint for rubber tree growth. The majority of the rubber plantations occur below 1,000 m.

Rubber plantation were so lucrative that many farmers would like to take risks to plant rubber on less than suitable lands that were once covered by forests. As low altitude land was being used, new plantations were moving upward in altitude. This was considered to be an extremely risky activity. Both the government and many scholars have sent warnings on the potential disaster for establishing plantations at higher altitudes, but in recent years farmers have not slowed the pace of expansion. It was estimated 10% of plantations were located at a risky altitude level for rubber plantation success.

5. CONCLUSION

From 1976 to 2007, forest land has decreased from 69 to 43.6%, while rubber and tea plantation areas have increased from 1.3 to 11.8%, and from 0.23 to 2.14%, respectively. The shrub land has increased from 11.7 to 23.6% (Li et al. 2008). The major impact of land-use changes in Xishuangbanna has been an increase in tropical crops and monoculture rubber tree plantations at the expense of lowland tropical seasonal rain forest (Qiu 2009). This has had a profound impact on the environment.

The case of Xishuangbanna illustrates changes in land use and land cover, attributed from primarily local factors to outside factors such as public policies and globalization and market forces. Attention should be directed to such change so policies can be designed to address issues relating to land use and deforestation.

The changing situation imposes new threats to tropical forests. For example, the lucrative profits drive farmers to take risks to plant a large amount of rubber in protected areas, and local government pays less attention to rainforest protection. The changing situation also provides new opportunities to conserve the remaining tropical forests. While the market force drives some land to be converted from forest land into rubber plantations, it also brings in capital and information. Income from rubber plantations ease the concerns of local farmers who depend on farming for food production. The accompanied road expansion has led to urbanization and a population concentration, and has brought about changes in culture and life styles. For example, shifting cultivation is rarely practiced now. Currently, it is profitability instead of food demand that drives the move of population and capital. That means some land is used more intensively for rubber plantations, while some other lands are able to be protected.

ACKNOWLEDGMENTS

We appreciate two anonymous reviewers for their comments and suggestion.

REFERENCES

Alcorn, J. (1990). Indigenous agroforestry strategies meeting farmers' needs. In A. Anderson (ed.) Alternatives to Deforestation: *Steps Toward Sustainable Use of the Amazon Rain Forest.* New York: Columbia University Press.

Brown J. C., Koeppe M., Coles B., & Price K. P. (2005). Soybean production and conversion of tropical forest in the Brazilian Amazon: *The case of Vilhena, Rondonia.* AMBIO 34, 462–469.

Butler R A., & Laurance W.F. (2008). New strategies for conserving tropical forests: *Trends of Ecological Evolution.* 23(9), 469-72.

Cai, K. (1997). Relationship changes between lowlanders and hill tribes in Xishuangbanna, P. R. China. *Asia Pacific Viewpoint.* 38 (2), 161-167.

Cao, M, Zhang, J. (1997). Tree species diversity of tropical forest vegetation in Xishuangbanna, SW China. *Biodiversity Conservation.* 6, 995-1006.

Cao, M, Zhang, J., Feng, Z., Deng J., & Deng, X. (1996). Tree species composition of a seasonal rain forest in Xishuangbanna, Southwest China. *Tropical Ecology.* 37, 183–192.

Carr, D. L. (2004). Proximate population factors and deforestation in tropical agricultural frontiers *Population and Environment*. 25(6), 585–612.

Coward, W. C. E., Jr. (2002). Tai politics and the uplands. *Working Paper.* Kunming, China: Center for Biodiversity and Indigenous Knowledge.

Fu, Y., Guo, H., Chen, A., & Cui, J. (2007). Rubber development by smallholder farmers in Xishuangbanna, Yunnan, China, a case study from Daka and Baka. Kunming, China: XTBG Working paper.

Grau, H. R, Gaspari, N. I, & Aide, T. M. (2005). Agriculture expansion and deforestation in seasonally dry forests of north-west Argentina. *Environmental Conservation.* 32, 140–148

Guo, H., Padoch, P., Coffey, K, Chen, A., & Fu, Y. (2002). Economic development, land use and biodiversity change in the tropical mountains of Xishuangbanna, Yunnan, Southwest China. *Environ Science & Policy.* 5, 471-479.

http://info.rubber.hc360.com: Xishuangbanna rubber industry a three-pronged approach to deal with financial crisis: Posted 2/10/2009, last accessed by 2/15/2009: http://info.rubber.hc360.com/2009/02/1109516353.shtml

Li, H. M., Aide, T. M., Ma, Y. X., Liu, W. J., & Cao, M. (2007). Demand for rubber is causing the loss of high diversity rain forest in SW China. *Biodiversity Conservation.* 16, 1731-1745.

Li, Z., Ma, Y., Li, H., Peng, M., Liu, W. (2008). Relation of land use and cover change to topography in Xishuangbanna, Southwest China. *Journal of Plant Ecology.* 32(5), 1091-1103.

Li, X. 2002. Green Giant: Xishuangbanna and Rubber. In X. Yin, & Y. Fukao (Eds.), *Rainforest and Rubber.* (pp. 260-302). Kunming, China: Yunnan Education Publisher.

Liu, W. 2000. Impact of climate variation on net primary productivity of natural vegetation in Xishuangbanna. *Journal of Mountain Science.* 18 (4), 296-300.

Jia, K. (2006). *Study on growth change of the rubber plantation along an altitudinal gradient in Xishuangbanna, Southwest China*. Master thesis. Kunming, China: Chinese Academy of Sciences.

Marcoux, A. (2000). Population and the environment: a review and concepts for population programmes. Part III: Population and deforestation. Rome, Italy: FAO.

Moran E. M. (1993). Deforestation and land use in the Brazilian Amazon. *Human Ecology.* 21(1), 1-21.

Moran, E. F., & Ostrom, E. (Eds.). (2005). *Seeing the Forest and the Trees: Human Environment Interactions in Forest Ecosystems.* Cambridge, MA: MIT Press.

Qiu, J. (2009). Where the rubber meets the garden. *Nature.* 457, 246-247.

Repetto, E., and M. Gillis. (Eds.). (1988). *Public Policies and the Misuse of Forest Resources*. New York: Cambridge University Press

Stone, R. (2008). Showdown looms over a biological treasure trove. *Science.* 319, 1604.

Uusivuori, J., Lehto, E. & Palo, M. (2002). Population, income and ecological conditions as determinants of forest area variation in the tropics. *Global Environmental Change.* 12, 313-323.

Wu, Z., Zhu, Y, & Jiang, H. (1987). The vegetation of Yunnan. Beijing: Science Press.

Xu, J., Fox, J., Vogler, J., & Zhang, P., Fu, Y., Yang, L., Qian, J. & Leis, S. (2005). Land-use and land-cover change and framer vulnerability in Xishuangbanna? Prefecture in Southwestern China. *Environmental Management.* 36(3), 404-413.

Zhang, J., & Cao, M. (1995). Tropical forest vegetation of Xishuangbanna, SW China and its secondary changes, with special reference to some problems in local nature conservation. *Biological Conservation.* 73, 229-238.

Zhang, K. (1988). The climatic dividing line between SW and SE monsoons and their differences in climatology and ecology in Yunnan Province of China. *Climatol. Notes.* 38, 197–207.

Zhang, Y., Uusivuori, J., & Kuuluvainen, J. (2000). Econometric Analysis of the Causes of Land Use Change in Hainan, China. *Canadian Journal of Forest Research.* 30, 1913-1921.

Zhang, Y., Zhang, K., Ma, Y., Liu, Y., & Liu, W. (1997). Runoff characteristics of different vegetation covers in tropical region of Xishuangbanna, Yunnan. *Journal of Soil Erosion Soil Water Conservation.* 3, 25-30.

In: Global Change and Forestry
Editors: J. Gan, S. Grado and I. A. Munn
ISBN: 978-1-60876-262-0

Chapter 7

ECONOMICS OF COGONGRASS CONTROL FOR NON-INDUSTRIAL PRIVATE FOREST LANDOWNERS IN MISSISSIPPI

Donald L. Grebner*[*1], *Gregory S. Amacher*[2], *Jon D. Prevost*[3], *Stephen C. Grado*[4] and *Jeanne C. Jones*[5]

[1] Department of Forestry, Mississippi State University, MS, USA
[2] Department of Forestry, Virginia Polytechnic Institute and State University, VA, USA
[3] Delta Wildlife/Delta F.A.R.M., Stoneville, MS, USA
[4] Department of Forestry, Mississippi State University, MS, USA
[5] Department of Wildlife and Fisheries, Mississippi State University, MS, USA

ABSTRACT

Cogongrass (*Imperata cylindrica* (L.) Beauv.) is an invasive plant species that arrived in the United States early in the last century. It is an aggressive plant species that can dominate a site and greatly inhibit the establishment and growth of native plant species. We evaluate alternative management control regimes and the reduction in soil expectation values that can occur on infested sites when control is not implemented. A simulation approach is used to evaluate six management control regimes using either aerial or ground-based applicators with pine afforestation using Mississippi as a case study. Results show that spraying 14.7 oz Accord Concentrate, 1.7 oz Arsenal AC, with 7.4 oz of surfactant and afforesting to pine yields the highest land expectation value and lowest reduction in soil expectation value.

Keywords: Invasive species; Cogongrass (*Imperata cylindrica* (L.) Beauv.); Financial analysis; Herbicide control.

---

* E-mail: dgrebner@cfr.msstate.edu, Department of Forestry, Forest and Wildlife Research Center, Mississippi State University, Box 9681, Mississippi State, MS 39762.

1. INTRODUCTION

In 2007, the world population grew to 6.7 billion persons and is expected to reach 9.1 billion persons by 2050 (UN 2008). As the global population increases in size, it is realistic to expect growing demands for food and wood resources that stretch the productive capacity of our various landscapes. While the world population expanded, trade barriers were being simultaneously loosened across the globe. This opening up of international boundaries may have played an important role in exposing countries to invasive pests that could threaten the productive capacity of our landscapes. Although there are many invasive plants species within the United States, this study will focus on Cogongrass (*Imperata cylindrica* (L.) Beauv.), which arrived in the United States in the early 1900s and has become a well known invasive plant species endemic to forests throughout the southern United States.

Cogongrass is a warm-seasoned perennial grass native to Asia and infests 500 million hectares worldwide (Holms et al. 1977, Dozier et al. 1998). The leaves contain silica crystals which make forage opportunities undesirable (Dozier et al. 1998). Where cogongrass is found, it crowds out native vegetation, forms single species communities, and inhibits natural ecosystem functions (Dozier et al. 1998). The patches form an extensive rhizome root system where the sharply pointed roots can pierce the roots of competing species such as tree seedlings (Prevost 2007). Work by Daneshgar et al. (2008) presented results that sites infested with cogongrass lead to lower survival rates for pine seedlings. It thrives in direct sunlight but has been known to tolerate ambient light levels as low as 1% as well as temperatures down to 6.8^{o}F (Wilcut et al. 1988). These attributes present many challenges to foresters and land managers.

Since the accidental introduction of cogongrass in Alabama and later in experimental forage crop trials, it has spread across the United States, inflicting biological and financial losses to private and public landowners (Tabor 1949, 1952). Cogongrass requires bare soil for germination. It is most often spread through seed dispersal or translocation of leaf and root fragments that follow forest stand disturbances caused by harvesting, site preparation, burning, and thinning operations (Dickens & Moore 1974, Holm et al. 1977, Shilling et al. 1997). Rhizomes are additionally spread by tires, grapples, and blades of cultivation and planting machinery used near existing populations (Willard 1988; Dozier et al. 1998). Once established, cogongrass forms dense patches that exclude most native vegetation other than large trees and dense shrubs which reduces biodiversity and inhibits forest stand regeneration (Lippincott 1997).

In this chapter, we study afforestation of forest sites under the assumption of certain cogongrass invasion that characterizes Loblolly pine (*Pinus taeda*) silviculture in the United States' South. We use a defensive expenditures approach that reveals the decrease in land rents from several afforestation regimes forest landowners can adopt to establish forests when cogongrass is endemic. These alternative regimes are developed with guidance from the forest biology literature and differ according to herbicide and mechanical site preparation. The decrease in land rents under each regime relative to the rent obtained under the assumption of no invasion is one measure of the reduction in soil expectation values from cogongrass invasion. The fact that common financial criteria are used across regimes allows us to determine losses from invasive species holding market factors constant.

Reduction in soil expectation value estimation is an important focus for two reasons. First, given there is some variation in recommended afforestation regimes in the literature, we will identify the most efficient treatment, i.e., the one that reduces rents the least when compared to a scenario where cogongrass is not present. This information will be useful to non-industrial private forest (NIPF) landowners seeking to establish plantations with limited budgets. Second, the reduction in soil expectation values can be compared to current cost sharing programs to determine whether these instruments bridge the gap in returns between control and no control cases. If, for example, cost sharing is greater or equal to the decrease in rents, then landowners are at least as well off after invasion as they would be if they did not need to use controls. If, however, current cost sharing is lower than the gaps we identify, then it is likely that land use will be affected by continued cogongrass invasion, possibly reducing productive forest cover in endemic regions.

The study of invasive species such as cogongrass has been an active and growing area of research in forestry for the past decade. Increasingly many studies are including some aspect of economic analysis within their research focus on this topic. Mehta et al. (2007) developed a model to assess the importance of uncertainties in detecting invasive species for eventual policy targeting. Prestemon et al. (2006) considered how the arrival of a United States invasive species is likely to affect global trade in forest products. Alavalapati et al. (2007) evaluated control measures for cogongrass in slash pine forests. Their results suggest positive returns to landowners when control is implemented. Matta et al. (2007) incorporated invasive species control in their assessment of the economic impacts of adopting conservation enhancing forest practices. Langholtz et al. (2007) also acknowledged that costs for controlling invasive species such as cogongrass on reclaiming mine sites may require greater expenditures for establishing short-rotation woody crops in Florida. In general, few studies have focused on landowner rents and varying management strategies for controlling invasive species on public or private lands in the southern United States.[1]

2. COGONGRASS CONTROL REGIMES AND COSTS

In this section, we discuss literature concerning afforestation regimes relevant to our landowner rent analysis. Controlling cogongrass requires an active application of integrated pest management principles (Johnson 1999, Faircloth 2004). Typically, a combination of multiple herbicide applications and mechanical disturbance is necessary to gain complete control of the grass. Single herbicide applications are normally insufficient in completely killing cogongrass because of the large amount of rhizome biomass present (Willard et al. 1997, Ramsey et al. 2003). However, using sufficient amounts of water during the coverage of target vegetation may allow for sufficient short-term control for afforesting a site (Prevost 2007).

Numerous studies have reported imazapyr and glyphosate based herbicides to be the most effective control agents (Gaffney 1996, Willard et al. 1997, Dozier et al. 1998, Miller 2000). Another study by Ramsey et al. (2003) applied Arsenal AC (imazapyr) and Accord (glyphosate) at rates up to 32 and 189.5 ounces (oz) per acre as single or split applications to

[1] One study by Grebner, Ezell, Prevost, and Gaddis (In Press) evaluates alternative forest management regimes for controlling kudzu (*Pueraria montana var. lobata (Willd.)*) and afforesting to pine or oak plantations.

cogongrass on both a cutover site and a 17-year-old loblolly pine stand. While all herbicide treatments reduced above-ground biomass in this cutover study, herbicides had little effect on rhizome biomass, but mid-rotation reductions in foliar and rhizome biomass were observed. A split application of Arsenal AC was found most effective even though the cogongrass had begun to recover at the end of 24 months. Ramsey et al. (2003) also incorporated post-treatment disking into their cutover study and found that, while foliar re-growth was not affected, a decrease in rhizome growth was observed. When comparing the cutover site to the mid-rotation site, shade and competition from the 17-year-old pines did not have a significant effect on cogongrass populations.

Another study by Miller (2000) recommended spraying Arsenal at 32 oz per acre at 25 gallons per acre (gpa) volume, Accord at 4 quarts per acre (qpa) at 5 to 10 gpa volume, or a tank mixture of 16 oz of Arsenal plus 4 quarts of Accord applied at 10 to 40 gpa volume. Re-treatment will be required, as Miller observed a 10 to 80% loss of control in cogongrass re-growth between 1 and 2 years after treatment. Miller (2000) also found that September was the best time to apply these treatments. Willard et al. (1997) reported that a tank mix of glyphosate and imazapyr provided the best control when sequential applications were evaluated and suggested a rate of 72 and 32 oz per acre. They also found that control with imazapyr alone increased from 20 to 40% as volume of diluent was increased from 5 to 25 gpa. Faircloth (2004) found that Arsenal AC, applied at 20 oz per acre, and a tank mix of GlyPro Plus (glyphosate) 25.4 oz per acre and Arsenal AC 8.6 oz per acre both provided the highest levels of control.

Johnson (1999) recommended an integrated mechanical-herbicide treatment with late spring mowing, followed by deep disking after 80% of the cogongrass has reached heights of 6 to 12 inches. After re-growth has begun, herbicide treatment should begin, preferably late summer to early fall before the first frost. Johnson (1999) suggested that treatment sites should be revisited and regrowth treated with herbicide as necessary. Burnell (2005) suggested that mowing alone can be an effective method of cogongrass control. In a field experiment conducted between 1999 and 2003 it was found that weekly mowing (40 mows per year) reduced dried root weights, total nonstructural carbohydrate of roots, and shoot density 70, 50, and 86%, respectively. Faircloth (2004) found that bi-weekly mowing for one summer followed by an application of glyphosate the following spring resulted in complete above-ground control. However, re-growth was noted at the end of the second year. Fire is also beneficial when used in an integrated approach. Applying herbicides after burning a cogongrass patch can achieve greater results because new growth absorbs and translocates herbicides at an accelerated rate (Lippincott 1997, Dozier et al. 1998). Willard et al. (1997) recommended a combination of imazapyr and glyphosate to provide the greatest level of control while still allowing for re-vegetation.

A recent study by Prevost (2007) evaluated the use of lower herbicide combinations of Arsenal and Accord plus a surfactant. This study evaluated field plots where five alternative herbicide treatments were applied. The study results indicated that one year after herbicide applications surviving cogongrass tillers covered from 2% to 44% of treated sites across the alternative herbicide control regimes. This study consistently used alternative control measures having 14.7 oz of Accord Concentrate but varied mixtures of Arsenal AC and surfactant. The first treatment applied a mixture of 3.5 oz of Arsenal AC with 11.8 oz of surfactant which gained 87% control after one year. The second treatment used 1.7 oz of Arsenal AC with 14.7 oz of surfactant garnered 69% control after one year. The third treatment maintained the same level of Arsenal AC but used 10.3 oz of surfactant and yielded 62% control. The fourth treatment had the same mix as the previous one but used only 7.4 oz of surfactant which had 56% control. A key difference between this study's application rate and the previous literature is the greater volume of water used in the tank mix. The greater use of water allowed for better coverage of the targeted plant species. In addition, Prevost (2007) evaluated the financial returns from the control regimes used in his study and from the literature.

Using this literature as a foundation, we estimated the cost of several management regimes using herbicides to control cogongrass with pine afforestation. Several of these are based on work by Prevost (2007). The per-acre values of these costs were calculated using container size and price and adjusted by application rates to determine per-acre control costs. Average costs per acre per activity for Mississippi in 2006 dollars are reported in table 1.[2] The various treatment regimes are represented by six combinations of Accord Concentrate and Arsenal AC with surfactant plus a control which assumed no cogongrass invasion.

Table 1. Costs for Herbicide, Application, Site Preparation, Planting, and Ownership Costs Used in Financial Comparison of Six Cogongrass Treatments in 2006 Dollars

Costs	$ per acre
Chemicals (per gallon)	
Arsenal AC	460.00
Accord Concentrate	26.00
Surfactant	7.00
Dye	39.25
Chemical Cost per Treatment[a]	
(1) 14.7 oz Accord Concentrate, 3.5 oz Arsenal AC, 11.8 oz surfactant sssssusurfasurfsurfactassufsurfactant[1]	22.95
(2) 14.7 oz Accord Concentrate, 1.7 oz Arsenal AC, 14.7 oz surfactant	16.83
(3) 14.7 oz Accord Concentrate, 1.7 oz Arsenal AC, 10.3 oz surfactant	16.59
(4) 14.7 oz Accord Concentrate, 1.7 oz Arsenal AC, 7.4 oz surfactant	16.43
(5) 15.7 oz Arsenal AC, 7.8 oz surfactant	57.09
(6) 188.3 oz Accord, 94.2 oz surfactant	43.66
Application	
Ground-based (20 gpa)	45.00
Aerial-based (20 gpa)	35.00
Ground-based (35 gpa)	65.00
Aerial-based (35 gpa)	55.00
Mechanical Site Preparation	
Ripping	60.00
Planting	
Seedlings (726 per acre)	43.56

[2] In some cases, information from personal communication with herbicide vendors was used. Details are found in Prevost (2007).

Table 1. (Continued).

Costs	$ per acre
Labor	36.30
Ownership Fees	
Land Use Tax[b]	4.46
Management Fees[c]	5.00

[a] Treatments 1-4 were applied once March 2004 in a volume of 35 gallons per acre on the John C. Stennis Space Center (Prevost 2007). Treatment 5 was applied in November 1999 and 2000, volume of 25 gallons per acre assumed (Ramsey et al. 2003). Treatment 6 was applied once in November 1999, 25 gallons per acre assumed (Ramsey et al. 2003).

[b] Personal communication with Ian A. Munn, Mississippi State University.

[c] Assumed miscellaneous annual expenses.

3. Growth and Yield, Harvesting, and Price Assumptions

Loblolly pine plantations were modeled under the different control treatments given in table 1. PTAEDA3 was used to project growth for potential rotation ages. Given that cogongrass primarily affects establishment but not growth of mature trees, our growth projections were made under the assumption that the plantation had been successfully established. We assumed a planting density of 726 trees per acre, which is reasonable for the Mississippi coastal plain. The forest site was assumed to be well-drained with a base age 25 site index of 65. Growth was projected in tons, and top wood was assumed available for sale as pulpwood. Product classes available for sale at harvest time included sawtimber, chip-n-saw, and pulpwood. Price information for these different product classes was taken from 2006 information reported in Timber Mart South using Mississippi state wide data averaged over four quarters (TMS 2006). Table 2 summarizes this information.

Table 2. Price Information for Standing Timber in Mississippi (2006)[a]

Species	Prices ($ per ton)
Pine sawtimber	41.28
Pine chip-n-saw	22.26
Pine pulpwood	6.76

[a] Timber prices from TimberMart South.

4. Landowner Management Regimes and Rents

We used six alternative management regimes for computation of landowner rents and a control (table 3). The regimes reflected recommendations in the silviculture literature and unit cost assumptions discussed earlier. The regimes differ mainly in the assumptions concerning herbicide applications and chemical dispersal equipment used prior to seedling planting. We also assumed that all herbicide applications occurred at the beginning of year zero, while seedling planting and soil ripping operations occurred in time period one for the purposes of computing landowner returns.

Table 3. Description of Alternative Management Regimes by Ground-Based and Aerial Applicators for Controlling Cogongrass and Afforesting to Loblolly Pine[a]

Management Regime -Chemical Rates Per Acre[b]	Dispersal Type
(1) 14.7 oz Accord Concentrate, 3.5 oz Arsenal AC, 11.8 oz surfactant	(T) Ground-based (H) Aerial
(2) 14.7 oz Accord Concentrate, 1.7 oz Arsenal AC, 14.7 oz surfactant	(T) Ground-based (H) Aerial
(3) 14.7 oz Accord Concentrate, 1.7 oz Arsenal AC, 10.3 oz surfactant	(T) Ground-based (H) Aerial
(4) 14.7 oz Accord Concentrate, 1.7 oz Arsenal AC, 7.4 oz surfactant	(T) Ground-based (H) Aerial
(5) 15.7 oz Arsenal AC, 7.8 oz surfactant	(T) Ground-based (H) Aerial
(6) 188.3 oz Accord, 94.2 oz surfactant	(T) Ground-based (H) Aerial

[a] For example, a regime coded 1T refers to a cogongrass being treated with 14.7 oz Accord Concentrate, 3.5 oz Arsenal AC, 11.8 oz with surfactant using an aerial applicator and afforested to pine.

[b] All herbicide treatments occur at the beginning of year zero.

Landowner rents were defined as the maximum land expectation value that could be attained under each regime in table 3 (Porterfield et al. 1977, Bullard & Straka 1998, Grebner et al. 2003). Land expectation value (LEV), or landowner rents as we define them, was computed for each regime j by choosing rotation age to maximize the following expression,

$$LEV_j = \frac{NPV_j(1+i)^n}{(1+i)^n - 1} \tag{1}$$

where n is the rotation age and i is the real interest rate. Equation (1) corrects for discounting over an infinite sequence of rotations that are all identical given our price and cost assumptions, and the numerator represents the net future value of a single rotation, i.e.,

$$NPV_j = \sum_{t=0}^{n} \frac{R_t}{(1+i)^t} - \sum_{t=0}^{n} \frac{C_{jt}}{(1+i)^t}, \tag{2}$$

where R_t is revenue that the landowner captures by harvesting in year t, with a final merchandised harvest at time n, and C_{jt} is cost incurred at time t and includes planting, establishment, and cogongrass control strategies specific to each afforestation regime given in table 3. In the results that follow, we assume a low thinning operation in year 17 which reduced the basal area to 65 square feet per acre. If we let j = 0 define the control scenario, then reduction in soil expectation values is defined as $LEV_0 - LEV_j > 0$ for all j afforestation regimes. A 6% real interest rate was assumed throughout the analysis in computing (1), (2), and reduction in soil expectation values.

5. Results and Discussion

As stated in the methods section, an optimal rotation age for each management regime was found by maximizing the LEV under the different cogongrass control scenarios. The optimal rotation age for these management regimes was 34 years for implementing the final harvest. Table 4 shows that the optimal LEV for each of the management regimes simulated to control a cogongrass infestation using either an aerial or ground-based spray applicator. The control regime expectedly yields the highest LEV given the lower costs of establishing seedlings due to the assumed lack of cogongrass invasion. Referring to the results by regime, we see that only management regime five was not financially feasible, resulting in negative LEV for applying herbicides by either aerial or ground-based spray applicators. Regime four yields the highest LEV for both applicators. We also find that aerial applicators lead to higher computed LEVs than ground-based applicators.

Table 4. Per-Acre Before-Tax LEV of Cogongrass Control Regimes Based on Ground (T) or Aerial (H) Applications

Regime	LEV ($)
Control	$233.37
1T	$140.53
2T	$147.62
3T	$147.90
4T	$148.09
5T	($24.01)
6T	$116.50
1H	$152.13
2H	$159.22
3H	$159.50
4H	$159.69
5H	($1.47)
6H	$128.10

Table 5 shows the per-acre reduction in soil expectation values for controlling cogongrass in loblolly pine plantations. Although the per acre losses appear relatively small in magnitude, the potential decrease in returns across large areas of existing cogongrass invasions would be very significant according to our results, even with landowners choosing the regime with the highest LEV. Referring to the table, management regime five has the highest reduction in soil expectation value of $257.37 per acre when an aerial spray applicator is employed and $234.83 per acre when a ground-based applicator is employed. The management regimes with the lowest reduction in soil expectation values are those for ground-based spray application and regimes three and four. These losses are around 30% of the loss from using management five, the least financially feasible one.

Table 5. Per-Acre Reduction in Soil Expectation Values of Cogongrass Control Regimes Based on Ground (T) or Aerial (H) Applications

Regime	LEV ($)
1T	$92.83
2T	$85.74
3T	$85.46
4T	$85.27
5T	$257.37
6T	$116.86
1H	$81.23
2H	$74.14
3H	$73.86
4H	$73.67
5H	$234.83
6H	$105.26

Our results thus far demonstrate that implementing management regimes to control cogongrass can be a costly endeavor for private landowners in terms of forgone returns. In Mississippi, there are some government-sponsored programs that provide cost-share assistance to private landowners for control of invasive species. If a non-industrial private forest landowner can gain access to these cost sharing programs, then this will improve LEV and lower reduction in soil expectation values under all of the cogongrass control regimes shown above. Our reduction in soil expectation value results are in fact a useful indicator of how close these cost sharing programs come to making the landowner as well off as they would be if cogongrass was not present and did not need to be controlled when regenerating loblolly pine sites.

A state funded cost-share program that private nonindustrial landowners in Mississippi are eligible to participate in is sponsored by the Mississippi Bureau of Plant Industry. This particular program supplies free herbicides to control cogongrass but is limited to areas of only 25 acres per landowner. A federally sponsored cost-share program, the Environmental Quality Incentives Program, is a well-funded program sponsored by the Natural Resources Conservation Service (NRCS). NRCS provides financial and technical assistance for conservation management on agricultural and forested lands. Mississippi counties receive annual state formula funding from the federal government that addresses several factors such as previous funding demand and performance and priority resource concerns. United States Department of Agriculture local working groups then distribute funds at the county level to non-industrial private forestland owners. In general, forestland owners are eligible for up to 75% cost-share up to a total of $450,000, but small forestland owners may be eligible for up to a total of $10,000 where they receive a 90% cost-share.

Since these cost sharing programs provide relief for the portion of LEV that is paid in year zero and only apply to the invasion-based regimes, it is easy to see how the programs would affect the reduction in soil expectation values reported in table 5. For example, a 75% cost share per acre would lower reduction in soil expectation values anywhere from $64.09

per acre for the most financially feasible regime to just over $223.88 per acre for management regime 5. A 90% cost share would lower reduction in soil expectation values between $76.91 per acre to $268.69 per acre depending on the cogongrass control regime employed by the landowner. Over large acreages that characterize loblolly pine in the region, minimizing the reduction in soil expectation values would be substantial and represent an important benefit of such programs for landowners in the state. According to Dr. John Byrd from Mississippi State University (Personal communication), 57 of 82 counties in Mississippi have various levels of cogongrass invasion. The Invasive Plant Atlas of the MidSouth suggests that several hundred thousand acres are infested across the southeastern United States and up to 1.2 billion acres across the globe.[3]

6. Conclusions

As stated earlier in this study, the primary goal of this work was to evaluate alternative management control regimes to combat cogongrass and the reduction in soil expectation values that occur in the presence of this invasive plant species. Positive returns are found for various cogongrass control options. Spraying 14.7 oz Accord Concentrate, 1.7 oz Arsenal AC, with 7.4 oz of surfactant and afforesting to pine yields the highest land expectation value and lowest reduction in soil expectation values. In addition, reduction in soil expectation values can be reduced by governmental agencies providing adequate cost-share levels to forest landowners to effectively control infested areas.

Although several alternative management control regimes were evaluated in this study, additional research will be necessary as new brands of herbicides are developed and field tested. In addition, future efforts that evaluate the effectiveness of government programs to combat invasive plant species would greatly aid the broader natural resources community.

References

Alavalapatti, J. R. R., Jose, S., Stainback, G. A., Matta, J. R., & Carter, D. R. (2007). Economics of cogongrass control in slash pine forests. *Journal of Agricultural and Applied Economics, 39*(October 2007), 61-68.

Bullard, S. H. & Straka, T. J. (1998). *Basic concepts in forest valuation and investment analysis.* Edition 2.1. Auburn, Alabama: Preceda Education and Training.

Burnell, K. D. (2005). Biology, detection, and management of cogongrass [*Imperata cylindrica* (L.) Beauv.] in Mississippi. Ph.D. dissertation, Plant and Soil Sciences Department, Mississippi State, Mississippi State University, 186 pp.

Daneshgar, P., Jose, S., Collins, A., and Ramsey, C. (2008). Cogongrass (*Imperata cylindrical*), an alien invasive grass, reduces survival and productivity of an establishing pine forest. *Forest Science*, *54*(6), 579-587.

Dickens, R. & Moore, G. M. (1974). Effects of light, temperature, KNO_3, and storage on germination of cogongrass. *Agronomy Journal, 66*, 187-188.

---

[3] Source is http://www.gri.msstate.edu/ipams/Species.php?SName=&CName=Cogongrass. Retrieved on February

Dozier, H., Gaffney, J. F., McDonald, S. K., Johnson, E. R., & Shilling, D. G. (1998). Cogongrass in the United States: history, ecology, impacts, and management. *Weed Technology, 12*, 737-743.

Faircloth, W. H. (2004). Integrated vegetation management of cogongrass in Alabama. Ph.D. dissertation, Agronomy and Soils Department, Auburn University, Auburn, Alabama, 182 pp.

Gaffney, J. F. (1996). Ecophysiological and technological factors influencing the management of cogongrass (*Imperata cylindrica*). Ph.D. dissertation, Agronomy Department, University of Florida, Gainesville, Florida, 114 pp.

Grebner, D.L., Ezell, A.W., Gaddis, D., & Bullard, S.H. (2003). Impacts of southern oak seedling survival on investment returns in Mississippi. *Journal of Sustainable Forestry, 17*(3), 1-19.

Grebner, D. L., Ezell, A. W., Prevost, J. D., & Gaddis, D. A. (In Press). Kudzu control and impact on monetary returns to non-industrial private forest landowners in Mississippi. Submitted to the *Journal of Sustainable Forestry* on December 17, 2007; Accepted June 8, 2008.

Holm, L. G., Pucknett, D. L., Pancho, J. B., & Herberger, J.P. (1977). *The World's Worst Weeds; Distribution and Biology.* University Press of Hawaii, Honolulu, Hawaii.

Johnson, E. R. R. L. (1999). Management of the non-native invasive weed cogongrass (*Imperata cylindrica* (L.) Beauv.) utilizing an integrated management program. M.Sc. thesis, Agronomy Department, University of Florida, Gainesville, Florida, 87 pp.

Langholtz, M., Carter, D. R., Rockwood, D. L., & Alavalapati, J. R. R. (2007). The economic feasibility of reclaiming phosphate mind lands with short-rotation woody crops in Florida. *Journal of Forest Economics*, *12*(4), 237-249.

Lippincott, C. L. (1997). Ecological consequences of *Imperata cylindrica* (cogongrass) invasion in Florida sandhill. Ph.D. dissertation, Botany Department, University of Florida, Gainesville, Florida, 165 pp.

Matta, J., Alavalapati, J., & Tanner, G. (2007). A framework for developing marked-based policies to further biodiversity on non-industrial private forests (NIPF). *Forest Policy and Economics*, *9*, 779-788.

Mehta, S., Haight, R., Homans, F., Polasky, S., & Venette, R. (2007). Optimal detection and control strategies for invasive species management. *Ecological Economics, 61*, 237–245.

Miller, J. H. (2000). Refining rates and treatment sequences for cogongrass (*Imperata cylindrica*) control with imazapyr and glyphosate. Proceedings Southern Weed Science Society *53*, 131-132.

Porterfield, R., Utz, K., & Balmer, W. (1977). Financial evaluation of artificial vs. natural regeneration of bottomland hardwoods. Paper presented at the Second Symposium on Southeastern Hardwoods, Dothan, Alabama.

Prestemon, J., Zhu, S., Turner, J., Buongiorno, J., & Rhuong, R. (2006). Forest product trade impacts of an invasive species: modeling structure and intervention trade-offs. *Agricultural and Resource Economics Review, 35*, 128–143

Prevost, J. D. (2007). Evaluation of forest protection practices on cogongrass (Imperata cylindrical (L.) BEAUV.) on the Mississippi Gulf Coast. M.Sc. thesis, Mississippi State University. 108pp.

9, 2009.

Ramsey, C. L., Shibu, J., Miller, D. L., Cox, J., Portier, K. M., Shilling, D. G., & Merritt, S. (2003). Cogongrass [*Imperata cylindrica* (L.) Beauv.] response to herbicides and disking on a cutover site and in a mid-rotation pine plantation in southern USA. *Forest Ecology and Management.* (179): 195-207.

Shilling, D. G., Beckwick, T. A. Gaffney, J. F., McDonald, S. K., Chase, C. A., & Johnson, E.R.R.L. (1997). Ecology, physiology, and management of cogongrass (*Imperata cylindrica*). Final Report. Florida Institute of Phosphate Research, Bartow, Florida.

Tabor, P. (1949). Cogongrass *[Imperata cylindrica (L.) Beauv.]* in the southeastern United States. *Agronomy Journal*, *41*, 270.

Tabor, P. (1952). Comments on cogon and torpedo grasses: a challenge to weed workers. *Weeds,* 1, 374-375.

Timber Mart-South (TMS). (2006). Timber Mart-South Market News Quarterly Reports. Timber Mart-South, Inc. Volumes *31*(1)-*31*(4).

United Nations (UN). (2008). *Development and globalization: facts and figures.* United Nations Publication, Geneva, Switzerland.

Wilcut, J. W., Dute, R. R., Truelove, B., & Davis, D. E. (1988). Factors limiting the distribution of cogongrass, *Imperata cylindrical*, and torpedograss, *Panicum repens*. *Weed Science*, *36*, 577-582.

Willard, T. R. (1988). Biology, ecology and management of cogongrass (*Imperata cylindrica* (L.) Beauv.). Ph.D. dissertation, Agronomy Department, University of Florida, Gainesville, Florida, 113 pp.

Willard, T. R., Gaffney, J. F., & Shilling, D. G. (1997). Influence of herbicide combinations and application technology on cogongrass (*Imperata cylindrica)* control. *Weed Technology, 11*, 76-80.

PART II. FOREST PRODUCTS MARKETS AND TRADE

In: Global Change and Forestry
Editors: J. Gan, S. Grado and I. A. Munn
ISBN: 978-1-60876-262-0

Chapter 8

STATUS AND TRENDS OF U.S. PULPWOOD MARKET

Peter J. Ince*
U.S. Forest Service, Forest Products Laboratory, WI, USA

ABSTRACT

Global trends in pulp, paper, and paperboard have deeply affected U.S. pulpwood markets since the late 1990s. Global trends included a shift of growth in paper and paperboard production from North America to Europe and Asia over the past decade. This trend was associated with generally slower growth in U.S. industrial production and displacement of growth in print media advertising expenditures by electronic media. Wood pulp consumption in the United States was also offset by increased paper recycling since the late 1980s. The net impact of these trends was a contraction of growth in U.S. pulpwood demand and generally declining real prices for pulpwood since the late 1990s. The current status and trends of the U.S. pulpwood market contrast sharply to decades prior to the 1990s, when growth in U.S. pulpwood demand was more robust. Future growth in U.S. pulpwood demand is likely to be relatively subdued for the near term. Total pulpwood demand for wood pulp, wood panel products, and export has potential to increase in the U.S. South but only gradually in the decades ahead, while likely to remain flat or declining in other U.S. regions. Biofuels and bioenergy represent emerging demands for wood biomass, and potential future demands for pulpwood, but sustained higher oil prices or further technical advances plus new capital investments will be needed before biofuels create strong competing demands for pulpwood. Pulpwood still has much higher value-added potential in production of wood pulp than in production of biofuels such as cellulosic ethanol. In the near future, consumption of pulpwood in the emerging cellulosic biofuel industry is likely to remain a small fraction of total U.S. pulpwood supply and demand, although long-range future wood use for biofuel could be potentially quite large, depending on future oil prices and technological developments. The current depressed status of the U.S. pulpwood market presents a near-term policy challenge in forest management, whether to provide incentives to sustain or expand

* E-mail: pince@fs.fed.us, U.S. Forest Service, Forest Products Laboratory, Madison, WI 53726-2398.

pulpwood supply in current times of lean demand growth when there may be much larger foreseeable needs, but only in the long run.

Keywords: Pulp and paper products; Market outlook; Production and demand shifts; Trade.

1. Introduction—Global Trends in Production

A significant global shift of growth in paper and paperboard production has occurred since the late 1990s, as growth shifted toward Asia and Europe and away from North America. Figure 1 illustrates historical changes in the tonnage of paper and paperboard production during three recent decades in North America, Europe, and Asia, based on data from the Food and Agriculture Organization of the United Nations (2009). Over the three decades, combined output of North America, Europe, and Asia accounted steadily for over 90% of global paper and paperboard output (world output reached 383.6 million metric tons in 2007). In each of the three major regions, growth in output each decade typically amounted to tens of millions of tons, but during the most recent decade (1997–2007), growth in production abruptly receded in North America while continuing to expand in Europe and Asia. In the past decade, China was the world's most rapidly expanding producer and consumer of paper and paperboard, as production and consumption of paper and paperboard receded in the United States. U.S. output peaked in 1999, whereas output in Canada peaked in 2000. In 1997, North America produced about 20 million tons more than either Europe or Asia, but by 2007 production in Asia (at 142 million tons) was 40 million tons higher than in North America (at 102 million), while Europe (at 115 million tons) was 13 million tons higher.

The shift in the global distribution of growth affected trends in pulpwood prices, both globally and in the United States. Global expansion of pulp, paper, and paperboard production put record demand pressures on pulpwood prices globally in the past decade, whereas lower production levels in the United States resulted in an easing of demand pressures relative to the late 1990s. Average global pulpwood prices rose steadily from 2002 to 2008, reaching historical record levels in early 2008 according to Håkan Ekström, president of Wood Resources International (Wood Resources International 2008). On the other hand, in the United States, the nominal pulpwood price index peaked in the late 1990s (shortly before paper and paperboard production peaked), and the price index averaged about 20% below that peak throughout the current decade from 2000 to 2009 (U.S. Bureau of Labor Statistics 2008). Figure 2 shows the trend in the U.S. pulpwood price index since the late 1990s. The price index experienced some gains from 2002 to 2008, after bottoming out during the 2001–2002 economic recession, but the index never recovered to its 1998 peak, and the index was falling again amid the latest economic recession in 2008–2009.

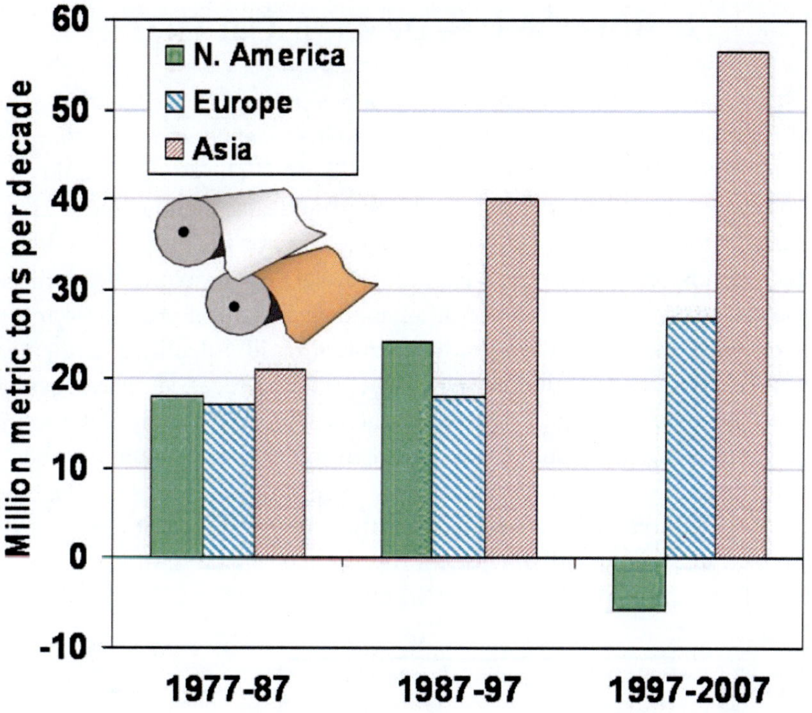

Figure 1. Changes in paper and paperboard production by decade for North America, Europe, and Asia, in millions of metric tons per decade (Food and Agriculture Organization of the United Nations 2009).

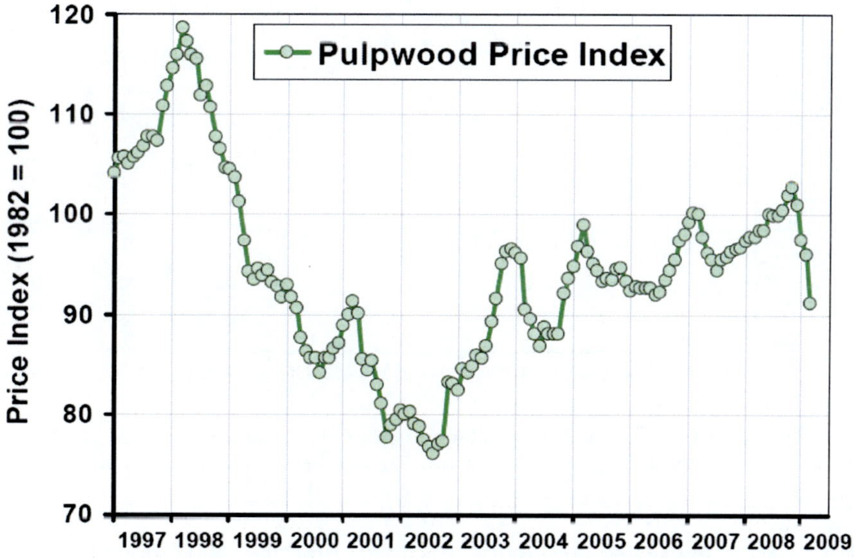

Figure 2. Trend in U.S. nominal price index for delivered pulpwood, monthly data, 1997–2009 (U.S. Bureau of Labor Statistics 2008).

2. Shifts in U.S. Demand and Trade

United States paper and paperboard demands peaked nearly a decade ago, around 1999. Purchases of paper and paperboard in the United States peaked and then sharply declined in 2000–2001, along with U.S. industrial production. Figure 3 shows the monthly U.S. index of industrial production (U.S. Federal Reserve 2008a) and also annualized U.S. purchases of paper and paperboard each month (the sum of purchases for the latest 12 months). The chart shows that U.S. purchases of paper and paperboard were approaching 94 million metric tons per year at the end of 1999 and remained at an annualized rate of around 94 million tons into the first half of 2000. However, the rate of purchases rapidly declined in the second half of 2000 and through 2001, as U.S. industrial production also receded. The recession of 2000–2001 followed the Asian financial crisis of the late 1990s and coincided with a period when the exchange value of the U.S. dollar was increasing, boosting consumer goods imports and negatively affecting U.S. manufacturing output. The largest U.S. demands for paper and paperboard are in the areas of packaging and print advertising, both historically linked to industrial production, so U.S. demands for paper and paperboard generally declined as manufacturing of consumer goods shifted overseas.

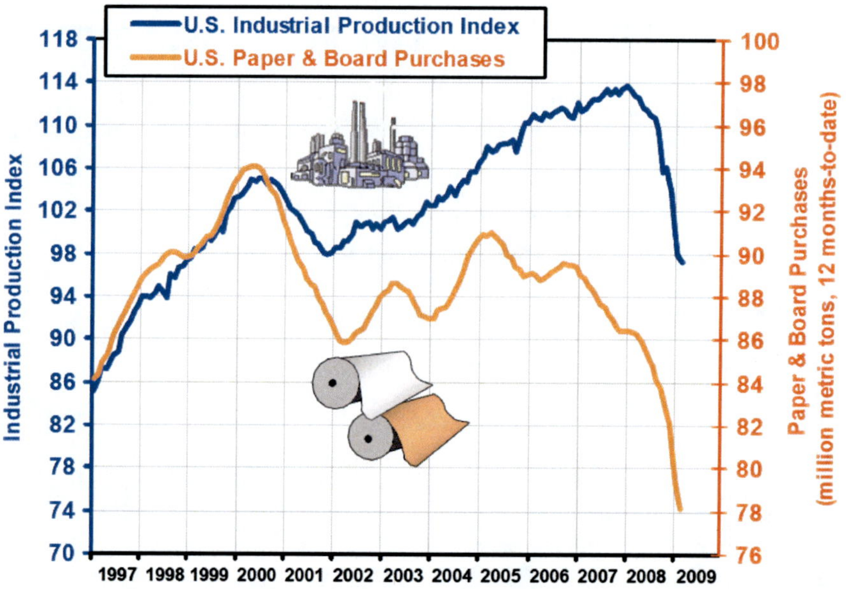

Figure 3. Monthly trends in U.S. industrial production index (U.S. Federal Reserve 2008a) and U.S. paper and paperboard purchases, 12-month-totals-to-date (American Forest & Paper Association 2008), 1997–2009.

Growth in U.S. industrial production resumed after the 2001 downturn, but average growth from 2001 to 2008 was *less than half* the average growth rate for industrial production in the latter half of the 20th century (U.S. Federal Reserve 2008a). The decline of growth in U.S. industrial output reflects expansion of manufactured goods imports, which have soared to record levels over the past decade. U.S. demands for newsprint and printing papers were also negatively affected as growth in advertising expenditures shifted from print media to electronic media. Slower growth in print advertising expenditures and other structural changes in paper and paperboard use resulted in a gradual divergence over the past decade between the industrial production trend and purchases of paper and paperboard. The recession of 2008 resulted in further steep declines in U.S. industrial production and in purchases of paper and paperboard (figure 3), but this time the downturn was due to sharp declines in retail sales and consumer spending, which were adversely affected by a credit crisis and rising unemployment. U.S. retail sales, for example, dropped by 12.4% in just six months from June to December of 2008 (U.S. Census Bureau), by far the largest and most rapid decline in decades.

In contrast to declining domestic demands for paper and paperboard, the U.S. trade gap in pulp, paper, and paperboard widened for a period in the past decade but then narrowed as the trade-weighted value of the U.S. dollar peaked (in 2002) and then declined. As shown in figure 4, the U.S. trade gap in total tonnage of pulp, paper, and paperboard products (imports minus exports) peaked at around 8 million metric tons in 2002–2003. Rising imports and weaker domestic demands led to industry downsizing and consolidation. For example, according to industry sources, 98 U.S. paper mills were closed between 1998 and 2003 (American Forest & Paper Association 2005). Consolidation resulted in job losses, but the industry achieved higher average labor and capital productivity. Higher productivity and a weaker dollar improved the competitiveness of U.S. producers, and by 2008 the U.S. trade gap as measured by tonnage of pulp, paper, and paperboard products was finally closed (American Forest & Paper Association 2008). Although not shown in figure 4, the trade-weighted dollar index increased in early 2009, as the value of foreign currencies declined amid the global economic recession. A higher dollar could adversely affect the U.S. pulp, paper, and paperboard trade balance.

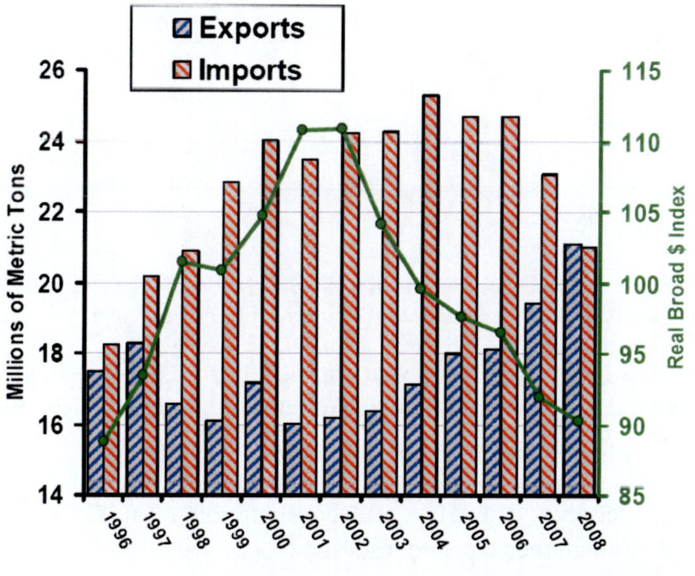

Figure 4. U.S. exports and imports of all pulp, paper, and paperboard products (American Forest & Paper Association 2007, 2008) and real broad trade-weighted dollar index (U.S. Federal Reserve 2008b), 1996–2008.

3. Relationship to GDP Growth

Another way to visualize the structural change that occurred in the past decade is to compare recent and historical growth trends for paper and paperboard to real Gross Domestic Product (GDP) trends. Throughout most of the 20th century growth in U.S. paper and paperboard consumption was fairly consistent with GDP growth. Figure 5 shows trends since 1975 in U.S. paper and paperboard apparent consumption along with U.S. real GDP. The two trends were not directly proportionate, as the average rate of growth in GDP was higher than the average growth rate for paper and paperboard consumption, but clearly both trends were growing in tandem up until 1999, at which point the trends diverged rather sharply. This divergence between GDP growth and U.S. consumption of paper and paperboard is a clear reflection of structural changes that have occurred in paper and paperboard demands, primarily a result of higher imports of manufactured goods and slower growth in U.S. industrial production and shifts in advertising expenditures from print media to electronic media.

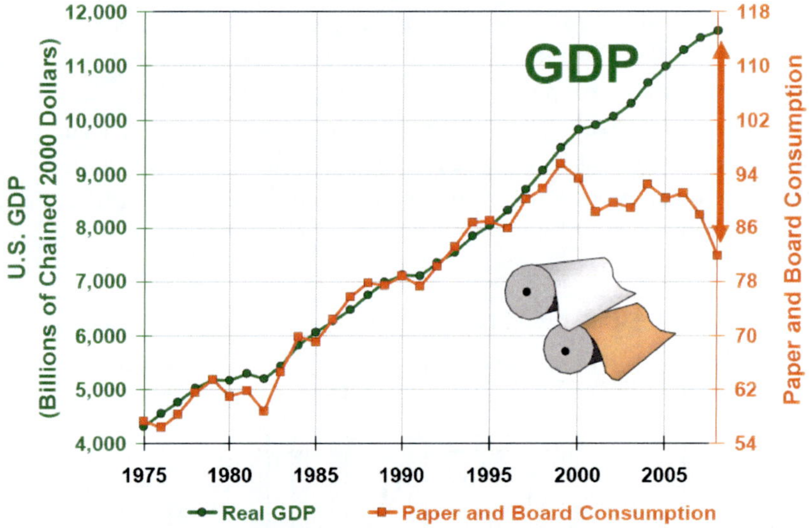

Figure 5. Divergence since 1999 between trends in U.S. real GDP (U.S. Bureau of Economic Analysis 2008) and U.S. paper and paperboard consumption (American Forest & Paper Association 2007, 2008).

The year 1999 was the historic peak year for U.S. paper and paperboard purchases, and from that point forward U.S. paper and paperboard consumption obviously diverged from real GDP. By 2008, the divergence from GDP suggests a loss of more than 30 million metric tons of U.S. market growth (figure 4) relative to demands that might have been reached if historical correlation between real GDP and consumption of paper and paperboard had continued from 1999 through the present decade.

The widening gap between U.S. paper and paperboard growth and GDP growth was offset in production only to a small extent by expansion of U.S. paper and paperboard exports and decreased imports in recent years. Most U.S. production still serves domestic markets, and exports are only a fraction of U.S. paper and paperboard production (about 15%), so U.S. paper and paperboard production also diverged from U.S. GDP growth, beginning around

1999, as shown in figure 6. Although increased exports have somewhat narrowed the gap between GDP and production trends, there was nevertheless a loss of well over 25 million metric tons of growth in U.S. production over the past decade relative to the historical GDP relationship.

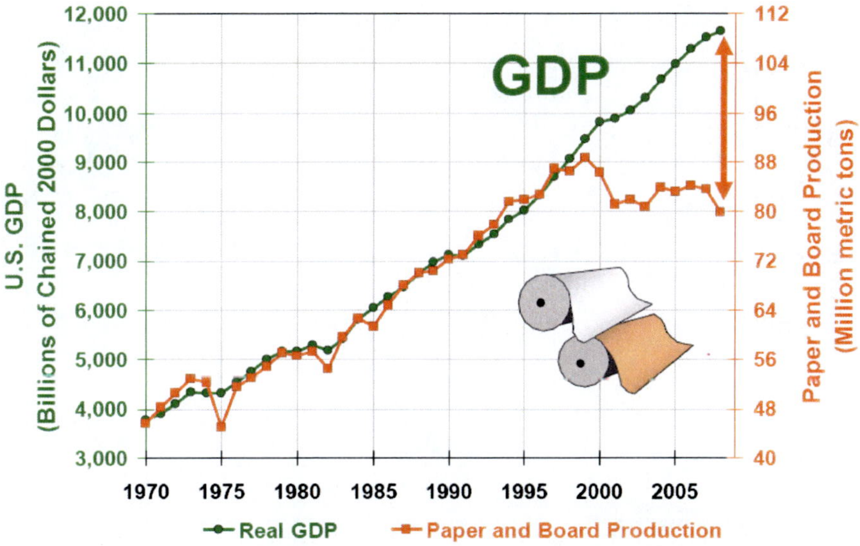

Figure 6. Divergence since 1999 between trends in U.S. real GDP (U.S. Bureau of Economic Analysis 2008) and U.S. paper and paperboard production (American Forest & Paper Association 2007, 2008).

Whether looking at U.S. paper and paperboard consumption (figure 5) or production (figure 6), it is obvious that there was a structural change in growth trends relative to overall economic growth or GDP, and the onset of the structural change was coincident with the Asian financial crisis of the late 1990s, the subsequent rise in value of the U.S. dollar, and consequent downturn in industrial production and recession of 2000–2001. Average historical growth rates prior to 1999 were well above 2% per year for U.S. production and consumption of paper and paperboard, but growth on average has been negligible from 2001 to 2008. The annual rate of growth in production from 2001 to 2008, for example, was just under 0% per year, obviously much less than the average annual growth rate in U.S. production from 1970 to 1999 (2.4% per year).

4. Impact on U.S. Wood Pulp Production

The trends in U.S. paper and paperboard demand and output since the late 1990s, along with earlier increases in paper recycling, had profound impacts on the historic trend in U.S. wood pulp production. For most of the 20th century, U.S. wood pulp production was increasing, but production peaked in the mid-1990s (figure 7). Growth was offset by increased paper recycling from the late 1980s to 1990s. However, the utilization rate of recycled fiber in U.S. paper and paperboard products has leveled out since the late 1990s (American Forest & Paper Association 2007). Increased paper recovery for recycling since then has gone mostly for export (mainly to China). Thus, since the late 1990s, shifts in U.S.

wood pulp output were determined primarily by trends in U.S. paper and paperboard output. Therefore, not surprisingly, growth in U.S. wood pulp output has been fairly negligible since 2001, although there were some gains in exports relative to imports. Compared with growth rates of earlier decades in the 20th century, growth in pulpwood consumption at U.S. wood pulp mills in this century has been relatively subdued.

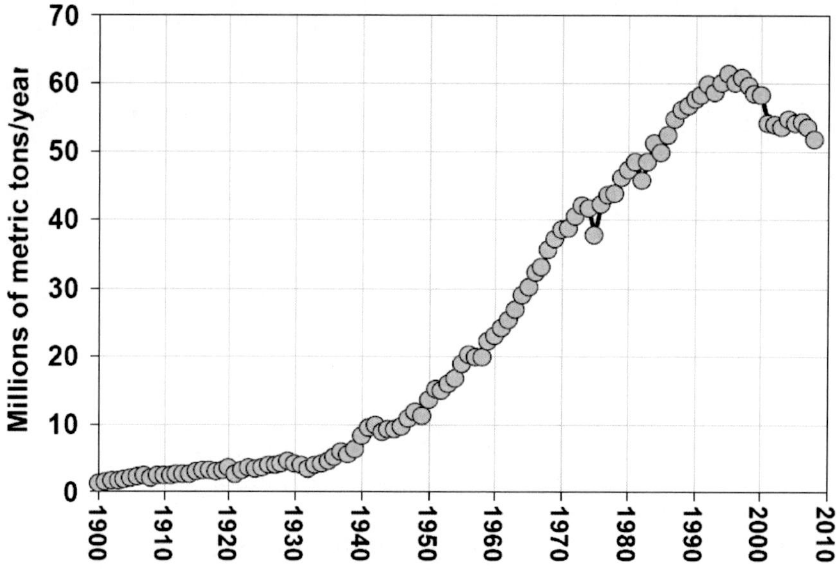

Figure 7. Annual U.S. wood pulp production, 1900–2008 (Department of Commerce, Forest Service, and American Forest & Paper Association 2007).

5. Status and Trends of Pulpwood Market

Shifts in U.S. pulp, paper, and paperboard production appear to have deeply affected the status and trends of the U.S. pulpwood market, as reflected by substantially lower real prices for pulpwood since the late 1990s. Figure 8 shows real price indexes for delivered pulpwood in the United States, along with real price indexes for paper and paperboard (1982 = 100). Nominal price indexes as reported by the U.S. Bureau of Labor Statistics (2008) were deflated using the all-commodity producer price index. As shown in figure 8, the real price of pulpwood generally declined over the past 25 years, but there was a transitory period in the 1990s when the real price of pulpwood was on the increase. However, after 1998 the real price of pulpwood declined sharply as U.S. wood pulp production peaked and declined (figure 7). In the decade between 1998 and 2008, the real price of pulpwood in the United States dropped by 42%. Consolidation in the paper industry helped avoid a similar collapse in real paper and paperboard prices, but real prices for paper and paperboard have also been generally declining with weaker demand since the late 1990s. It can be noted that nominal U.S. paper and paperboard prices actually increased from 2002 to 2008, but that was partly attributable to rising energy prices and higher costs for fuel, transport, and chemicals. After adjusting for inflation, the real prices of paper and paperboard (relative to other commodities) were in fact declining, as shown in figure 8.

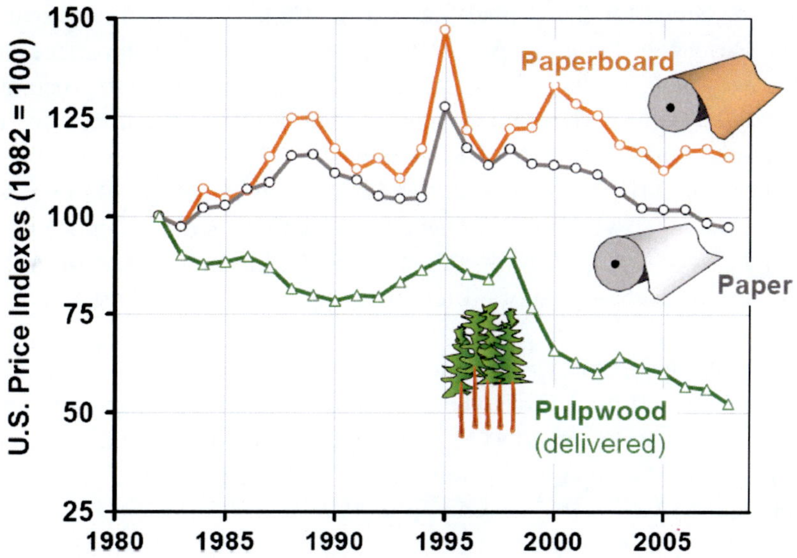

Figure 8. U.S. real price indexes for paper, paperboard, and pulpwood, 1982–2008 (BLS producer price indexes (U.S. Bureau of Labor Statistics 2008) deflated using all-commodity producer price index).

Several market implications are suggested by the real price trends. First, there is the implication that pulpwood supply in the United States generally increased more than demand over the past 25 years, as indicated by generally declining real prices, although there was a period from 1990 to 1997 when the real price trend was increasing, partly a result of reductions in National Forest timber harvests in the late 1980s and early 1990s. However, expansion in the area and productivity of Southern pine plantations no doubt contributed to overall expansion of U.S. pulpwood supply, as did other supply-enhancing developments such as improved pulpwood harvesting technology.

Second, demands for paper and paperboard were sufficient to sustain elevated real prices for paper and paperboard through the 1990s, but declining real prices imply that market demands have weakened for paper and paperboard since the late 1990s. It can be seen that real prices for paper and paperboard were generally increasing prior to the late 1990s, when paper and paperboard production and consumption were still increasing in line with GDP growth, but not since the late 1990s.

Lastly, an implication is that weaker paper and paperboard demands contributed to declines in the real price of pulpwood since the late 1990s. Although the nominal price index for delivered pulpwood actually increased in recent years with higher energy costs (which contributed to higher harvest and transport costs), the real price of pulpwood (after adjusting for inflation) has declined relative to other commodities.

6. PULPWOOD DEMAND OUTLOOK

Future pulpwood demands depend on projections of wood pulp production, which in turn depend on projections of paper and paperboard production, along with projections of other pulpwood demands such as demands for pulpwood in production of oriented strandboard

(OSB). Such projections for the United States as a whole have been developed in periodic Forest Service Resources Planning Act (RPA) assessment studies, most recently the 2005 RPA Timber Assessment Update (Haynes et al. 2007). Implications of structural changes in demand should be considered when viewing the RPA projections. In addition, an unprecedented temporary spike in global oil prices in 2005–2008 elevated policy concerns about future energy security and led to renewed interest in expansion of biomass energy and use of biomass (including wood) for biofuels. Such biofuels could introduce new competing demands for pulpwood in the future, although the timing and extent of that outcome is uncertain, because the global price of oil fell off abruptly from peak levels in the second half of 2008.

The 2005 RPA Update projected much lower growth rates for U.S. pulp, paper, and paperboard production in the decades ahead, more in line with growth rates observed since 2001. Figure 9 illustrates historical U.S. paper and paperboard production and wood pulp production and shows the RPA projections to 2030. Average annual growth in paper and paperboard production was 2.4% per year from 1970 to 1999 (the peak year for paper and board production). Average annual growth in production from 2001 to 2030 was projected to be just 0.8% per year for paper and paperboard, and also for wood pulp, or only about one-third of the pre-1999 growth rate for paper and paperboard output.

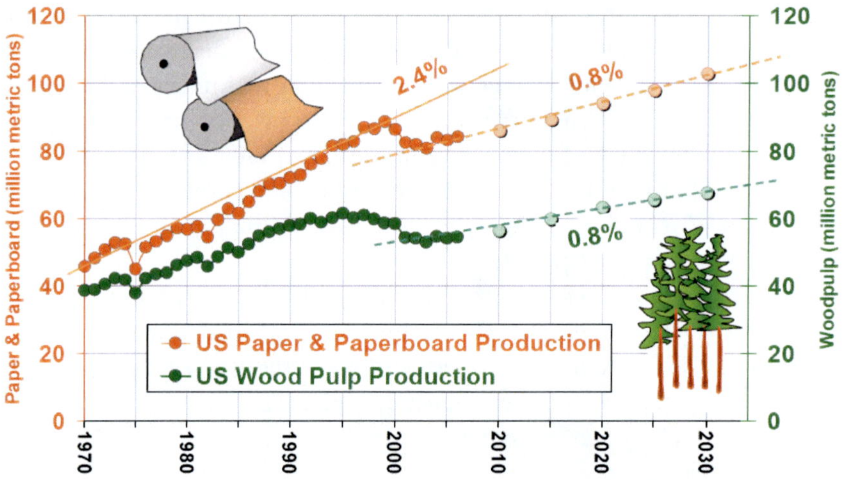

Figure 9. Annual U.S. production of paper and paperboard and wood pulp, 1970–2006 (American Forest & Paper Association 2007), with 2005 RPA projections, 2010–2030 (Haynes et al. 2007).

The 2005 RPA Update (Haynes et al. 2007) and earlier 2000 RPA Timber Assessment (Haynes 2003) recognized that global trends were resulting in much slower growth in U.S. paper and paperboard demands, wood pulp production, and corresponding pulpwood consumption than projected in earlier RPA assessment studies. The last RPA assessment to project a continuation of pre-1990s growth trends in U.S. pulpwood consumption was the 1989 RPA Timber Assessment (Haynes 1990). The 1993 RPA Update (Haynes et al. 1995) recognized that inroads made by increased paper recycling in the 1990s would result in somewhat lower levels of future pulpwood consumption. Figure 10 shows historical pulpwood receipts at U.S. wood pulp mills from 1960 to 2005 (Forest Resource Association

2006) and shows also for comparison projections of pulpwood receipts from the four most recent RPA timber assessments (1989, 1993, 2000, and 2005).

For pulpwood and timber markets, the implication of global change was a substantial lowering of projected U.S. pulpwood consumption relative to earlier RPA assessments, and hence the more recent RPA projections of U.S. pulpwood receipts at wood pulp mills are generally much lower than projections of earlier assessments, as shown in figure 10. Looking at the historical data, it is apparent that after decades of robust expansion, U.S. pulpwood receipts peaked in the mid-1990s (along with U.S. wood pulp production). Growth resumed after the 2000–2001 recession but at a much slower growth rate. According to the 2000 and 2005 RPA timber assessments, pulpwood receipts at wood pulp mills are projected to take decades to fully recover, even with steady GDP growth, as assumed in the RPA analyses.

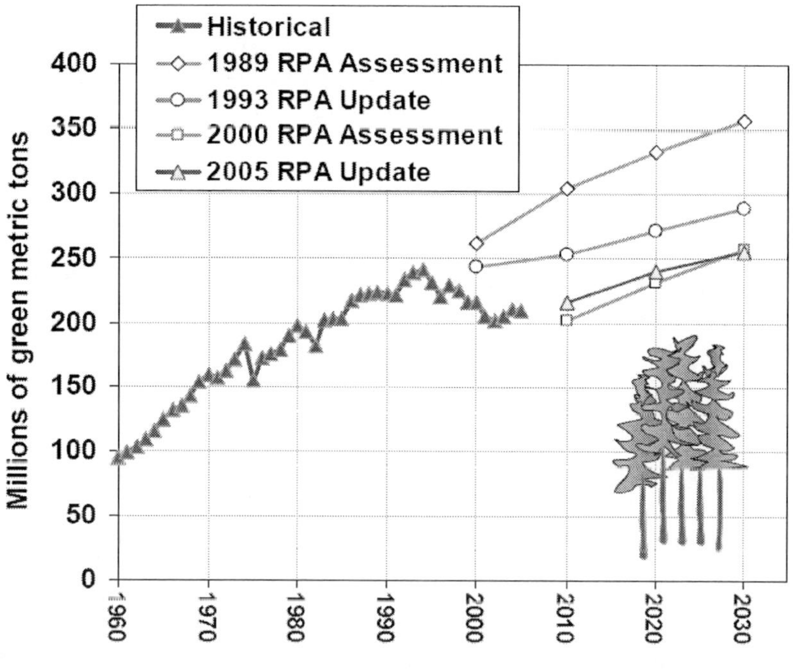

Figure 10. Annual U.S. pulpwood receipts at wood pulp mills, 1960–2005 (Forest Resource Association 2006) and projected receipts from last four RPA studies (Haynes 1990, Haynes et al. 1995, Haynes 2003, Haynes et al. 2007).

7. OTHER IMPLICATIONS

Several other implications for forestry and forest products are related to the status and trends of the U.S. pulpwood market. Those implications include industry divestiture of timberlands and resulting shifts in financial objectives of land management; potential expansion of pulpwood consumption in uses other than wood pulp, such as for other wood products or potentially for wood energy; and emergence of the U.S. South as the only U.S. region with much potential for growth in pulpwood supply or demand, but with generally much slower demand growth than prior to the 1990s.

Since the mid-1990s, coincident with receding U.S. pulpwood demand, there was substantial divestiture of timberland ownership in the United States by integrated forest product firms, primarily large pulp and paper companies. As recently as the early 1990s, the industry owned over 25 million hectares of U.S. timberland, but since then most of that timberland changed ownership and is no longer under direct control of forest product firms (Clutter et al. 2005, Hagen et al. 2005). Timberland investment organizations and real estate investment trusts now own the largest share of the former forest industry land. Their financial objectives reflect fiduciary responsibilities to an investor clientele, and in many cases they actively seek higher investment returns by conversion of forest land into so-called "higher and better uses" (such as residential development). Research has shown that expansion of housing development on America's private forests has serious implications for the condition and management of affected private forests and the watersheds in which they occur (Stein et al. 2005).

The leading alternative use for pulpwood in the United States is oriented strandboard (OSB), a structural wood panel product made primarily from pulpwood. OSB output expanded rapidly since it was introduced commercially in the United States in 1979. OSB is used primarily in housing construction. Thus, North American OSB demand was affected by the recent downturn in U.S. housing construction (2006–2009), but higher U.S. exports of OSB helped avert a more significant collapse in U.S. production. U.S. OSB exports were boosted by generally lower OSB prices, a weaker U.S. dollar, and excess capacity. On the other hand, Canada lost roughly half of its OSB output just since 2006, as the housing slump and a strong Canadian dollar reduced demand for Canadian OSB exports. Figure 11 shows historical trends since 1999 in U.S. and Canadian OSB production.

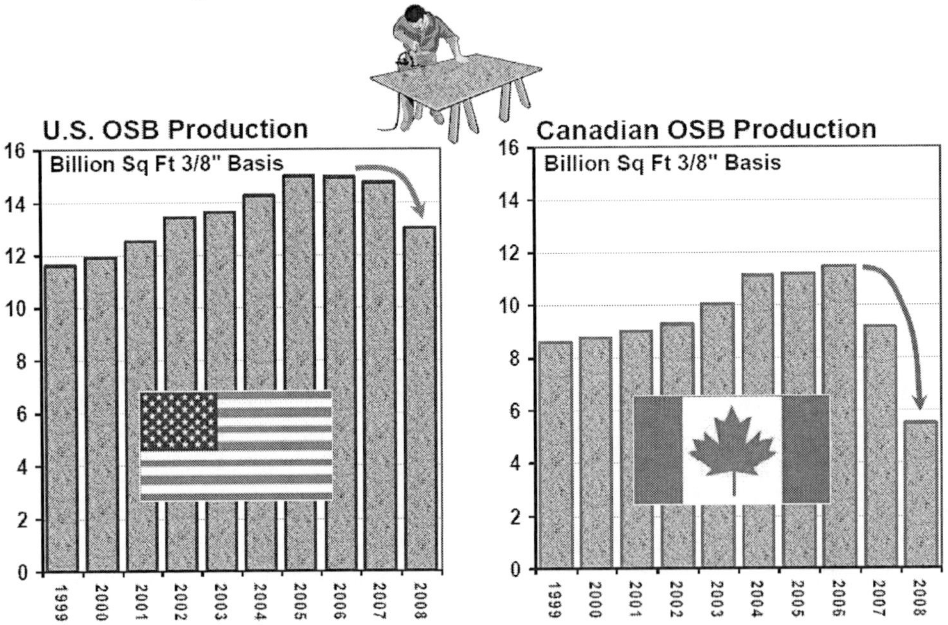

Figure 11. U.S. and Canadian OSB production, 1999–2008 (APA—The Engineered Wood Association 2008); 2008 estimate based on data through 3rd quarter.

According to the 2005 RPA Update, OSB is expected to be a growing element of pulpwood consumption in the decades ahead, but wood receipts at OSB mills will remain relatively small in comparison to wood receipts at wood pulp mills. Figure 12 shows historical data (to 2000) on pulpwood receipts along with RPA projections, including roundwood pulpwood and wood residue receipts at wood pulp mills and roundwood pulpwood receipts at OSB and other wood panel mills. The RPA projections should be viewed in light of the current economic recession, where the downturn in housing is affecting OSB demand and the downturn in industrial production is affecting U.S. output of paper, paperboard, and wood pulp. However, in any case, wood pulp is likely to remain the dominant use of pulpwood in the United States for the foreseeable future.

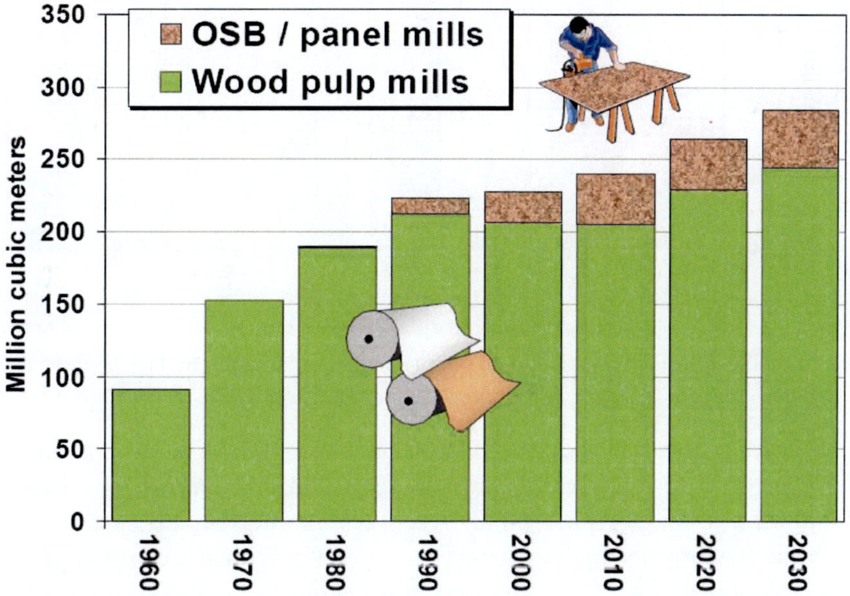

Figure 12. Historical and projected annual pulpwood receipts at U.S. wood pulp mills and at OSB and wood panel mills, 1960–2030 (Haynes et al. 2007).

According to the regional timber supply and demand analysis that was included in the 2005 RPA Timber Assessment Update, the U.S. South was the only U.S. region with much projected expansion of pulpwood supply or demand in the decades ahead, but with much slower overall growth in quantity than that experienced in the late 20th century. Figure 13 shows historical and projected total pulpwood supply quantity by U.S. region, including pulpwood roundwood and residue supplies to wood pulp mills, roundwood pulpwood supplies to OSB and wood panel mills, and pulpwood exports.

There are also some other noteworthy caveats about the current status and trends of the U.S. pulpwood market. The 2005 RPA projections were based on a view that expansion of electronic media in advertising would continue to offset future demand for print advertising, and thus despite projected expansion in U.S. population and GDP, the projections assumed declining demand for pulpwood in newsprint production and gradually declining demand for pulpwood in production of printing and writing paper. Projected expansion in pulpwood consumption is thus limited to wood pulp used for other paper and paperboard products (chiefly packaging paper and paperboard, and tissue and sanitary paper products) and OSB products. Demands for many of those products, particularly OSB and packaging paper and paperboard, are being adversely affected by the current economic recession.

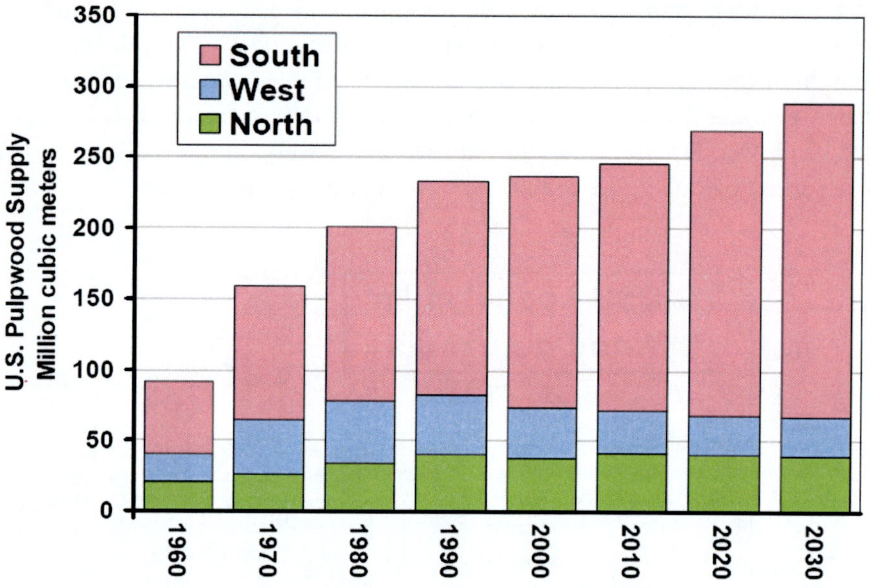

Figure 13. Historical and projected total annual pulpwood supply quantity by U.S. region, 1960–2030 (Haynes et al. 2007).

The 2005 RPA analysis included projections of conventional fuelwood demand but did not include projections of large-scale conversion of wood or biomass to biofuels. With the recent spike in oil prices and related policy responses, biofuels would appear to represent a potential emerging future demand for wood that could potentially compete for pulpwood. Nevertheless, technology for efficiently converting cellulosic materials such as wood into biofuels is yet to be demonstrated, and large-scale investment and R&D efforts would be needed to create competitive demands for pulpwood from biofuels. Also, the precipitous drop in oil price that occurred in the second half of 2008 suggests less impetus for rapid commercial expansion of biofuel production in the near term.

Based on recent average historical prices for various product alternatives, production of wood pulp still offers much higher revenue potential for pulpwood than does production of biofuels such as cellulosic ethanol. For example, in production of bleached kraft market pulp, the yield of pulp in dry tons per dry ton of pulpwood input is 45% to 50%, meaning that about 0.45 to 0.5 dry metric tons of wood pulp output can be produced per dry metric ton of pulpwood input. The market value of that pulp in 2007 was $350 to $440, with market pulp prices ranging from $700 to $800 per metric tonne air-dry (10% moisture). Cellulosic ethanol production from wood via hydrolysis and fermentation is generally thought to have a yield potential of 80 to 90 gallons per dry metric ton of wood input, and with ethanol prices recently around $1.75 per gallon, the market value of the potential ethanol output from one metric ton of wood is $140 to $160 (or less than half the revenue from production of wood pulp). Taking into account total costs of production, the profit margins would be generally lower for conversion of pulpwood into ethanol than for production of wood pulp. Of course, prices and costs vary over time for ethanol, wood pulp, and pulpwood. Higher future market prices for ethanol or other biofuels or improved production cost efficiency could increase competitiveness of biofuel production from wood. Also joint production of wood pulp and biofuels at integrated forest product biorefineries is a potentially more lucrative opportunity that is being explored by the forest industry.

The next RPA forest assessment (in 2010) will explore future timber supply and demand implications of scenarios that involve global expansion of wood use for energy. Although long-range future wood use for biofuel could be potentially quite large, for the next 5 to 10 years, consumption of pulpwood in the emerging biofuel industry is likely to remain just a small fraction of total U.S. pulpwood demand. This presents an economic and policy challenge in forest resource management, whether to sustain or expand pulpwood supplies in current times of relatively subdued timber demand when there may be much larger foreseeable needs, but only in the long run. This challenge must also be viewed in the context of ongoing structural changes in timberland ownership and shifts in land management objectives that are now being driven by other fiduciary interests (such as land development) in addition to timber production.

8. Summary

Change occurred in the global distribution of growth for paper and paperboard production since the late 1990s, as growth shifted toward Asia and Europe and away from North America. U.S. paper and paperboard purchases peaked in 1999 and then sharply declined along with U.S. industrial production in 2000–2001. Comparatively modest growth was experienced from 2002 to 2007 (averaging less than 1% per year), and yet another downturn in industrial production and paper and paperboard demand occurred in 2007–2008 along with a more generalized global recession. The recent downturn was associated with declining retail sales and consumer spending.

Average historical growth rates before 1999 were well above 2% per year for U.S. production and consumption of paper and paperboard, but post-2001 growth rates have averaged 0% per year, and projected future growth rates are well below 1% per year. The status and trends of the U.S. pulpwood market are reflected by declining real prices for pulpwood in the United States over the past 25 years, and especially since the late 1990s, as pulpwood demand has receded while supply has expanded. The general implication is that growth in pulpwood supply has exceeded growth in demand. OSB has appeared as a growing demand for pulpwood in recent decades, but wood use at U.S. OSB mills remains relatively small compared with pulpwood consumption at wood pulp mills. The recent downturn in housing construction has also negatively affected OSB demand. On the positive side, improved competitiveness and a weaker U.S. dollar helped erase the deficit in net export tonnage of pulp, paper, and paperboard products, and OSB exports have also surged in the past couple years. According to the 2005 RPA Timber Assessment Update, the U.S. South is the only U.S. region with any projected expansion of pulpwood supply and demand in the decades ahead, but at a much slower overall growth rate than experienced in the late 20th century. The next RPA forest assessment (in 2010) will explore timber supply and demand implications of scenarios that involve large-scale future expansion of wood use for energy, but for the next 5 to 10 years, consumption of pulpwood in the emerging biofuel industry is likely to remain a small fraction of total U.S. pulpwood demand. A challenge for forest management and policy is whether to sustain or expand pulpwood supplies in current times of relatively lean demand growth when there may be much larger foreseeable needs (such as for wood energy) but only in the long run.

REFERENCES

American Forest & Paper Association. (2005). U.S. Forest Products Industry—Competitive Challenges in a Global Marketplace. http://bipac.net/page.asp?g=afpa&content=White_Paper_re_Competitiveness_Summary&parent=AFPA.

American Forest & Paper Association (2007). Statistics of Paper, Paperboard & Wood Pulp. 45th and earlier editions. Washington, DC: American Forest & Paper Association.

American Forest & Paper Association. (2008). Paper, Paperboard & Wood Pulp, Monthly Statistical Summary. Volume 86, No. 11 and earlier editions. Washington, DC: American Forest & Paper Association.

APA—The Engineered Wood Association. (2008). *Engineered Wood Statistics*. Third Quarter 2008 (and earlier editions). Tacoma, WA: APA—The Engineered Wood Association.

Clutter, M., Mendell, B.C., & Newman, D.H. (2005). The changing landscape of industrial timberland ownership in the South. *Carolina Forestry Magazine*, *November*, 25-27.

Food and Agriculture Organization of the United Nations. (2009). FAOSTAT. http://faostat.fao.org/site/626/default.aspx#ancor.

Forest Resource Association (2006). *Annual Pulpwood Statistics Summary Report, 2001-2005.* Rockville, MD: Forest Resources Association.

Hagen, J.M., Irland, L.C., & Whitman, A.A. (2005). Changing Timberland Ownership in the Northern Forest and Implications for Biodiversity. Report # MCCS-FCP-2005-1. Brunswick, ME: Manomet Center for Conservation Sciences.

Haynes, R.W. (1990). *An Analysis of the Timber Situation in the United States: 1989-2040.* Gen. Tech. Rep. RM-199. Fort Collins, CO: U.S. Department of Agriculture, Forest Service, Rocky Mountain Forest and Range Experiment Station.

Haynes, R.W. (technical coordinator) (2003). *An Analysis of the Timber Situation in the United States: 1952 to 2050.* Gen. Tech. Rep. PNW-GTR-560. Portland, OR: U.S. Department of Agriculture, Forest Service, Pacific Northwest Research Station.

Haynes, R.W., Adams, D.M., & Mills, J.R. (1995). *The 1993 RPA Timber Assessment Update.* Gen. Tech. Rep. RM-259. Fort Collins, CO: U.S. Department of Agriculture, Forest Service, Rocky Mountain Forest and Range Experiment Station.

Haynes, R.W., Adams, D.M., Alig, R.J., Ince, P.J., Mills, J.R., & Zhou, X. (2007). *The 2005 RPA Timber Assessment Update*. Gen. Tech. Rep. PNW-GTR-699. Portland, OR: U.S. Department of Agriculture, Forest Service, Pacific Northwest Research Station.

Stein, S.M., McRoberts, R.E., Alig, R.J., Nelson, M.D., Theobald, D.M., Eley, M., Dechter, M., & Carr, M. (2005). *Forests on the Edge: Housing Development on America's Private Forests.* Gen. Tech. Rep. PNW-GTR-636. Portland, OR: U.S. Department of Agriculture, Forest Service, Pacific Northwest Research Station.

U.S. Bureau of Economic Analysis. (2008). Gross Domestic Product (GDP). http://www.bea.gov/national/index.htm#gdp.

U.S. Bureau of Labor Statistics. (2008). Producer Price Index Commodity Data. http://data.bls.gov/PDQ/outside.jsp?survey=wp.

U.S. Federal Reserve. (2008a). Statistical Release G.17. Industrial Production and Capacity Utilization. http://www.federalreserve.gov/releases/G17/table1_2.htm.

U.S. Federal Reserve. (2008b). Statistical Release H.10. Foreign Exchange Rates. http://www.federalreserve.gov/releases/H10/summary/.

Wood Resources Quarterly. (2008). Wood Resources International (Press Release, August 23, 2008). http://www.woodprices.com (Contact: Håkan Ekström).

In: Global Change and Forestry
Editors: J. Gan, S. Grado and I. A. Munn
ISBN: 978-1-60876-262-0

Chapter 9

A Forest Products/Bioenergy Mill Location and Decision Support System Based on a County-Level Forest Inventory and Geo-Spatial Information

T. Luke Jones[1], Emily B. Schultz[1*], Thomas G. Matney[1], Donald L. Grebner[1], David L. Evans[1], Curtis A. Collins[2] and Patrick A. Glass[3]

[1] Forest and Wildlife Research Center, Mississippi State University, MS, USA
[2] Larson and McGowin, Inc., AL, USA
[3] Mississippi Institute for Forest Inventory, MS, USA

Abstract

The forest products industry is a major component of the economic base for many states in the southeastern United States. Forest inventories that are precise within a 50- to 80-mile mill working circle (i.e., procurement area) and the availability of decision support tools for locating mills are of primary importance in attracting and sustaining the industry. This chapter reviews the current status of the State of Mississippi's efforts to provide forest inventory information to attract forest industry and balance potentially increased utilization due to new markets with resource sustainability. A pilot study is described that integrated a decision support system (DSS) with a geo-spatially based county-level forest inventory to determine the feasibility and optimal location of forest products mills. The DSS aids economic development decisions for state planners, forest industries, and forest and wildlife managers. Geographic Information System (GIS) layers were constructed for a 15-county southwest Mississippi study area for forest type, age, ownership, and volume. Growth and drain ratios were calculated from volume-age

* Email: eschultz@cfr.msstate.edu, Forest and Wildlife Research Center, Mississippi State University, Mississippi State, MS.

relations and historical Landsat multi-spectral images. Raw material and finished product haul distances and costs were determined from a geo-spatial transportation network of the primary, secondary, and county roads. The user enters mill specific input parameters through a Microsoft Windows® interface, and a linear programming (LP) model minimizes costs of procuring wood to the potential mill site. Net present values (NPV) were calculated to assess financial feasibility of placing a mill at the selected location. Future versions will incorporate minimization of transporting finished products to market.

Keywords: Mill location; Forest products; Cost minimization; Linear programming.

1. INTRODUCTION

Selecting an optimal location for a forest-based industry is a data intensive and costly exercise typically conducted by corporate and government decision-makers. The process involves the manual integration of analyses from numerous consultants and experts and can be error prone. Mill location studies are very costly and can take months or years to complete. Since such a study takes a long period to complete, it also has to keep pace with changes that might affect the outcome of such an analysis. Included in the information that potential investors must know are: 1) the quantity, location, and type (i.e., species, size class) of woody feedstock available, 2) transportation networks and delivery costs of raw material to the mill and of finished product to the market, 3) sustainability of feedstocks (e.g., future growth, utilization), 4) competing utilization for raw materials, 5) product demand, and 6) various types of socio-economic data on education and the workforce. Without this information, investors will not be capable of making sound minimum risk investment decisions. This analysis is mandatory not only to establishing new facilities but also for planning modification or expansion of existing facilities. Computer-based decision support systems (DSS) are needed that efficiently coalesce geo-spatial information, forest inventory, transportation costs, and socio-economic factors to allow forest-based investors to rapidly assess economic viability and sustainability of proposed plant locations and capacities. These same tools can also be used to evaluate proposed plant expansions or modifications and predict future raw material availability and evaluate the impact of competing resource consumers (e.g., biofuels, particle-based panels and lumber, plywood, sawtimber, poles, veneer, pulpwood, reconstituted solid wood products, exports). The State of Mississippi has a unique opportunity to develop a DSS for selecting optimal mill locations and monitoring forest resource sustainability because of its current county-level forest inventory system, and its construction is the focus of this chapter.

A variety of industries compete for a limited forest resource that meets specific manufacturing requirements. Each product requires a minimum log diameter size class and quality. Any one forest-dependent plant must compete with all other product manufacturers and buyers within a targeted procurement area. Traditional forest products or bioenergy plants must find sufficient volume/biomass among competing uses that is obtainable at a reasonable cost within a required product size class. The price and availability of feedstocks will depend upon supply and existing demand within the procurement area of a proposed plant. Multi-product forest product companies direct the flow of products to their various mills based on tree diameter, length, and quality. For example, representative pine merchantability specifications are 3.6 to 7.5 inches diameter at breast height (dbh) for pulpwood, 7.6 to 8.5 inches for chip-and-saw, 8.6 to 11.5 inches for peelers, and 11.6 inches and greater for sawlogs. Higher quality pulpwood sized trees are also utilized for particle-based panels and lumber [e.g., oriented strand board (OSB), engineered solid wood products]. Biofuels will most likely compete directly with mill and forest residues (e.g., logging slash) and pulpwood-sized trees, which can also be considered sawtimber growing stock.

Thirteen southeastern states contain nearly one-third of the forest inventory and almost one-half of the timber harvested in the United States (U.S.). Forest industries in these 13 states produce 45% of the softwood lumber, 56% of total paper production capacity, and 72% of total wood pulp production capacity (Sun and Zhang 2001). The forest products industry is a major component of Mississippi's economic base, with timber one of the most valuable agricultural crops accounting for more than $1 billion of harvested forest products annually (Munn and Tilley 2005). In 2001, the amount of pine and hardwood stumpage utilized resulted in $801 million in payments to Mississippi landowners. The total (i.e., direct and indirect) output for aggregated forest-related sectors was approximately $13.4 billion with $5.3 billion of value-added (Munn and Tilley 2005). The early indications are that increased utilization of forest thinnings and logging residues for biofuels promises to add more value and importance to Mississippi's forest industry (Grebner et al. 2008). Mississippi has enough woody biomass residues to produce 280 to 400 million gallons of ethanol annually utilizing a range of currently available conversion technology or an equivalent of 6,500 to 9,300 MW of electricity. The development of a biofuel industry could translate into additional jobs, wages, expenditures, and regional production (Grebner et al. 2008). Forest lands, like those in Mississippi, provide an array of sources for woody biomass that could potentially be utilized for biofuels production. These sources include standing woody biomass, forest residues, and manufacturing residues. Growing demand for woody biomass from biofuel markets heightens the importance of obtaining forest inventory information and DSS tools on a scale fine enough for attracting and maintaining the industry and sustaining the resource.

This chapter reviews the current status of the State of Mississippi's efforts to provide forest inventory information to attract forest industry and balance potentially increased utilization due to new markets with resource sustainability. A pilot study is described that integrates a DSS with a geo-spatially based county-level forest inventory to determine the feasibility and optimal location of forest products mills. The DSS was developed in phases, and background information important to the current work in optimal mill location and feasibility, forest resources assessment, transportation, and mill requirements is reviewed. The next DSS phase of the study will include the addition of socio-economic data to the spatial database and the minimization of transporting finished products to market.

2. BACKGROUND

2.1. Optimal Mill Location

Optimal mill location and feasibility studies have been traditionally carried out by forest industry and, therefore, are usually proprietary and never published. As a result, there are few examples in the literature. An optimal mill location can be defined as a site that minimizes the sum of procurement and shipment costs from a processing plant to market (McCauley and Caulfield 1990). Distribution and market costs may also be considered (Lin et al. 1996). Important factors involved in an optimal location study include stumpage availability, furnish mix, wood costs, market distance, market penetration, species type, capital availability, timber supply, and product market demand (Lin et al. 1996; McCauley and Caulfield 1990).

McCauley and Caulfield (1990) divided Alabama into supply and demand points that consisted of 10 regions based on topography, species, and supply characteristics. Potential plant site locations were designated as each region's geographic center. A linear programming model was used with a base-level set of parameters to find an optimal plant location. The model's objective function value minimized the costs of procuring timber to potential sites and transporting finished products to demand points. Model variables included potential plant sites, supply regions, market locations, cost of procuring pine and hardwood timber, cost of transporting final product to market, amount of pine and hardwood available, demand for the final product, amount of wood required by the mill, amount of mill furnish consisting of hardwoods, amount of final product produced, number of plants located within the region, amount of pine and hardwood transported to the mill, amount of the final product transported to market from a plant site, and total amount of wood transported to a potential plant site. Supply information was obtained from the U. S. Department of Agriculture Forest Service (FS) Southern and Southeastern Experiment Station analyses that calculated growth minus natural mortality and pulpwood annual removals. Calculated surplus was assumed to be the only source of wood available. Timber cost was estimated from the sum of stumpage, logger profits, and harvesting and transportation costs. Transportation and harvesting costs were derived from prices of pine and hardwood stumpage provided by *Timber Mart-South*. To assess how changes in assumptions would influence the optimal location decision, a sensitivity analysis was conducted. While McCauley and Caulfield (1990) addressed a complex management problem, they separated capacity planning and location planning.

Lin et al. (1996) considered capacity and location simultaneously while including operating costs by using OSB/LOCATION software, a decision support program with financial and mixed integer linear programming (MILP) models for locating OSB mills in eastern Alabama. MILP determined the optimal location and size based on the most cost efficient set of options and considered factors such as capital availability, timber supply, and product market demand. Economic desirability was evaluated with a financial model developed under expenditure and capital investment assumptions and analyzed by discounted cash flows. The objective function of the Lin et al. (1996) MILP problem minimized the sum of costs for mill establishment, procuring timber for the potential mill, transporting products to markets, and mill operation. Model constraints were mill input, mill output, mill site and size, target market, timber supply, and budget. Input data required for the OSB/LOCATION program fell into six categories: model parameters, timber supply information, OSB market information, transportation data, financial data, and operating cost data. Model parameters accounted for the number of potential plant locations and sizes, timber supply regions, and product markets. These factors determined the size of the MILP model. Timber supply information was derived from FS Forest Inventory and Analysis (FIA) data and delivered mill prices were input by region. OSB market information was the estimated demand quantities, predicted selling prices, and target market shares. Transportation data were distances and costs from plant locations to various markets, and financial data included the working capital of plants, the initial capital investment, and a minimum rate of return. The OSB/LOCATION program returned optimal plant location and size, timber resource use information, production-distribution cost information, investment return information, and marketing information (Lin et al. 1996).

2.2. Mill Feasibility

In addition to determining the optimal location for a forest products-based mill, financial feasibility should be examined by calculating net present values (NPV) and internal rates of return (IRR). Blinn et al. (1986) calculated after tax NPV to determine economic and market feasibility of a proposed 1986 yellow-poplar [*Liriodendron tulipifera* L.] OSB mill producing 120 million ft^2 in a preselected area. Labor supply, geographical distribution of markets, resource quality and availability, and raw material costs were factors considered. Quantity of available resource and influences of landowner attitudes and accessibility were based on FS data. Assumptions for simulating a prototype OSB plant included capital requirements and costs for production, maintenance, engineering, energy, residue, supplies, and other fixed costs such as taxes, insurance, depreciation, and administrative salaries. A specified plant operating schedule was also assumed. Risk was included in the cash flow projection by inputting derived data in a computer-based Forest Products Investment Model (FPIM) that performed a Monte Carlo simulation and generated NPV probability distributions (Chambers et al. 1986). Other estimates and assumptions were made for depreciation and investment tax credits, gross revenues, and investment period. FS data estimated excess yellow-poplar growth and availability within a 75-mile radius of the plant site. A market analysis estimated the percentage of market penetration necessary to sell plant products within 300- and 500-mile radii (Blinn et al. 1986).

2.3. Forest Inventory, Growth, and Drain

All mill feasibility and optimal location studies require an accurate estimate of wood resources available to the mill. Several factors influence availability including land ownership, current growing stock, and projected yields. In Mississippi, 65% of the land is owned by nonindustrial private forest (NIPF) landowners, 18% by forest industry, and 17% by the state and federal government (Arano and Munn 2004). NIPF landowners generally own small and fragmented tracts of land and may have less capital to support management activities. Future increases in demand, like that from a new biofuel market, could necessitate less traditional or more intensive forest management practices to avoid timber scarcities (Arano and Munn 2004). Since NIPF landowners control a large proportion of Mississippi forest land and employ fewer management practices, there could be serious implications for future timber availability. Information on landowner attitudes and ownership patterns is important in estimating the proportion of forest resource availability to a mill and for planning sustainability of a potential mill. Ownership statistics for Mississippi have been collected by the Mississippi Institute for Forest Inventory (MIFI), and another study is currently being conducted at Mississippi State University to assess the willingness of different landowner groups to manage and sell timber.

Increased transportation costs of delivering raw materials to a mill have decreased procurement working circle radii and created a greater need for forest inventories that are precise for areas as small as a county. In 2002, MIFI was created to fulfill the need for a county-level forest inventory. This inventory together with a Web-based DSS tool is being used to assist statewide forest resource planning and mitigate the risk associated with decisions made in selecting a forest product-based mill location. Before MIFI inventory data were available, forest resource volume estimates were primarily based on FS FIA large area inventory data, which is currently sufficiently precise at only a 10 county or more area basis (USDA FS 2005). For FIA data, even at a 10+ county area, the percent sampling error at a confidence level of 95% is around 67%, and this sampling error is too large to secure funding from most financial institutions. Conversely, a five-county or larger area is typically larger than the maximum working circle required by most forest industries. Industry will be reluctant to invest in new facilities unless mill location analyses support the ability of the resource in a targeted area to sustain the mill throughout its expected life. The only way that reliability numbers can be placed on estimates is through an integrated inventory growth and harvest projection system derived from a geo-spatial database.

The MIFI inventory facilitates the estimation of growth and drain on a much smaller acreage basis than the FIA inventory. The State has been divided into five regions (i.e., north, central, Delta, southeast, southwest) with one region inventoried each year on a rotating basis. The inventory was spatially-based with sufficient sampling to be reliable at the county-level. Targeted sampling error levels are ± 15% at the 95% confidence level (Parker et al. 2005). Inventory plots were allocated according to an optimal stratified random sampling scheme based on satellite imagery and forest GIS strata (i.e., pine, hardwood, mixed pine-hardwood) within a county.

Inventory data and spatial information were incorporated into the MIFI Inventory Dynamic Reporter DSS (Matney and Schultz 2008). This Microsoft Windows® based interface allows users to obtain geo-spatial estimates of current and expected future forest volume and biomass within specified counties, working circles, or polygon bounded regions for combinations of nine forest and political selection attributes. Historical Landsat scenes were used to determine forested areas and change detection (i.e., afforestation, harvest detection, deforestation), and growth was obtained through stand table projection procedures.

2.4. Transportation

Transportation distances in acquiring wood, delivering products, and related costs are important factors in determining the feasibility and optimal location of a mill. McCauley and Caulfield [1990] measured transportation distance from the supply region to a potential mill site as the straight line distance between two points on a road map (i.e., the center point of a region). The disadvantage of this method was that it underestimated the transportation distance and, thus, the transportation cost. Current geo-spatial road databases and GISs now present the opportunity for making more accurate distance calculations.

Forest industries and local economic development organizations need reliable and precise forest inventory estimates based on a 50- to 80-mile mill working circle. This is critical to determining where forest manufacturing facilities can be located, as they need to be centrally positioned in areas where raw material supplies can be sustained at reasonable transportation costs over the manufacturing facility life. Since transportation costs are an ever increasing factor for a facility to operate profitably, the mill must be located in an area that has a transportation network sufficiently small to make the facility operation feasible.

2.5. Feedstock Requirements and Capacities

Mill production capacities and annual wood, species, and age class requirements are important input factors in feasibility and optimal location studies. Blinn et al. [1986] reported on the production capacity for an OSB mill in the Appalachians that was 120 million ft^2 on a 3/8-inch basis per year. Yellow-poplar was the required species because of its similar properties to the preferred species of aspen (Populas spp.). Annual solid wood requirements for the mill consisted of 7.5 million ft^3 based on industry data and manufacturer's specifications. Timber supplied to the mill was in 16-foot lengths and ranged from 4 to 20 inches in diameter (Blinn et al.1986). The McCauley and Caulfield [1990] OSB plant study required 110,000 cords of wood annually to produce 160 million ft^2 of 3/8-inch OSB. Species mixture included sweetgum (*Liquidambar styraciflua* L.), yellow-poplar, and southern pine (Pinus spp.). Current production statistics are confidential by nature within the industry and are not generally available. Operationally, however, an integrated DSS for evaluating optimal mill locations and their feasibility would allow industry users to input their own production statistics without fear of disclosure.

3. Mill Location: A Mississippi Decision Support System

A mill location and feasibility case study for the southwest region of Mississippi was carried out based on MIFI inventory data and the Dynamic Reporter Interface DSS developed at the Forest and Wildlife Research Center (FWRC), Mississippi State University. Enhancements were made to the existing MIFI interface DSS to allow users to choose options that determine the feasibility and optimal location of a forest products based mill (figure 1). User inputs included selecting the inventory region of Mississippi, type of mill and requirements, resource area size surrounding the proposed mill location, timber type and size class, and ownership class. The final products were an objective function value and the NPV of the proposed mill. An optimal mill location is determined by comparing objective function values and NPVs among proposed sites.

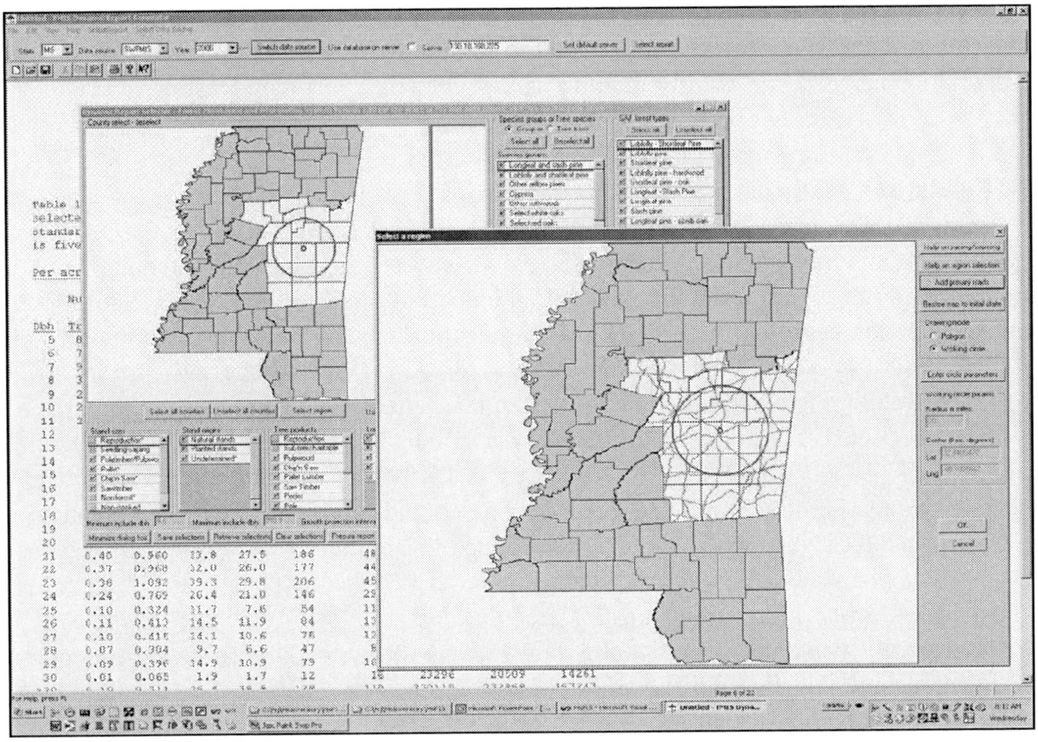

Figure 1. The pilot version Mississippi Institute for Forest Inventory (MIFI) interface decision support system (DSS) for determining optimal forest product-based mill locations in Mississippi.

3.1. Methods

The pilot version of the mill location DSS utilized the 15-county 2004 - 2005 MIFI southwest forest inventory, where an average of 150, 0.2-acre fixed radius plots per county were allocated (figure 2). Inventory data and spatial information were used to develop GIS layers for forest type and age (Collins et al. 2005), ownership, and volume (figure 3).

Volumes for pine, mixed pine-hardwood, and hardwood forest types (in 11 different volume units) were estimated from the regression of the natural log of volume on the inverse of age (Equation 1) and allocated to each image pixel in a user selected area.

$$\ln(vol) = b_0 + b_1(1/age) \quad (1)$$

where ln = natural logarithm, vol = volume, and b_0 and b_1= regression coefficients.

ERDAS Imagine's® spatial modeler was used to create volume layers from the appropriate regression equations (Leica Geosystems 2003). Growth (i.e., the forest resource growing at the present rate projected to a specified time) was calculated from growth curves based on the volume-age relation in Equation 1. Drain (i.e., the forest resource removed or harvested within a given time period) was calculated by comparing 2004 and 2006 Landsat images (figure 4) to identify harvested areas and predicted removals. Removals were randomly allocated to procurement areas that matched volume, age, forest type, and ownership classes. The resulting growth to drain ratio indicated the sustainability of forest resources under current and future harvest demands.

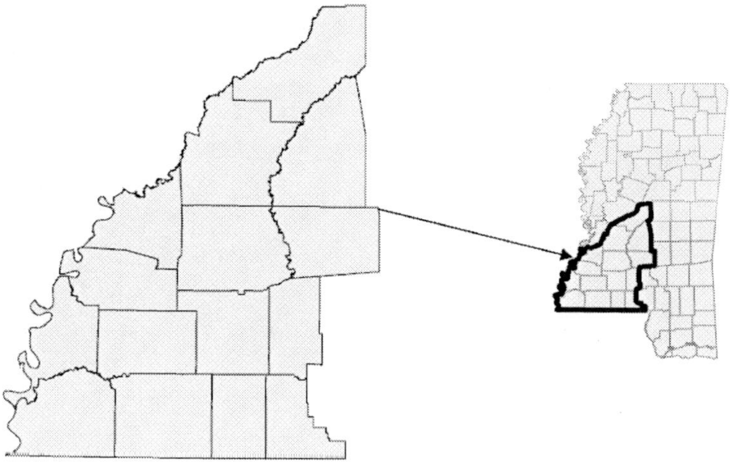

Figure 2. Fifteen-county southwest Mississippi study area inventoried by Mississippi Institute for Forest Inventory (MIFI) in 2004 and 2005.

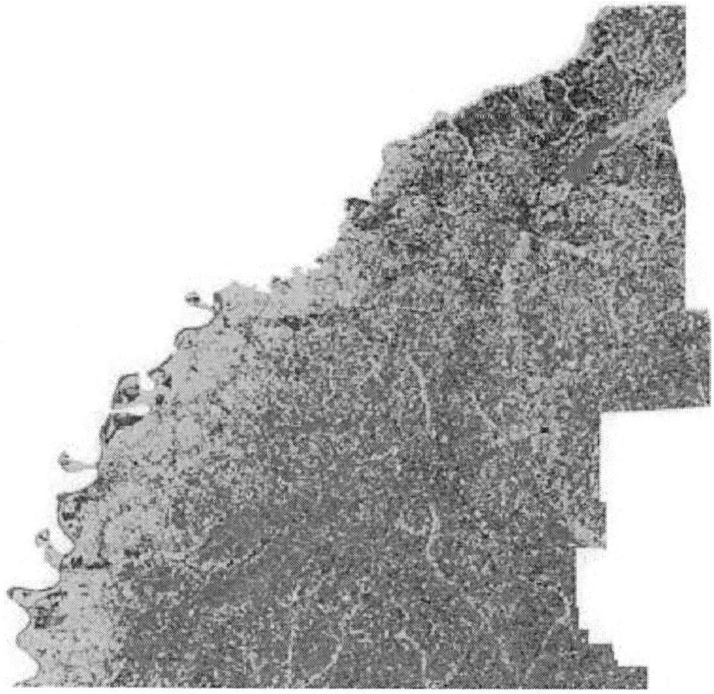

Figure 3. The 2004 Geographic Information System (GIS) forest type layer for the southwest Mississippi Institute for Forest Inventory (MIFI) inventory region developed from Landsat data and used as decision support system (DSS) input for calculating growth and drain volumes by forest type.

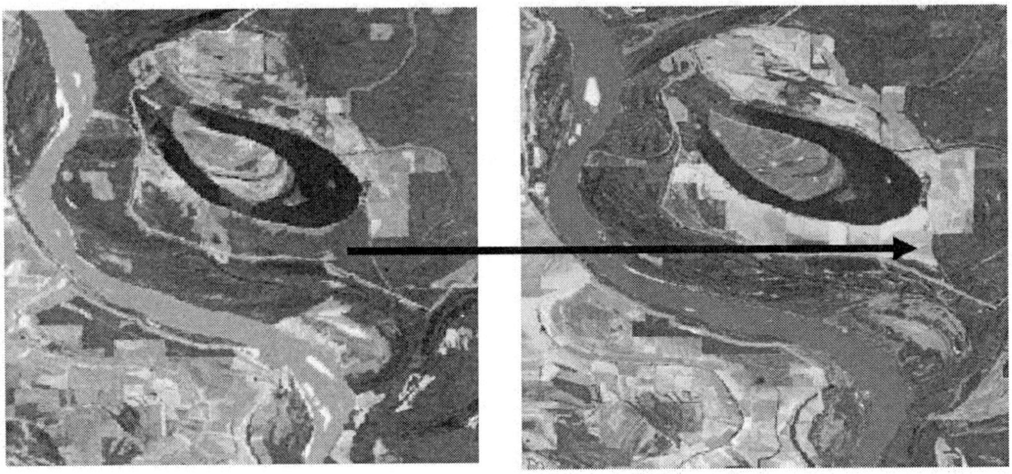

Figure 4. Historical Landsat scenes of the Eagle Lake area located near Vicksburg, Mississippi, in 1972 (left) and 2004 (right) as an example of change detection techniques used to determine forested area and derive growth and drain ratios.

Geo-spatial data for Mississippi's road network were obtained from the Mississippi Automated Resource Information System (MARIS) (MARIS 2006), and transportation network nodes and segment lengths were constructed from the primary, secondary, and county roads (figure 5) using ArcGIS® software (ESRI 2005). The resulting network provided input to a Microsoft Visual C++® (Microsoft 2005) minimum route algorithm that accumulated volumes and weights by GIS forest type and calculated minimum route distances from each forested pixel in the selected area to the proposed mill location.

Procurement and transportation costs were input parameters to a linear programming (LP) model that minimized costs for a potential mill site. Total procurement cost for a user defined procurement area was based on *Timber Mart-South*'s (Timber Mart-South 2008) dollar/volume unit price for mills in Mississippi. Transportation costs were obtained from Route Chaser software (Stuart and Grace 2008). This program allows the user to enter scheduling, equipment, and cost components. The scheduling component includes work days per year, hours per day, allowable extra hours, and overtime. The equipment component includes gross vehicle loaded weight (lbs) and vehicle tare weight (lbs). The cost component includes equipment purchase price, financing, labor, fuel, repair and maintenance, insurance, and overhead costs. The Route Chaser program outputs per ton cost for each mile traveled (Stuart and Grace 2008).

Once the growth and drain ratio and procurement and transportation costs were determined, a LP model was constructed with C-whiz software (KMS 2003). The LP objective function minimized costs of procuring wood for a potential mill site. Input constraints and variables (tables 1 and 2, respectively) were similar to those used by McCauley and Caulfield (1990). Sensitivity analyses identified LP factors with the greatest influence on mill location, such as stumpage volume, timber type, size class, and ownership class. Mill feasibility was assessed by calculating NPVs. Present value of costs incurred by a specified mill were subtracted from the present value of the mill's annual production revenue. The DSS accumulates procurement and transportation costs in the selected procurement area while the user specifies other costs such as labor, utilities, and plant maintenance. The user also specifies expected annual production revenue. If the resulting revenues minus costs produce an NPV that is zero or greater, it is deemed feasible to locate the specified mill in the selected procurement area. Ongoing work will seamlessly combine all optimal mill location and suitability components into the current MIFI interface (see http://www.mifi.ms.gov/mission.htm) (figure 1).

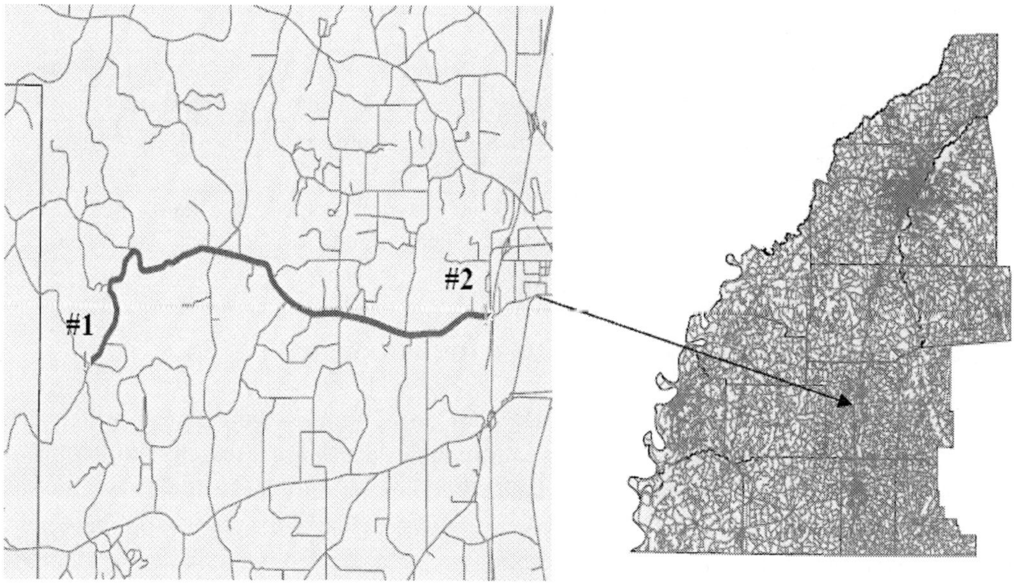

Figure 5. Example transportation network constructed for 15 counties in the southwest inventory region of Mississippi that calculated the shortest distance between a procurement area/pixel (#1) and a potential mill site (#2).

Table 1. Constraints used in the C-whiz linear programming (LP) model that minimizes the cost of procuring wood to a potential mill site (adapted from McCauley and Caulfield 1990)

• Annual mill wood volume requirement
• Mill annual mill volume requirement does not exceed the amount of timber available in the selected procurement area
• Mill annual production requirement
• Mill GIS forest type (pine/hardwood/pine-hardwood mixed) requirement proportion limit
• Non-negativity constraint

Table 2. Input variables used in the C-whiz linear programming (LP) model that minimizes the cost of procuring wood to a potential mill site (adapted from McCauley and Caulfield 1990)

• Potential plant sites and timber supply regions
• Cost of procuring a given volume of pine/hardwood/pine-hardwood mix from the selected procurement area to the potential mill location
• Volume of pine/hardwood/pine-hardwood mix available to a mill from the selected procurement area
• Volume wood required by the mill
• Mill GIS forest type (pine/hardwood/pine-hardwood mixed) requirement mix
• Targeted minimum mill unit production of finished product
• Binary variable for selecting or not selecting a site
• Volume of pine/hardwood/pine-hardwood mixed GIS forest type transported to proposed mill site from the selected procurement area

3.2. Challenges, Weaknesses, and Future Improvements

Many challenges exist for the development of an automated geo-spatial mill location DSS, and enhancements will continue as additional data and research results become available. Challenges and weaknesses include: 1) The inability to estimate volumes by product class directly from satellite imagery necessitates that surrogate variables such as forest age must be used to partition the aggregate volumes from an image into product classes. This means that the proportional partitioning of pixel volume must be taken from a growth and yield model or other inventory source of comparable age and timber type to that of the pixel. If in the future methods are developed to accurately estimate pixel density in both trees and basal area, in addition to change detection age, then a more accurate partitioning of the pixel volume could be obtained from a growth and yield model. These estimates could also be greatly improved if a measure of pixel average tree height were available from a secondary source of data such as Light Detection and Ranging (LiDAR). 2) The pilot DSS cannot estimate volumes in adjacent state areas because of the lack of county-level inventory information. If mills are proposed for locations near state borders, then methods must be developed to estimate forest growth/drain ratios in those areas. Models are currently under development at Mississippi State University that allow volume and weight estimates by GIS forest type from satellite imagery of surrounding states. 3) DSS information for ownership category and associated willingness to sell timber was not complete and was essential for adjusting estimates of total volume available to a mill. MIFI is currently collecting ownership category information, and an associated project is evaluating points at which NIPF landowners are willing to sell their timber by merchandizing class. 4) Estimates of the errors associated with image interpretation propagated through the inventory must be added to the sampling error of the inventory to obtain a true estimate of error. Another FWRC project is investigating the propagation of errors associated with perturbations in image interpretation which will help analyze classification errors in imagery and, in turn, help develop new classification methods. Good estimates of total error are required so investors can assess the true risk associated with locating a proposed mill.

Plans have been made to overlay socio-economic data such as demographics, education, income, housing, retailing, migration, and other economic metrics on MIFI's spatial inventory. These data are an important component of the industrial decision-making process where inadequate resources, such as educational facilities, can be the "deal killer" for local and state economic developers. Future DSS versions will include modules to access MIFI data, do mill location and feasibility studies, and provide raw growth and drain information. The transportation model required for the mill location and feasibility analyses is in place, and the growth and drain module is near completion. Making these modules operational will require the preparation of age and GIS forest type layers for every year across the entire inventory since its operational beginning in 2004. Improved information on road and bridge weight limits for hauling wood is needed to obtain a better estimate of minimum transportation costs, and product demand information will allow the minimization of transporting finished products to market.

4. Conclusion

The MIFI DSS will fill a critical role in assessing the current and future availability and costs of procuring forest resources within user defined regions of Mississippi. Results of forecasting future raw material availability should be used to assess the need for forest landowner educational and incentive programs in reforestation, management, harvesting, and utilization, ensuring a sustainable supply and reducing the risk of mill failures. Lawmakers will have information upon which to base policy decisions guaranteeing the environmental and economic sustainability of the forest lands that support a major component of the State's economic base. The forest products industry will possess the best possible information to evaluate economic risk for new and existing mills. An effective, Web-based, easy-to-use presentation of complex models associated with the DSS will empower its use. Without these tools, it will be difficult to entice industry investors to build or expand existing mills. No matter how technologically advanced a mill is, unless it is placed in a location with sustainable feedstock sources, it will fail.

The final product will be a monumental step forward in automating the process of evaluating forest-based resource supply and mill feasibility. The DSS will incorporate the most recent and precise inventory information and expert knowledge making that knowledge readily available to users in an effective, low cost, and timely manner. The ultimate impacts are: 1) promotion of stand management and reforestation, 2) more efficient utilization of the resource, and 3) prevention of over-utilization of the resource and environmental degradation.

References

Arano, K. G. &. Munn, I. A. (2004). Nonindustrial private forest landowners' forest management activities and expenditures in Mississippi, 1998-2000 data. *Forest and Wildlife Research Center, Bulletin FO249,* Mississippi State University, Mississippi State, Mississippi, 13pp.

Blinn, C. R., Sinclair, S. A., Brown-Gallagher, L., Wengert, E. M., & Crist, J. B. (1986). Economic feasibility and market potential for producing yellow-poplar oriented strandboard in the Appalachians. *Forest Products Journal, 36(9)*, 40-44.

Chambers, P. C., Sinclair, S. A., Hassler, C. C., & Hansen, B. G. (1986). Forest Products Investment Model: a microcomputer tool for incorporating risk into capital budgeting. *Forest Products Journal, 36(1),* 64-68.

Collins, C. A., Wilkinson, D. W., & Evans, D. L. (2005). Multi-temporal analysis of landsat data to determine forest age classes for the Mississippi statewide forest inventory-preliminary results. *2005 International Workshop on the Analysis of Multi-Temporal Remote Sensing Images.* Biloxi, Mississippi, May 16-18, 2005, p. 10-14 (CD-ROM).

Environmental Systems Research Institute (ESRI). (2005). ArcGIS® Software, Version 9.1. Redlands, California.

Grebner, D. L., Perez-Verdin, G., Sun, C., Munn, I. A., Schultz, E. B. & Matney, T. G. (2008). Woody biomass availability for bioethanol conversion in Mississippi. SOFEW. *Global Change and Forestry: Economic and Policy Implications.*

Ketron Management Science (KMS), Optimization Products, The Bionetics Corporation. (2003). C-whiz, Version 4, Linear Programming Optimizer. Arlington, Virginia: Ketron Management Science User's Reference Manual, 126pp.

Leica Geosystems GIS and Mapping. (2003). ERDAS Imagine®, Version 8.7. Norcross, Georgia.

Lin, W., Carino, H. F. & Muehlenfeld, K. J. (1996). OSB/LOCATION: A computer model for determining optimal oriented strandboard plant location and size. *Forest Products Journal,* 46 (2), 71-78.

Matney, T.G. & Schultz, E.B. The MIFI Inventory Dynamic Reporter. (2008). Accessed on October 20, 2008. Available from: http://www.mifi.state.ms.

McCauley, K.C. & Caulfield, J.P. (1990). Using mixed-integer programming to determine the optimal location for an oriented strandboard plant in Alabama. *Forest Products Journal, 40(2),* 39-44.

Microsoft Corporation. 2005. Microsoft Visual Studio® (2005). Redmond, Washington.

Mississippi Automated Resource Information System (MARIS). (2006). Accessed on May 12, 2006. Primary, secondary, and county road files. Available from: http://www.maris.state.ms.us.

Munn, I.A. & Tilley, B.K. (2005). Forestry in Mississippi: The impact of the forest products industry on the Mississippi economy, an input output analysis. *Forest and Wildlife Research Center, Bulletin FO301,* Mississippi State University, Mississippi State, Mississippi, 27pp.

Parker, R.C., Glass, P.A., Londo, H.A, Evans, D.L., Belli, K.L., Matney, T.G. & Schultz, E.B. (2005). Mississippi's forest inventory pilot program: Use of computer and spatial technologies in large area inventories. *Forest and Wildlife Research Center, Bulletin FO274,* Mississippi State University, Mississippi State, Mississippi, 43pp.

Sun, C. & Zhang, D. (2001). Forest resources, government policy, and investment location decisions of the forest products industry in the southern United States. *Forest Science, 47(2),* 169-177.

Stuart, W.B. & Grace, L.A. (2008). Route Chaser Program. Accessed on October 22, 2008. Available from: http://www.wssg.cfr.msstate.edu.

Timber Mart-South. (2008). *Market News Quarterly*, Vol. 13, No.4. Center for Forest Business. Warnell School of Forest Resources. University of Georgia. Athens, Georgia.

USDA Forest Service (USDA FS). (2005). USDA Forest Service Forest Inventory and Analysis: Data collection and analysis and Forest Inventory and Analysis: Phase 2 and Phase 3: Ground measurements. Accessed on September 10, 2005. Available from: http://fia.fs.fed.us/library/fact-sheets.

In: Global Change and Forestry
Editors: J. Gan, S. Grado and I. A. Munn
ISBN: 978-1-60876-262-0

Chapter 10

ECONOMIC IMPACTS OF WOODY BIOMASS UTILIZATION FOR BIOENERGY IN MISSISSIPPI[1]

Gustavo Perez-Verdin, Donald L. Grebner**, Ian A. Munn, Changyou Sun and Stephen C. Grado
Department of Forestry, Mississippi State University, MS, USA

ABSTRACT

This study examined the economic impacts of woody biomass utilization for bioenergy conversion in Mississippi. Analysis of economic impacts was organized around three groups of events: (1) recovery of logging and thinning residues, (2) electricity generation from co-firing systems, and (3) construction and operation of biofuel facilities. Input-output analysis was used to simulate economic impacts in terms of gross output, value-added, and employment. Cost information and woody biomass inventories were obtained from the literature, a statewide forest inventory, and Impact Analysis for Planning (IMPLAN) database. Results showed that the single activity of recovery of all available logging and thinning residues would create a considerable number of jobs and stimulate the rural economy with more resources coming to local industries and households. Due to construction and operation costs, economic impacts of biofuels were higher than biopower. However, biofuels reported the lowest employment and value-added multipliers of all three groups. This may be due, in part, to equipment and materials manufactured outside of the state. It is expected that as technology, equipment, and human capital were gradually attracted to the area, the multiplicative effect and expenditure retention should increase. These results can help decision-makers evaluate Mississippi's potential for future bioenergy development.

Keywords: Biofuels; Biopower; IMPLAN; Economic multipliers; Logging residues.

---

[1] Reprinted with permission from Forest Products Journal 58(11): 75-83.

* E-mail: dgrebner@cfr.msstate.edu, Department of Forestry, Forest and Wildlife Research Center, Mississippi State University, Box 9681, Mississippi State, MS 39762.

1. INTRODUCTION

Concerns over energy security, environmental health, and economic development have generated increasing interest in the search for alternative sources of fuel supplies (Bartuska 2006). Woody biomass, the fibrous and generally inedible portions of stems and branches of plants, is one alternative energy supply source because it is renewable, abundant, and can reduce greenhouse gas emissions (Hamelinck et al. 2005, USDE 2006, Solomon et al. 2007). Cellulose and hemicellulose, the main compounds of woody biomass, are made of energy-rich sugars that can be converted to ethanol through hydrolysis and fermentation procedures (USDE 2006). Alternatively, biomass can be used as a feedstock to produce electricity replacing fossil fuel-fired power plants.

Woody biomass for bioenergy conversion is obtained from a diverse and widespread resource base which includes low-value forest products such as logging residues and thinning of overstocked stands, among other sources (Perlack et al. 2005). Frequently, this material is left on-site, piled, and burned at additional cost, or left on-site for decomposition and incorporation into soil nutrients. As pressure for green energy develops, there has been a strong interest in utilizing this material for bioenergy conversion. A significant number of studies have evaluated the potential of woody biomass for fuels, electricity, and heat generation (e.g., Giampietro et al. 1997, Graf and Koehler 2000, CEC 2001, Solomon et al. 2007). A question remains about the economic impacts that recovery of logging residues for use as a feedstock to bioenergy would have on a state's economy. These impacts, generally estimated in terms of expenditures, taxes, and employment, have not been well-documented due to the incipient market development and low-scale commercial production (CEC 2001, REMI 2006, Gan and Smith 2007).

The study objectives were to describe the main economic impacts of developing bioenergy and to specifically quantify the economic impacts on Mississippi's economy of logging residue recovery, electricity generation from woody biomass, and construction and operation of a biofuel facility. The paper begins with a brief review of basic concepts of bioenergy and input-output models, which are the basis for economic impacts simulations. It then presents a case study using data for the state of Mississippi to analyze preliminary economic impacts and closes with a summary and conclusions.

2. BACKGROUND

Bioenergy is solar energy stored via photosynthesis in organic matter. While this definition implies plant-converted energy, the concept also includes animal and food processed wastes (Perlack et al. 2005). Electricity (biopower), liquid fuels (biofuels), and heat generated from biomass are examples of bioenergy. The first two are largely considered for commercial scale development due to existing processing capacity and an increasing demand for residential and industrial uses (Cook and Beyea 2000, Gan and Smith 2006, Solomon et al. 2007). Electricity generated from biomass operates in a similar way as fossil fuel-based plants. Instead of fossil fuels, however, biomass is direct-fired in a boiler to produce high-pressure steam that rotates a turbine, which in turn is connected to an electric generator. The majority of power plants are direct-fired systems, but some plants use a combination of coal and biomass feedstocks, a process called co-firing (NREL 2007). Co-firing systems have demonstrated that combining coal and biomass for electricity generation increases boiler efficiency, reduces fuel costs, and significantly decreases emissions of nitrates and fossil carbon (Demirba 2003). Typically, the same power plants that generate renewable electricity also yield useful steam and heat in combined heat and power. It is expected that the combined share of biomass and other non-hydropower renewable electricity for the next 30 years will increase from 2.2 to 4.3 percent of total generation (EIA 2006).

Biofuels are liquid or gaseous fuels produced from biomass feedstocks. Bioethanol, biodiesel, biogas, and hydrogen fuel are some examples of biofuels. Bioethanol and hydrogen fuel are produced through microbial fermentation of sugar, starch, and cellulosic materials. Biodiesel is made by combining alcohol (usually methanol) with vegetable oil, animal fat, or recycled cooking grease. Biogas is produced through anaerobic digestion of volatile organic matter. Ethanol is the most common biofuel produced from cellulose and hemicellulose (NREL 2007). Ethanol has been in the center of political debates as it was announced as a serious alternative fuel to displace 30 percent of U.S. current gasoline use by 2030 (USDE 2006). Current technology, based on dilute acid pretreatment and fermentation, can produce up to 80 gallons of ethanol from one dry ton of woody biomass (Hamelinck et al. 2005). To meet the 2030 goals, it would be necessary to produce 60 billion gallons of cellulosic ethanol a year, which requires one billion dry tons of woody biomass in supplies. Nation-wide studies estimated that current woody biomass potential is about 1.3 billion dry tons per year, of which 73 percent comes from agricultural residuals (Perlack et al. 2005). This amount is enough to cover the 2030 energy goals.

Mississippi's forests cover approximately 20 million acres, and each year they generate over $1 billion worth of timber and related forest products (Munn and Tilley 2005). The forest products industry, which consists basically of four sectors: logging, solid wood products, pulp and paper, and wood furniture manufacturing, on average processes 1.3 billion cubic feet of roundwood and by-products per year (Howell et al. 2005, MAFES 2007). In 2001, all sectors combined were responsible for more than 54,000 direct jobs, roughly 3.7 percent of the State's total employment, and its impact on the economy was about $13 billion (Munn and Tilley 2005). Wood furniture contributed 44 percent of the direct jobs, followed by the solid wood products industry (28%), pulp and paper (13%), and logging and miscellaneous forest products (15%). Grebner et al. (2009) estimated the availability and production costs of woody biomass in Mississippi's 82 counties. Their results showed that about four million dry tons of woody biomass can be produced every year in the state (figure 1). The material consists of logging residues, small diameter trees, urban waste, and mill residues. They also indicated that logging residues, the largest feedstock, can be produced at $40 per dry ton and small-diameter trees at $49 per dry ton, for a 50-mile procurement radius.

To process these supplies, it is necessary to know where exactly feedstocks are located and how much it will cost to transport the material to a manufacturing facility. Region-specific studies are necessary to account for variation in cost and availability of woody biomass. Generally, the best facility location is conceived as the place which significantly reduces production costs (i.e., a site close to feedstocks). Once site location is identified and availability and production costs are estimated, a next step is to analyze the economic impacts on a state's economy of processing woody biomass. The economic impacts can be analyzed in terms of various socio-economic indicators such as gross output, value-added, employment, and taxes (CEC 2001, Gan and Smith 2007), and input-output models are often used in the appraisal. This type of evaluation would help decision-makers evaluate and develop strategic energy transformation centers in the state.

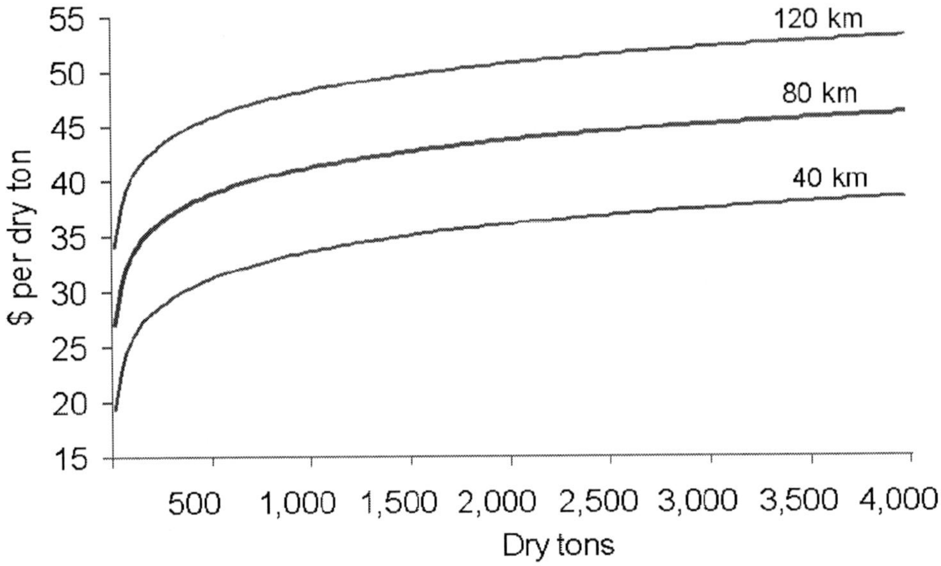

Figure 1. Adjusted cost curves of woody biomass feedstocks for three procurement distances in Mississippi (Source: Grebner et al. 2009).

3. METHODS

To estimate economic impacts of bioenergy development, this study used the Impact Analysis for Planning (IMPLAN) model developed by MIG, Inc. and the USDA Forest Service. The IMPLAN model, based on input-output (I-O) analysis, consists of a nationwide matrix and estimates of sector activity for final demand, final payments, industry output, and employment for each county in the Unites States. New databases, which include 509 economic sectors, are annually developed using U.S. Census Bureau, U.S. Department of Commerce, Bureau of Economic Analysis, and other government-based information.

3.1. INPUT-OUTPUT MODELS

I-O models trace commodity flows from producers to intermediate and final consumers (Schaffer 1999). Industries produce goods and services for final demand and purchase raw material from producers. Producers, in turn, purchase goods and services from other industries. This process continues until *leakages* from the region stop the cycle (Schaffer 1999). The total industry purchases of commodities, services, employee compensation, value-added, and imports are in the end equal to the value of the commodities produced. I-O models also provide multipliers that estimate the relationship between the initial effect of a change in final demand and the total effects of that change (Miller and Blair 1985). The effects, which can be direct, indirect, or induced, describe the change of output for each and every local industry caused by a one dollar change in final demand and, typically, they are estimated in terms of output, value-added, and employment (Rickman and Schwer 1995). Gross output represents the total value of production; value-added is total output minus costs of purchased inputs and represents the amount of money available for disbursement, either in the form of wages, owner compensation, or taxes; and employment includes the number of full- and part-time jobs in the sector (Munn and Tilley 2005).

Direct effects measure changes associated with the immediate impacts in the level of production of an economic activity. Indirect effects are the production changes from various rounds of inter-industry re-spending. For instance, a cellulosic ethanol plant requires a considerable amount of chemical and biological materials such as sulfuric acid and *cellulase* enzymes to process woody biomass. Changes in sales, jobs, and income in the cellulosic ethanol industry are direct effects. Changes in sales, jobs, and income in the chemical industry represent indirect effects of changes in ethanol production. The chemical industry, in turn, requires inputs from other industries that represent another round of indirect effects. Indirect effects disappear gradually until leakages stop the cycle. Induced effects are the changes in economic activity resulting from household spending of income earned directly or indirectly as a result of bioenergy conversion. Cellulosic ethanol and chemical industry employees spend their earned income for housing, food, transportation, insurance, health, and education. The sales, income, and jobs that result from household spending of added wage, salary, or proprietor's income are induced effects. The magnitude of this ripple effect is captured in multipliers (Tilley and Munn 2007).

Various types of multipliers have been developed to measure the effects (i.e., direct, indirect, and induced) and the economic activity analyzed (i.e., output, value-added, employment, and income) (Rickman and Schwer 1995). I-O multipliers include Type I, Type II, Type III, and social account matrices (SAM). Differences among all multipliers involve the type of effects being analyzed and the way that induced effects are computed. For instance, Type SAM multipliers are estimated by adding direct, indirect, and induced effects and dividing by the direct effects. Induced effects are based on information in the social account matrix. Type SAM multipliers usually are preferred to other types of multipliers because they account for social security and tax leakages, institutional savings, and interinstitutional transfers (Rickman and Schwer 1995). IMPLAN generates multipliers for output, value-added, employment, and income. Output multipliers measure total impact of selling industries of each dollar brought by purchasing industries. Value-added multipliers are a measure of economic output that eliminate inter-industry sales and, therefore, reflect the amount of output added by each industry. Employment multipliers are defined as the ratio of the number of employees over one million of dollars of output (Schaffer 1999). Income multipliers are the ratio of household income to output or total changes in output over total changes in income (Miller and Blair 1985, Schaffer 1999).

Few studies have analyzed the economic impacts of woody biomass utilization in terms of employment, output, and value-added since bioenergy development is relatively new. Gan and Smith (2007) estimated co-benefits of woody biomass for electricity generation in East Texas. They found that about 1.3 million dry tons could be used to generate 2,400 MWh of electricity in coal-fired power plants. They also reported that about $352 million would be generated as total output and 1,340 jobs would be created. Their study, however, did not provide specific information about costs and industry sectors actually included in the simulations. In a conjoint study developed by Michigan Tech University and Regional

Economic Models, Inc. (REMI) in the Great Lakes Region, it was found that bioethanol can bring considerable economic impacts, including population migration and value-added, due to construction and operation of processing facilities (REMI 2006). This study, based on dynamic input-output models, simulated three scenarios for ethanol production as a function of plant capacity. The greatest impacts were found for the construction and operation of a 52-million gallon ethanol plant (REMI 2006).

3.2. Classification of Bioenergy Impacts

The analysis of economic impacts of woody biomass for bioenergy conversion was organized around three groups of events: (1) recovery of logging and thinning residues, (2) electricity generation from co-firing systems, and (3) construction and operation of biofuel facilities. The first group, recovery of logging residues and thinning, involved the procurement of wood in the form of tree tops and branches, and the chipping, loading, and transporting of raw material to a processing facility. Transportation costs were estimated for a 50-mile supply radius, which included two miles of in-woods hauling. The total cost of recovery of logging residues and thinning was estimated at \$151.7 millions per year (Grebner et al. In press). This number was used in IMPLAN simulations. To avoid conflicts in woody biomass allocation to any type of bioenergy, the effects of collecting all woody biomass supplies were simulated independently of the other two groups.

The second group involved operation of existing processing facilities to convert woody biomass into biopower. Due to technical considerations, derived from incipient markets, it was difficult to estimate investments and input requirements for modifications, operation, and maintenance of biopower plants. Following Gan and Smith (2007), it was assumed that biopower was somewhat similar to electricity generation from fossil fuel power plants. Though possibly unrealistic, this assumption was based on the feasibility that existing power plants can also be used for co-firing systems and on similarities of per-unit production costs of initial investments between biomass-fired systems and coal or other fossil fuel-fired plants (Hughes 2000, Gan and Smith 2007). Thus, the analysis of biopower did not consider new plant construction and the impact analysis was conducted by replacing coal as the primary feedstock with woody biomass. As of 2006, according to the Energy Information Administration (EIA), there were 17 power plants operating in Mississippi that may use coal and wood for electricity generation with a production capacity of 2,800 MW (17% of the state capacity) (EIA 2007). As expected, not all woody biomass supplies would be sufficient to meet power plant demands, especially if part of the material was diverted for biofuels conversion. Woody biomass was perceived as an alternative to reduce problems associated with fossil fuels consumption, but it was not expected to completely solve fossil-fuel dependency.

A 100-MW electrical generating plant was considered to determine the economic impacts of electricity generation. This power plant required 240,000 tons of coal at 47 percent efficiency (Beér 2007) or 430,000 tons of woody biomass at 35 percent of efficiency (Gan and Smith 2007) to generate 800,000 MWh of electricity per year. Production costs of fuels were estimated at \$10.3 million (EIA 2007) for coal and \$17.6 million for wood. According to EIA, the delivery price for each MWh of electricity generated in Mississippi was \$90 (EIA 2007), which gave \$72 million of total output. Since the plant was already in operation, fixed and variable costs were assumed constant and fuel costs accounted for most differences (table 1)[2]. The economic impacts of electricity generation from co-firing systems were simulated for various combinations of wood and coal use. Simulations started with 100 percent of wood and 0 percent of coal, then 75-25, 50-50, 25-75, 10-90, 5-95, and 0 percent wood and 100 percent of coal use. To compare effects, results were standardized in terms of Type SAM multipliers, which were the ratio of total impacts to direct impacts.

Table 1. Operation costs for a 100 MW power plant[a]

IMPLAN sector	Cost category	Million $	$/MWh
014	Biomass cost	17.63	21.99
032	Water	0.30	0.38
037	Modification cost	1.89	2.35
269	Annualized charge	14.89	18.58
043	Maintenance	2.66	3.32
427	Insurance	1.77	2.21
460	Ash transport	0.79	0.98
460	Ash disposal	0.31	0.38
10004	Payroll	2.76	3.45
10007	Other property income	6.45	8.05
11003	Federal taxes	10.75	13.41
12003	State taxes	5.16	6.44
	Total	65.36	81.54

[a] Based on a fluid bed combustor, steam turbine cycle, and 100% wood feedstock.
Source: Caputo et al. (2005), Hughes (2000), and Caywood (1972).

A different approach was applied to the third group (i.e., biofuels). The effects of construction and operation of new manufacturing facilities were estimated because there were no cellulosic ethanol processing facilities in existence. Plant construction impacts included employment for the construction industry, and equipment and material purchases. We assumed a plant capacity of 52 million gallons of ethanol per year that would require about 700,000 dry tons of woody biomass annually. Information regarding construction and operation costs of manufacturing facilities was obtained from various sources including California Energy Commission (2001), REMI (2006), Gan and Smith (2007), and Solomon et al. (2007). A plant processing 52 million gallons of cellulosic ethanol per year would need an estimated capital investment of $224 million (table 2) plus other costs to cover feedstocks, enzymes, utilities, maintenance, labor, and administrative fees during the plant lifetime.

To perform IMPLAN simulations of biofuel impacts (e.g., on a year-by-year basis), construction costs were annualized to produce an annual charge over the economic lifetime of the plant. This was done by multiplying the total construction cost by a capital recovery factor. The capital recovery factor is a function of variables such as project life, cost of capital, tax rate, depreciation methods, and equipment type (Axelsson et al. 2003). In this analysis, the economic lifetime of the processing facility was established at 15 years and assumed a 10 percent after-tax rate of return on capital investment. Following Solomon et al.

[2] See Caputo et al. (2005) for details in construction and operation of wood-fired power plants.

(2007), capital recovery factor was set at 20 percent. Thus, the annualized capital charge was estimated at $44.8 million. This amount was distributed into three IMPLAN sectors: manufacturing and industrial buildings (42%), miscellaneous fabricated metal product manufacturing (32%), and other industrial machinery and equipment manufacturing (26%). The spreading of percentages across industries was based on a demand study of plant construction and cellulosic ethanol production in the Great Lakes Region of the U.S. conducted by REMI (2006). Operating costs and annualized capital charges were shown in Table 3 including cost categories under IMPLAN sectoring.

Table 2. Estimated construction costs for a 52-million-gallon cellulosic ethanol manufacturing facility

Cost category	Million $
Feedstock handling and equipment	11.4
Pretreatment	37.5
Xylose fermentation	9.7
Cellulase production	4.5
Saccharification and fermentation	33.2
Ethanol recovery	6.4
Off-site tankage	6.4
Environmental systems	6.3
Miscellaneous	7.6
Utilities (steam, electricity, water)	80.7
Start-up costs and working capital	20.3
Total capital investment	224.0

Source: Adapted from Solomon et al. (2007), p. 421.

Table 3. Estimated operation costs and annualized capital charges for a 52-million gallon cellulosic ethanol manufacturing facility

IMPLAN sector	Cost category / industry	Million $/year	$/gallon
Operation costs			
014[a]	Feedstock procurement	29.04	0.56
030[a]	Electricity	-4.94[b]	-0.10
032[a]	Water	0.22	0.00
043	Maintenance	7.75	0.15
151	Enzymes	11.57	0.22
154	Other chemical materials (sulfuric acid, lime, nutrients)	5.73	0.11
427[a]	Insurance and property taxes	3.93	0.08
444[a]	Consulting services	2.36	0.05
455[a]	Overhead	8.54	0.16
460[a]	Gypsum disposal	0.56	0.01
Annualized capital charges[c]			
037	Manufacturing and industrial buildings	18.80	0.36
255	Miscellaneous fabricated metal product manufacturing	14.40	0.28

Table 3. (Continued)

IMPLAN sector	Cost category / industry	Million $/year	$/gallon
269	Other industrial machinery and equipment manufacturing	11.59	0.22
Labor, profit, and taxes			
10005 [a]	Payroll	2.85	0.05
10007 [a]	Other property income	18.20	0.35
11003 [a]	Federal taxes	2.97 [d]	0.06
12003 [a]	State taxes	14.15	0.27
	Total production cost	147.72	2.83

Source: Adapted from Solomon et al. (2007), p. 422.

[a] Regional purchase coefficients (RPC) for these events were assumed to be 100%. All other cost categories used IMPLAN default estimates. An RPC of 100% means that additional demands are fully met by local production.

[b] Industry by-products (e.g., lignin) are recycled for electricity generation.

[c] Based on a capital recovery factor of 20%; economic lifetime of 15 years; and 10% after-tax rate of return on capital investment.

[d] Federal taxes include a $0.51/gallon discount (subsidy).

4. RESULTS

IMPLAN estimates of economic impacts of logging residue recovery, biopower, and biofuels were shown in table 4. The recovery of 4 million dry tons of woody biomass available each year in Mississippi would create 585 direct jobs and generate $152 million of gross output. Another 481 indirect jobs and 646 induced jobs are generated as a result of logging operations. The total value-added, which is the sum of employee compensation, proprietor income, other property income, and indirect business taxes, provides $37.3 million in direct impacts and $23.4 million in indirect impacts. The induced value-added impact (generated due to income earned by employees in direct and indirect business and industries and spent on goods and services) adds another $37 million. Employee compensation and other property income (or profit earned by loggers) accounted for 48 percent and 30 percent, respectively, within total value-added impacts. The multiplier effect indicates that for every dollar worth of initial stimulus in the recovery of logging residues results in an additional $1.62 of value-added. The highest multiplier within total value added is indirect business taxes (4.42) which include payments of excise and sales taxes by individuals to businesses. These taxes, collected during normal operation of businesses, do not include taxes on profit or income. Similarly, the multiplier for employment, the highest number of all three groups of bioenergy, shows that for every single job directly created, 1.92 jobs are generated elsewhere. The most affected sectors by recovery of logging residues are agriculture and forestry support activities, petroleum refineries, wholesale trade, owner-occupied dwellings, and food services and drinking places, among others.

Table 4. Economic impacts of three groups of bioenergy development in Mississippi

Group	Economic impacts				Type SAM multiplier [a]
	Direct	Indirect	Induced	Total	
1. Recovery of logging residues					
Gross output (M$)	151.75	68.46	62.28	282.49	1.86
Total value added (M$)	37.27	23.45	37.07	97.78	2.62
Employee compensation (M$)	13.28	12.52	21.16	46.95	3.54
Proprietors income (M$)	9.91	4.13	2.35	16.39	1.65
Other property income (M$)	12.87	5.46	10.73	29.07	2.26
Indirect business taxes (M$)	1.22	1.34	2.82	5.38	4.42
Employment (# of jobs)	585	481	646	1,712	2.92
2. Biopower (100-MW plant)					
Gross output (M$)	64.47	16.27	22.68	103.42	1.60
Value added (M$)	14.98	6.42	13.54	34.94	2.33
Employee compensation (M$)	7.95	3.41	7.73	19.09	2.40
Proprietors income (M$)	2.60	0.85	0.85	4.30	1.66
Other property income (M$)	3.74	1.69	3.92	9.35	2.50
Indirect business taxes (M$)	0.70	0.47	1.03	2.21	3.13
Employment (# of jobs)	281	115	236	631	2.25
3. Biofuels (52-million-gallon plant)					
Gross output (M$)	150.11	35.91	56.71	242.74	1.62
Value added (M$)	38.11	14.55	33.74	86.41	2.27
Employee compensation (M$)	23.11	7.69	18.97	49.77	2.15
Proprietors income (M$)	7.78	1.89	2.14	11.81	1.52
Other property income (M$)	6.14	3.91	9.98	20.03	3.26
Indirect business taxes (M$)	1.08	1.08	2.64	4.80	4.45
Employment (# of jobs)	908	261	586	1,756	1.93

[a] Type SAM multiplier is the ratio of total impact to direct impact.

The economic impacts of operating a 100-MW power plant using woody biomass as a replacement for coal were also shown in table 4. As expected, the impacts were not as high as biofuels because no construction costs were assessed. In this case, electricity generation from wood would directly support 281 direct jobs, produce direct gross output of $64.5 million, and $15 million of direct value-added. Employee compensation (55%) and other property income (27%) had the highest contribution within total value-added. Type SAM multipliers indicated that for every dollar of gross output generated by the power plant, $0.60 worth of indirect and induced output impacts was generated in other industries. Similarly, Type SAM multiplier for employment is 2.25, which indicated that for every job directly created in biopower, 1.25 jobs were created in other industries. The most affected sectors by biopower included logging, maintenance and repair of nonresidential buildings, insurance carriers, agriculture and forestry support, activities, waste management and remediation services, petroleum refineries, and owner-occupied dwellings, among others.

Electricity generation from co-firing systems creates larger economic impacts when wood was the primary feedstock. Overall, impact multipliers for output, employment, and value-added steadily increase as wood replaces coal in power plants. The multiplier for gross output is 1.4 when wood is not used and 1.6 when wood is fired alone. Smaller differences were found for employment and value added multipliers. The multiplier for employment was 2.1 when coal is the only feedstock and 2.2 when wood is the primary feedstock. The multiplier for value-added is 2.2 when wood is not used and 2.3 when wood is fired alone (figure 2).

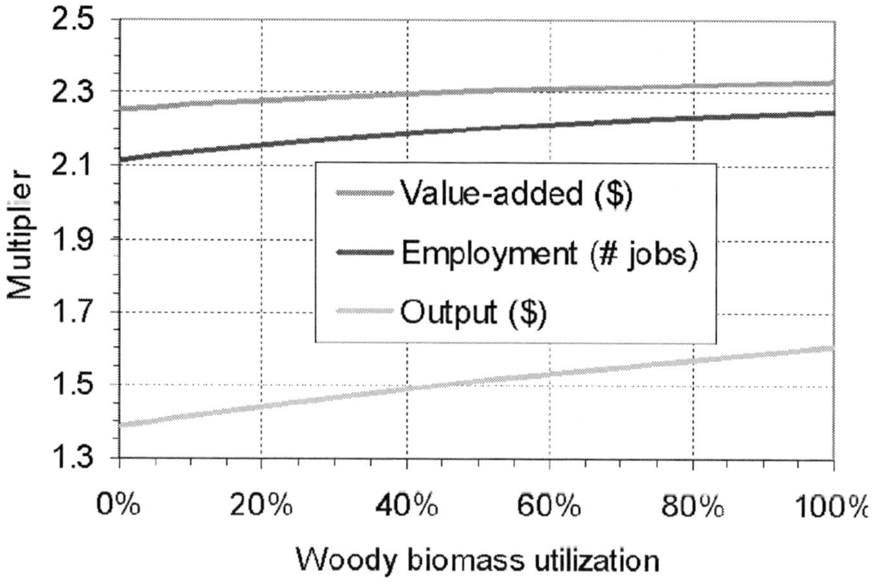

Figure 2. Economic impacts of co-firing woody biomass and coal in Mississippi. The x-axis shows woody biomass rate. Coal is assumed to be the difference to complete 100 percent utilization. For example, 20 percent of woody biomass means 80 percent of coal use.

The economic impacts of construction and operation of an ethanol plant were also shown in table 4. Considerable expenditures are immediately lost from the State, due to temporary workforces and the importation of equipment and services not currently available in Mississippi. Regional purchase coefficients (RPC) were determined based on the knowledge of industries acquiring resources within the area, but in some cases they were based on IMPLAN default estimates. These leakages represent about 21 percent of direct output. Overall, 1,756 jobs would be created from biofuels conversion (52-million-gallon plant) and would generate $242.7 million of total gross output. Total value-added was estimated at $86.4 million with employee compensation (58%) and other property income (23%) having the highest share. While the number of total jobs seems to be high, the employment multiplier of biofuels is the lowest of the three groups of bioenergy being analyzed. For every single job created in this industry, only 0.9 jobs will be created in other industries. The most affected sectors of biofuels were logging, manufacturing, and industrial buildings; domestic trade; business support services; other basic organic chemical manufacturing; maintenance and repair of nonresidential buildings; insurance carriers; building material and garden supply stores; and petroleum refineries. In total, these sectors accounted for 72 percent of total gross output.

5. Discussion

The economic impacts of recovery of logging residues for either biopower or biofuel conversion were examined in this study. The 4 million dry tons of woody biomass available each year in Mississippi can supply up to nine 100-MW power plants or five 52-million gallon plants of ethanol production. The economic impacts of biofuels and biopower were modeled for a certain plant capacity to be used as a baseline reference because results can be extrapolated to more plants depending on the proportion of woody biomass utilization. For example, a 40 percent distribution of available biomass for biopower would be enough to supply three plants and generate \$310 million of gross output. The remaining 60 percent would be enough to construct and operate three bioethanol plants, which would generate \$728 million of output (figure 3). Either form of bioenergy is compatible with the objectives of reducing non-renewable energy dependency and offsetting greenhouse gas emissions. Cellulosic ethanol is seen as a political, strategic, and economic source to control the price of corn and other agricultural feedstocks currently used for ethanol production (Solomon et al. 2007). However, ethanol development has been impeded by high construction costs, technological efficiencies, and financial constraints (Wyman 2003, Hamelinck et al. 2005, Solomon et al. 2007). Biopower has a slight advantage over biofuels due to existing plants that may use woody biomass and coal in co-firing systems. Electricity generated in co-firing systems requires no substantial investments in equipment and technology (Hughes 2000). This advantage represents a considerable savings in bioenergy development compared to facility building for biofuel production.

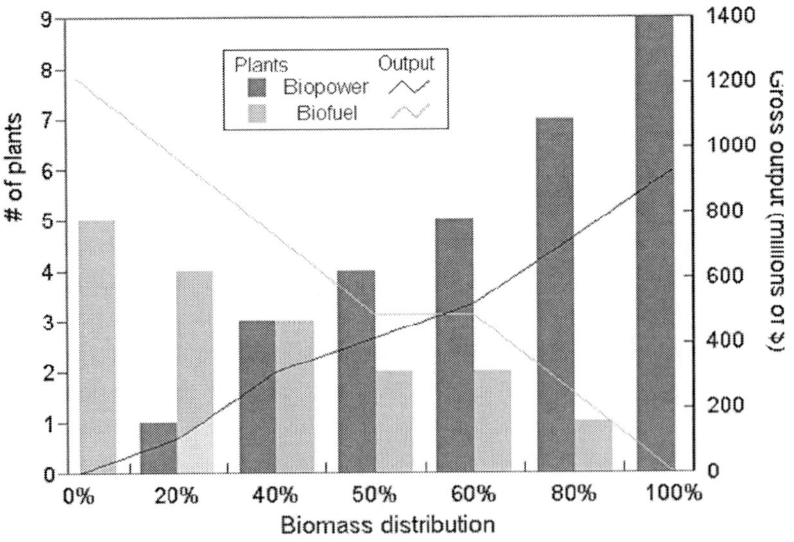

Figure 3. Trade-offs between biopower and biofuels development in Mississippi. The left y-axis represents number of plants whereas the right y-axis stands for gross output. The x-axis is referenced for biopower.

To differentiate economic impacts of biopower and biofuels on a per-unit basis, a standard energy unit called British thermal unit measured as millions of Btu (MMBtu) was used. In this case, one MWh is equivalent to 3.41 MMBtu (Gan and Smith 2006) whereas one gallon of ethanol is approximately 0.08 MMBtu. Per unit indicators of output and value-added were measured as dollars per MMBtu and employment as MMBtu per person. Accordingly, output, value-added, and employment indicators were greater for biofuels than biopower. Biofuel impacts were estimated at $55.3/MMBtu for output, $19.7/MMBtu for value-added, and 2,499 MMBtu/person for employment. Biopower impacts were estimated at $37.8/MMBtu for output, $12.8/MMBtu for value-added, and 4,328 MMBtu/person for employment. These results indicated that biofuels is more resource efficient that biopower. Biofuel yields more revenues and requires fewer jobs to produce one energy unit.

Replacing coal in power plants resulted in slightly more economic impacts for wood. Coal, Mississippi's leading energy source, has a higher energy content, is more efficient and, therefore, a lower tonnage is required to produce the same amount of electricity than wood (Hughes 2000). However, the majority of coal has to be imported from other areas, as far away as Colorado and Wyoming (EIA 2007), increasing its delivery price. Coal is estimated at $43/ton (EIA 2007) whereas wood has been estimated at $41/dry ton (Grebner et al. In press). In contrast, woody biomass is locally produced and all production inputs (i.e., labor, capital, equipment) are found in the area. Acquiring resources in the area make both impact multipliers and expenditure retention to be slightly higher than coal. Another major impact of woody biomass utilization for electricity generation stems from the amount of carbon dioxide (CO_2) that can be displaced if coal were the primary energy source. The 430,000 tons of woody biomass required to power a 100-MW plant would displace 340,000 tons of CO_2 emitted from a coal-fired plant (see Gan and Smith 2007 for conversion procedures). This is equivalent to 1.3 percent of the total CO_2 emissions generated in 2007 by all electric power industries in Mississippi. At the current price of CO_2 traded at the Chicago Climate Exchange ($4.08/ton CO_2, March 3, 2008), the value of carbon displaced is $1.4 million.

The economic impacts of each group of bioenergy to its respective aggregated sector were analyzed following IMPLAN production functions. The idea was to evaluate the role of each bioenergy group in Mississippi's economic sectors. The sectoring developed for IMPLAN was based on the former Standard Industrial Classification (SIC), now called North American Industry Classification System (NAICS), and the Bureau of Economic Analysis classification (IMPLAN 2004). Accordingly, impacts of recovery of logging residues were compared to the *logging* sector and biopower to the aggregated *power generation and supply* sector (see IMPLAN 2004, Appendix A, for details). Due to the nature of feedstocks and technology involved, no apparent sectors were identified for a biofuels analysis and comparison. The closest sector was *pulp and paper* industry, but differences in technology, type of feedstock, and efficiency made comparisons difficult. To establish a reference point, results were compared to this sector considering that the future vision of integrated biorefineries is to process all products for chemical feedstocks and energy as a byproduct of the pulping process (Coleman and Stanturf 2006). With 2006 data adjusted to the year 2008, total gross output of the logging sector was estimated at $1.6 billion, value-added at $409 million, and the number of full- and part-time jobs were estimated at 6,427. Comparing these numbers to our results (table 4), recovery of logging residues represents 17 percent of gross output, 24 percent of value-added, and 27 percent of employment of the aggregated sector.

The logging sector is largely dominated by roundwood for pulpwood production[3], but pulpwood's low cost contributed little to this sector's economy.

The economic impacts of biopower to the aggregated electricity sector represent only 4 percent, 2 percent, and 11 percent of gross output, value-added, and employment, respectively. Note that impacts of biopower and biofuels were estimated at a certain plant production capacity. These numbers could be higher if more plants were operated or capacity was increased. In contrast, biofuels seemed to narrow differences with the pulp and paper sector. The proportion of economic impacts of biofuels to the aggregated pulp and paper sector represents 25 percent, 11 percent, and 35 percent of gross output, value-added, and employment, respectively. This narrowing may be attributed to the construction factor that attracts considerable investments, expenditures, and employment (figure 4).

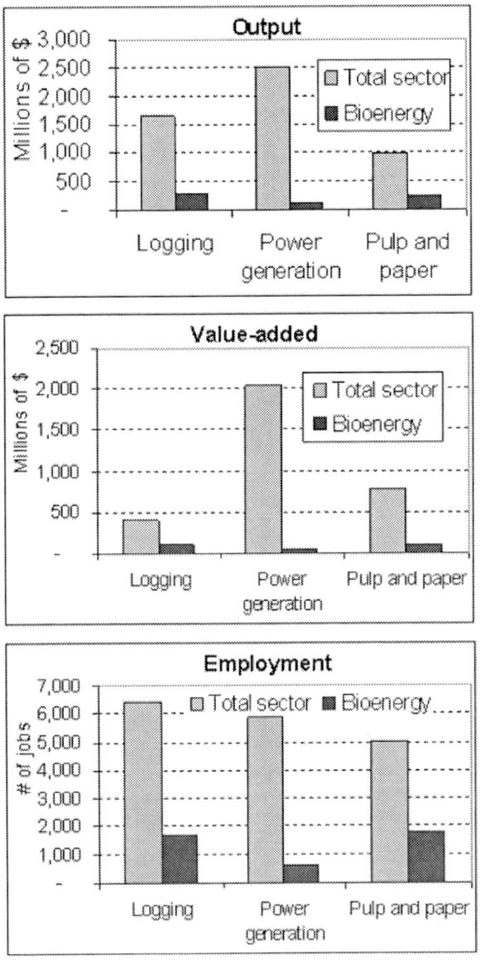

Figure 4. Comparison of three groups of bioenergy to aggregated sectors in Mississippi. The aggregated sectors were logging, power generation, and pulp and paper (see Tilley and Munn 2007, for details on pulp and paper's aggregation).

[3] In 2006, roundwood for pulpwood production accounted for 80% of total timber production in Mississippi (MAFES 2007).

Previous works have demonstrated the multiplicative impacts of bioenergy in other regions of the U.S. (Hjerpe 2006, REMI 2006, Gan and Smith 2007). Gan and Smith estimated Type SAM multipliers for recovery of logging residues of 2.0, 1.7, and 2.1 in terms of value-added, gross output, and employment, respectively. They estimated multipliers for electricity generation of 1.3, 1.3, and 5.2 for the same components as well. Though results of impact multipliers are difficult to compare across regions, because they depend on regionalized socioeconomic structures, what is important to note is that bioenergy can help diversify and improve regional economies in many rural areas of the country (Hjerpe 2006, Gan and Smith 2007). In particular for Mississippi, by looking at the contribution of each group to the aggregated sectors, employment generation represents the greatest impact of bioenergy development. It is expected that as technology, equipment, and human capital are gradually attracted to the area, the multiplicative impact and expenditure retention should also increase.

6. Summary and Conclusions

This study examined woody biomass utilization economic impacts for bioenergy conversion in Mississippi. Results showed that the single activity of recovery of logging and thinning residues would create a significant number of jobs and stimulate rural economies with more economic resources coming to local industries and households. Economic impacts of biopower, based on co-firing systems, showed some differences between wood and coal use. Wood registered slightly more economic impacts than coal due to higher regional expenditure retention and lower energy content, which requires more feedstock supplies and higher feedstock costs. About 240,000 tons of coal, compared to 430,000 tons of woody biomass, are required to generate 800,000 MWh of electricity per year. Production costs of wood were 41 percent greater than coal. This study suggested that wood brings more economic impacts, but it is not as efficient as coal. Other differences stem from environmental impacts such as carbon dioxide offset, as well as reduction of sulfur dioxide, nitrogen oxides, and other emissions. Wood has the unique advantage to offset emissions generated by the otherwise use of coal.

It was determined that employment generation is the greatest impact of bioenergy development in Mississippi. Due to construction and operation costs, the economic impacts of biofuels were higher than biopower. Despite that total value-added was higher and more jobs would be created in biofuels, the value-added and employment multipliers were lower than biopower. This difference can be due in part to equipment and material manufactured or produced in other areas generating leakages in Mississippi's economy. Gan and Smith (2007) and REMI (2006) found employment multipliers of 5.2 and 5.7 for biopower and biofuels, respectively. In these cases, it may be possible that most of the equipment and material were acquired within the study area. Mississippi has low regional purchase coefficients due to the lack of specialized industry and incipient market for bioenergy. However, it is expected that this change could be reversed because of the State's forest potential that could attract more investments into the region.

More research is necessary to assess the combined effect of biopower and biofuels in one industry (i.e., integrated biorefineries). Evaluation of the effects of land use changes as a result of increasing demand for biofuels or biopower is also necessary. Finally, more research is necessary to address social and ecological issues of biomass utilization to provide a more comprehensive analysis of bioenergy development

REFERENCES

Axelsson, H., Harvey, S., Asblad, A., and Berntsson, T. (2003). Potential for greenhouse gas reduction in industry through increased heat recovery and/or integration of combined heat and power. *Applied Thermal Engineering,* 23(1), 65–87.

Bartuska, A. (2006). Why biomass is important--The role of the USDA Forest Service in managing and using biomass for energy and other uses. Taken from speech at 25x'25 Summit II. Washington, D.C. [Available at http://www.fs.fed.us/research/pdf/biomass_importance.pdf], (Accessed 04-03-08).

Beér, J. M. (2007). High efficiency electric power generation: The environmental role. *Progress in Energy and Combustion Science,* 33(2), 107-134.

Caputo, A. C., Palumbo, M., Pelagagge, P. M., and Scacchia, F. (2005). Economics of biomass energy utilization in combustion and gasification plants: effects of logistic variables. *Biomass and Bioenergy,* 28(1), 35-51.

Caywood, R. E. (1972). Electric utility rate economics. New York, NY: McGraw-Hill Book Company, Inc.

California Energy Commission (CEC). (2001). Costs and benefits of a biomass-to-ethanol production industry in California. Report: P500-01-002. [Available at http://www.energy.ca.gov/reports/2001-04-03_500-01-002+002A.PDF] (Accessed 04-03-08).

Coleman, M. D., and Stanturf, J. A. (2006). Biomass feedstock production systems: Economic and environmental benefits. *Biomass and Bioenergy,* 30(8-9), 693-695.

Cook, J., and Beyea, J. (2000). Bioenergy in the United States: Progress and possibilities. *Biomass and Bioenergy,* 18(6), 441-455.

Demirba, A. (2003). Sustainable cofiring of biomass with coal. *Energy Conversion and Management,* 44(9), 1465-1479.

Energy Information Administration (EIA). (2006). Annual Energy Outlook 2006. National Energy Information Center. Report: DOE/EIA-0353(2005). Washington, D.C.

Energy Information Administration (EIA). (2007). State energy profiles. [Available at http://tonto.eia.doe.gov/state/state_energy_profiles.cfm?sid=MS], (Accessed 01-23-2008).

Gan, J., and Smith, C. T. (2006). Availability of logging residues and potential for electricity production and carbon displacement in the USA. *Biomass and Bioenergy,* 30(12), 1011-1020.

Gan, J., and Smith, C. T. (2007). Co-benefits of utilizing logging residues for bioenergy production: The case for East Texas, USA. Biomass and Bioenergy, 31(9), 623-630.

Giampietro, M., Ulgiati, S., and Pimentel, D. (1997). Feasibility of large-scale biofuel production. *Bioscience,* 47(9), 587-600.

Graf, A., and Koehler, T. (2000). Oregon cellulose-ethanol study. An evaluation of the potential for ethanol production in Oregon using cellulose-based feedstocks. Oregon

Office of Energy [Available at http://www.oregon.gov/ENERGY/RENEW/Biomass/docs/OCES/OCES.PDF], (Accessed 04-08-08).

Grebner, D. L., Perez-Verdin, G., Sun, C., Munn, I.A., Schultz, E. B., and Matney, T. G. (2009). Woody biomass feedstocks: a case study on availability, production costs, and implications for bioenergy conversion in Mississippi. *In* Renewable Energy From Forest Resources in the United States, eds. B. Solomon, and V. Luzadis, pp. 261-280. Taylor & Francis.

Hamelinck, C. N., van Hooijdonk, G., and Faaij, A. P. (2005). Ethanol from lignocellulosic biomass: Techno-economic performance in short, middle, and long term. *Biomass and Bioenergy,* 28(4), 384-410.

Hjerpe, E. E. (2006). Economics of forest restoration and fuels reduction programs in the Southwest. Dissertation. Northern Arizona University, School of Forestry. Flagstaff, AZ.

Howell, M., Johnson, T. G., and Bentley, J. W. (2005). Mississippi's timber industry - An assessment of timber product output and use, 2002. USDA Forest Service, Southern Research Station. Report: RB-SRS-102.

Hughes, E. (2000). Biomass cofiring: economics, policy and opportunities. *Biomass and Bioenergy,* 19(6), 457-465.

Impact Analysis for Planning (IMPLAN). (2004). IMPLAN Pro. User's guide, analysis guide, and data guide. Stillwater, MN: MIG, Inc.

Mississippi Agricultural and Forestry Experiment Station (MAFES). (2007). Economics: Harvest of forest products report. [Available at http://msucares.com/forestry/economics/reports/index.html], (Accessed 04-14-2008).

Miller, R. E., and Blair, P. D. (1985). Input-Output analysis, foundations and extensions. *Englewood Cliffs,* NJ: Prentice-Hall, Inc.

Munn, I. A., and Tilley, B. K. (2005). Forestry in Mississippi. The impact of the forest products industry on the Mississippi economy: An input-output analysis. Forest and Wildlife Research Center, Mississippi State University. Report: FO301.

National Renewable Energy Laboratory (NREL). (2007). Renewable Electric Plant Information System. [Available at http://www.nrel.gov/analysis/repis/], (Accessed 01-30-2008).

Perlack, R. D., Wright, L. L., Turhollow, A. F., and Graham, R. L. (2005). Biomass as feedstock for a bioenergy and bioproducts industry: The technical feasibility of a billion-ton annual supply. U.S. Department of Energy. Report: DOE/GO-102005-2135. Oak Ridge, TN.

Regional Economic Models, Inc. (REMI). (2006). An assessment of the economic and demographic impacts of developing, operating, and maintaining a lignocellulosic ethanol plant in the Great Lakes region. Prepared by REMI for Michigan Technological University. Report: Amherst, MA.

Rickman, D. S., and Schwer, R. K. (1995). A comparison of the multipliers of IMPLAN, REMI, and RIMS II: Benchmarking ready-made models for comparison. The Annals of Regional Science, 29(4), 363-374.

Schaffer, W. A. (1999). Regional impact models. Regional Research Institute, West Virginia University..

Solomon, B. D., Barnes, J. R., and Halvorsen, K. E. (2007). Grain and cellulosic ethanol: History, economies, and energy policy. *Biomass and Bioenergy,* 31(6), 416-425.

Tilley, B., and Munn, I. A. (2007). 2001 Economic impacts of the forest products industry in the South. *Southern Journal of Applied Forestry,* 31(4), 181-186.

U.S. Department of Energy (USDE). (2006). Breaking the biological barriers to cellulosic ethanol: A joint research agenda. Office of Science and Office of Energy Efficiency and Renewable Energy. Report: DOE/SC-0095. Rockville, MD.

Wyman, C. E. (2003). Potential synergies and challenges in refining cellulosic biomass to fuels, chemicals, and power. *Biotechnology Progress,* 19(2), 254-262.

In: Global Change and Forestry
Editors: J. Gan, S. Grado and I. A. Munn
ISBN: 978-1-60876-262-0

Chapter 11

Optimal Alternative Woody Biomass Management Regimes and Implications for Bioenergy Production in Mississippi

***Zhimei Guo*[1], *Donald L. Grebner*[*2],**
***Changyou Sun*[2] *and Stephen C. Grado*[2]**
[1] Department of Forestry, Wildlife and Fisheries,
The University of Tennessee, TN, USA
[2] Department of Forestry, Mississippi State University, MS, USA

Abstract

This paper evaluates and compares alternative management regimes of loblolly pine (*Pinus taeda*) plantations for joint production of timber and biomass and exclusively for biomass production in Mississippi. Alternative management regimes were mimicked via the PTAEDA 3.1 computerized yield simulator utilizing a variety of site preparation techniques, initial planting densities, and thinning activities. Evaluation criteria used to compare regimes included land expectation values (LEV) and mean annual increment (MAI) of stemwood biomass. Results indicate that culminating sustainable annual outputs of total stemwood biomass exclusively for bioenergy production using intensive site preparations were unprofitable due to the current low biomass price and high site preparation costs. Therefore, joint production of timber and logging residue biomass maximizing monetary return will be more favorable for private forest landowners. Sensitivity analyses indicate that as the relative prices of biomass to sawtimber increase, management strategies will gradually evolve to maximize total biomass with intensive site treatments and shorter rotation length. Increases in the discount rate reduced the intensity of site treatments and shortened rotation length. With regards to management implications, maximizing financial benefits may have more positive impacts on global climate change due to the potential of carbon sequestration, and provide more favorable wildlife habitats. Culminating biomass availability with intensive site treatments may result in negative ecological impacts, such as soil erosion and water run-off. Additional

* E-mail: dgrebner@cfr.msstate.edu, Department of Forestry, Mississippi State University, Mississippi State, MS 39762.

research is necessary to evaluate the possibilities of land use change associated with these management strategies.

Keywords: Woody biomass; Bioenergy; Optimal management regimes; Mississippi.

1. INTRODUCTION

Since the 1970s, policies and programs in the United States have been gradually enacted to foster the use of alternative renewable resources and decrease dependence of the national economy on imported fossil fuels. Various benefits associated with renewable energy utilization can be largely categorized as securing energy supplies, improving the environment, and increasing employment and economic development opportunities (Guo, Sun, & Grebner, 2007). Recently, more attention has been focused on forest biomass utilization due to potential problems of relying on grains for bioenergy production (NRC, 1999; Ford & Senauer, 2007). Though some bio-technologies using woody biomass are still in the development and demonstration stage, cellulosic ethanol plants are beginning to be built in the southern U.S., which will use wood wastes from pine and other biomass.

An incipient alternative market for woody biomass in the southern U.S. would increase small diameter wood utilization from private and public forest lands. Use of thinning materials from young southern pine stands, which have limited demand in solid wood markets because of the presence of high levels of juvenile wood, will increase forest management options for growing larger diameter wood and maintain forest health. Therefore, this new market will provide a substantial opportunity for forest management and rural economic development in Mississippi, where approximately 20 million acres of forest land comprise 66% of the total land base.

Given the growing interest in using woody biomass as a feedstock for bioenergy production, specific objectives addressed in this chapter were to: (1) evaluate and compare alternative loblolly pine (*Pinus taeda*) management regimes and biomass availability; (2) examine how optimal forest management strategies and biomass availability are subject to changes in the relative price of biomass to sawtimber and real discount rates; and (3) explore the management implications of these woody biomass management strategies for bioenergy production in Mississippi.

2. LITERATURE REVIEW

Previous studies have examined and compared land management alternatives for biomass (Schultz, 1997; Bjornstad & Skonhoft, 2002; Oregon Department of Energy, 2003; Mead, 2005). Bjornstad and Skonhoft (2002) compared the two main roles whereby forestry affects global climate by using biomass sources for energy production and carbon sequestration. Results suggested that accounting for CO_2 storage increased the optimal rotation length while managing the forest as a source of bioenergy shortened it. Oregon Department of Energy (2003) looked at three counties in eastern Oregon and indicated that modifying existing forest and agricultural practices was needed to make biomass available for local bio-production.

Mead (2005) examined alternative silvicultural regimes for enhancing productivity for biomass production on plantations and concluded that early site manipulation and fertilization generate the greatest potential gains along with appropriate species selection. Stocking and rotation length were also common factors increasing biomass production. Previous studies suggested that shorter rotations required a closer tree spacing to maximize productivity of unthinned loblolly pine plantations. Optimal spacing will become wider if management objectives target larger merchantable timber (Schultz, 1997; Mead, 2005).

Determining how various site preparation techniques impact loblolly pine plantation production in Mississippi is critical for evaluating alternative management regimes for growing biomass feedstocks. These site treatments enhance micro-site conditions for seedlings, control competing plants, remove obstructions, and consequently improve tree growth (Schultz, 1997). Schultz (1997) and Allen (2001) indicated that drainage treatments on wet flatwood sites in the southeastern U.S. has led to increases of site index (SI) by 10 to 33 feet which denoted a 30 to 100% increase in mean annual increment (MAI) over the rotation. Early biomass growth of loblolly pine plantations reached 240% and 370% in response to bedding and bedding plus mole-plowing (i.e., drainage improvement) treatments, respectively (Xu, Burger, & Aust, 2000). Fertilization at planting can greatly increase loblolly pine seedling growth on southern coastal plains. On phosphorous-deficient sites, it can improve plantation growth rates over the entire rotation when sufficient nutrients were applied (Allen, Dougherty, & Campbell, 1990; Schultz, 1997).

Biomass growth can also be enhanced by removing competing weed species (Lauer, Glover, & Gjerstad, 1993; Schultz, 1997; Martin & Shiver, 2002). Studies by Schultz (1997) and Martin and Shiver (2002) showed that woody weed control at year 7 increased stand volume at harvest by 25%, and control of both woody and herbaceous weeds led to increases of 39 to 45% in volume growth at age 12 depending on site conditions. Lauer et al. (1993) showed that early height gains of loblolly pine plantations from removal of grass and herbaceous weeds ranged from 1.3 to 5.2 feet, and volume response at year 9 averaged 4,682 and 7,357 board feet per acre with 1- and 2-year weed treatments, respectively. In addition, rotation length became shorter and a gain in MAI approached 14% (Lauer et al., 1993).

3. Methods

The main approach in this analysis was to compare alternative management regimes for growing woody biomass as a feedstock source for supporting renewable energy industries such as co-fired power-generation facilities or an incipient bio-fuels industry based on cellulosic materials. As in any analysis of this type, tree growth simulations are critical for evaluating long-term wood flow projections. In this chapter, volume growth of loblolly pine was projected for stemwood using the PTAEDA 3.1, a growth and yield simulator developed by the Virginia Tech Growth and Yield Cooperative in Blacksburg, VA. This simulator uses data collected from permanent plots scattered across every state in the southern U.S. PTAEDA 3.1 is an individual tree model that incorporates competition and mortality effects to project volume yields by various management scenarios (Bullock & Burkhart, 2003; Burkhart, Amateis, Westfall, & Daniels, 2004). The yield of total stemwood contained, by default, green weight of all trees greater than 1 inch diameter at breast height (DBH). The output of stemwood residue was defined as the difference of green weight between total stemwood and three mutually exclusive products: sawtimber, chip-n-saw, and pulpwood.

All management scenarios run within the PTAEDA 3.1 growth and yield simulator required specifying input factors such as physiographic region, drainage class, plant spacing, site index, and rotation age. This chapter assumed that the Gulf Coastal Plain region in Mississippi was appropriate for evaluating conditions since there was little to no piedmont lands. Due to the software's conservative projection of growth responses for alternative site preparation techniques, the study consulted with several relevant experts and adjusted the site index prior to each simulation instead of utilizing the site preparation option within the software (table 1). This approach allowed for the consideration of additional site preparation techniques such as sub-soiling and using three-in-one plowing. In addition, this chapter accepted the default merchandising limits offered by PTAEDA 3.1 for the sawtimber, and chip-n-saw and pulpwood product classes. Product class default dimensions for sawtimber are 12+ inches DBH to an 8-inch top diameter, for chip-n-saw 8 to11 inches DBH to a 6-inch top diameter, and for pulpwood 6 to 7 inches DBH to a 5-inch top diameter. Small diameter trees with a DBH less than 6 inches and topwood from all three product classes were considered as stem residue potential for bioenergy production.

Table 1. Hypothetical site preparation techniques and growth response* of planted loblolly pine (*Pinus taeda*) expressed as the improvement of site index (base age 25) in the Mississippi Gulf Coastal Plain

Site Preparation	Improvement of Site Index
D[1] (P&W)[2]	2
H[1] (P&W)	5
S[1] (P&W)	5
H+[3]S (P&W)	10
H+S+F[2] (P&W)	13
D+H (P&W)	7
D+H+F (P&W)	11
H+F (P&W)	10
B[1] (P)[2]	2
B+H (P)	7
B+H+F (P)	11
BDS[4] (P)	8
BDS+H (P)	13
BDS+H+F (P)	15
None[5] (P&W)	

[1] B, H, F, D, and S indicate site preparation technique: bedding, herbicide use, fertilization, disking, and subsoiling, respectively.

[2] P&W in brackets indicates the specific site preparation or combination of site preparation techniques is performed on both poor- and well-drained lands, whereas P&W is only conducted on poorly drained lands.

[3] '+' indicates a combination of different site preparation techniques.

[4] 'BDS' indicates site preparation with a three-in-one plow (i.e., a tractor-mounted setup that deep rips the subsoil, disks, and forms raised beds behind).

[5] 'None' indicates that there is no site preparation for control projection.

* Source: Review of literature (Fisher & Binkley, 2000; NCASI, 2004); Experts in growth and yield, R. Amateis (pers. commun., Oct. 2006); Experts in silviculture, Drs. A.W. Ezell and S.D. Roberts, and industry representatives (pers. commun., Oct. 2006).

3.1. Monetary and Biological Criteria

To simulate a decision-making environment that private forest landowners may undertake on their forest properties, this chapter utilized a land expectation value (LEV) model to evaluate financially optimal management practices for loblolly pine plantations on different sites in Mississippi. Chang (1998) considered this model to correctly incorporate the opportunity cost of capital and land. The basic LEV model was defined as:

$$LEV = \frac{\sum_{i=0}^{t} V_{(timber)i} e^{-rt} - \sum_{i=0}^{t} C_i e^{-rt}}{1-e^{-rt}} \quad (1)$$

where $V_{(timber)i}$ was the value of timber at time i (i.e., stumpage price times yield), C_i the stand establishment cost and management cost incurred at time i, r the real discount rate, and t the rotation age.

To incorporate stemwood residue revenue from thinnings and harvesting, the LEV model was modified as follows:

$$LEV = \frac{\sum_{i=0}^{t} \left[V_{(timber)i} + V_{(resi.)i}\right] e^{-rt} - \sum_{i=0}^{t} C_i e^{-rt}}{1-e^{-rt}} \quad (2)$$

where $V_{(timber)i}$ was the sum of the values of the three wood classes in year i (i.e., stumpage prices times their green weights predicated by PTAEDA 3.1), $V_{(resi.)i}$ the value of stemwood residues from thinnings and harvesting in year i (i.e., assumed biomass price times its green weight), C_i is the stand establishment and management costs incurred in year i, r the real discount rate, and t the rotation age. LEVs were calculated and compared using before-tax, a real annual discount rate of 5%, similar to a forest landowner's real rate of return.

To simulate another decision-making environment that a private forest landowner may employ, this chapter also employed the maximization of MAI of total stemwood biomass, as an alternative approach for finding the optimal management regime for producing woody biomass. This chapter defined it as the green weight of total stemwood divided by rotation age. This biological criterion was useful for identifying the maximum sustainable annual output of total stemwood biomass on a site, assuming that timberlands were managed only for bioenergy production. The management regime which culminated MAI on a specific site was considered biologically optimal.

3.2. Management Scenarios

This chapter developed management scenarios that represented different site conditions, site preparation techniques, planting densities, and alternative thinning ages. It also replicated simulation work by Cason (2006) which suggested that using site indices of 50, 60, and 70 (base age 25) across poorly-drained and well-drained classes were appropriate for representing common forestry sites in Mississippi. Silvicultural practices on poorly-drained sites have required techniques that employ bedding, disking, herbicide, fertilization, sub-soiling, and three-in-one plowing as pertinent management options. Similar site preparation techniques were employed on well-drained sites with the exception of bedding. Six different planting densities represented the common range of possible tree spacings of 10×10, 9×10, 8×10, 7×10, 6×10, and 5×10 feet. This chapter followed a recommendation by Sharma, Burkhart, and Amateis (2002) that rectangular spacing was appropriate for using mechanized equipment between rows and minimizing adverse impacts on diameter and height growth.

Table 1 highlights factors used in developing the management scenarios evaluated in this chapter, which evaluated 226,512 scenarios on poorly-drained sites and 169,884 on well-drained sites. All management scenarios employed one low thin, which was a combination of selection and row thinning. Residual basal area was 70% in all cases. Thinning age was pre-set to occur only between years 15 and 25. No thinning occurred 5 years prior to final harvest.

3.3. Price and Cost Data

Timber prices and cost data used in this analysis came from various sources. Table 2 presents costs associated with silvicultural practices acquired from interviews with experts and published sources such as Smidt, Dubois, and Folegatti (2005). Timber prices came from Timber Mart-South 2006 yearly state-wide average stumpage prices for Mississippi. Assumptions were made in regard to estimates for biomass prices. Daniels (1999) indicated that many mills in Mississippi were paying $1 to $2 less per green ton for 'juvenile pine pulpwood' specified as thinnings of age 17 years or less. This chapter assumed that $1.50 per green ton would represent the most likely price deduction for small diameter trees and topwood.

Table 2. Costs of forestry management practices (in 2006 dollars) for pine plantations in the Mississippi Gulf Coastal Plain

Activity	Cost ($/acre)
Administration	2.00
Bedding	45.00
Disking	45.00
Herbicide	76.68
Sub-soiling	91.94
Fertilization	54.47
Three-in-one plow	120.47
Hand planting	0.064 ($/seedling)
Seedlings	0.043 ($/seedling)

Source: Dr. A.W. Ezell, pers. commun., Oct. 2006; and Smidt et al. (2005)

3.4. Sensitivity Analyses

A sensitivity analysis was employed to evaluate the impact of changes in discount rates and relative biomass prices on simulated biomass management regimes in Mississippi. The real annual discount rates, 3, 7, and 9%, were also used for evaluation and comparison of management alternatives. This chapter used relative biomass prices expressed as a percentage of the sawtimber price to illustrate how changes will influence a landowner's optimal management strategies and availability of forest biomass for bioenergy production (Henderson & Munn, 2004). According to Timber Mart-South price reports (from 1994 to present), average prices for chip-n-saw and pulpwood were around 70% and 22% of the sawtimber price in Mississippi, respectively. Therefore, it was assumed that when the potential biomass price increased to 25% of the sawtimber price, forest landowners will sell pulpwood for bioenergy production. When biomass prices increased to 70% or 100% of the sawtimber price, even chip-n-saw or sawtimber can be considered for bioenergy production. When biomass prices fall at or below 10% of the sawtimber price, only stemwood residues would be available for bioenergy production.

4. Results

Both financially and biologically optimal management regimes and resulting average annual biomass yields were presented and compared (table 3). The financially optimal site treatment, thinning year, and rotation length varied in response to differences in site productivity. Application of three-in-one plowing was optimal site preparation on SI 50 and 60 poorly-drained lands, and the combination of disking and herbicide use appeared optimal on SI 50 and 60 well-drained lands. Optimal site preparation on SI 70 was applying herbicides and fertilization regardless of soil drainage. As SI increased from 50 to 70, optimal rotation lengths declined from 35 to 27 years. In general, the optimal thinning year also decreased from 23 to 15 years. The initial tree spacing, 10×10 feet, was financially optimal for all site combinations. LEVs ranged from \$644 to \$1,338 per acre over site indices between 50 and 70. Results indicated that higher site quality yields more biomass. Average annual yields of stemwood residues potentially available for bioenergy production ranged from 0.86 to 1.20 green tons per acre across sites with SI 50 to 70. In line with the limited demand for small diameter wood in the forest product industry (Neilson & Flynn, 2006), it was possible to include pulpwood for bioenergy production. The sum of annual yields of stemwood residues and pulpwood almost doubled biomass availability, ranging from 1.64 to 2.38 green tons per acre.

Mean annual biomass output from biologically optimal management regimes ranged from 5.28 to 7.73 green tons per acre for SI 50 to 70 (table 3). In comparison with financially optimal management regimes for timber and biomass joint production, culminating MAI of total stemwood biomass exclusively for bioenergy production involved highly intensive management activities. The most intensive site preparation considered, the application of three-in-one plowing, herbicide use and fertilization on poor-drained land; and the combination of herbicide use, sub-soiling, and fertilization on well-drained land, became optimal. Close initial tree spacings appeared optimal, varying from 623 to 872 trees per acre on different sites. No thinning and shorter rotation length became optimal, with rotation age declining from 22 to 18 years in response to the rise of SI. These biologically optimal management regimes were, however, unprofitable on all sites. Biomass availability more than tripled compared with financially optimal management regimes using stem residue and pulpwood for bioenergy production. The trade-offs between these two optimal management strategies ranged from $848 to $1,419 from SI 50 to 70.

Table 3. Comparison of financially [maximizing land expectation value (LEV) at a 5% discount rate] and biologically [culminating mean annual increment (MAI) of total biomass] optimal management regimes of loblolly pine (*Pinus taeda*) plantation, available average annual biomass yields (in green tons), and resulting LEVs by site index (SI) and drainage class (2006 dollars)

Site		Max. LEV Regime[5]	Residues tons/ac/yr	Resi.& Pulp. tons/ac/yr	LEV $/ac	Max. MAI Regime	Biomass tons/ac/yr	LEV $/ac
50[1]	Poor	BDS[2]-23[3]-**35**[3]	0.86	1.64	666	BDSHF[2]-*623*[3]-none[4]-**21**	5.45	-200
	Well	DH[2]-17-**35**	0.86	1.60	644	HSF[2]-*872*-none-**22**	5.28	-204
60	Poor	BDS-17-**31**	0.98	2.08	914	BDSHF-*727*-none-**19**	6.48	-161
	Well	DH-18-**34**	0.99	1.94	900	HSF-*727*-none-**19**	6.27	-127
70	Poor	HF[2]-16-**27**	1.16	2.34	1,268	BDSHF-*872*-none-**18**	7.73	-110
	Well	HF-15-**27**	1.20	2.38	1,338	HSF-*872*-none-**18**	7.43	-81

[1] Site index (base age 25).

[2] BDS, DH, HF, BDSHF, and HSF indicate site preparation techniques three-in-one plowing, combination of disking and herbicide use, combination of herbicide use and fertilization, combination of three-in-one plow, herbicide use and fertilization, and combination of herbicide use, subsoiling and fertilization, respectively.

[3] Numbers in italic, underlined, and in bold indicate initial planting density, thinning year, and rotation length, respectively.

[4] 'none' indicates that no thinning is required for this optimal management regime.

[5] The optimal planting density for Max LEV regimes is 436 trees per acre for all sites.

5. SENSITIVITY ANALYSIS

5.1. Changes in Relative Biomass Price

Financially optimal management strategies (at a 5% real discount rate) changed substantially, as biomass prices increased from 10% to 100% of the sawtimber price (tables 4-5). The intensity of site preparation was increasing on almost all sites. On poorly-drained lands, optimal site treatment techniques were three-in-one plowing (on SI 50 and 60) or application of herbicides and fertilizers (on SI 70) at low relative biomass prices. When the biomass price equaled the sawtimber price, highly intensive site preparations became optimal (i.e., application of three-in-one plowing and herbicide on SI 50, and use of three-in-one plowing, herbicide, and fertilization on SI 60 and 70). On well-drained lands, optimal site preparation practices gradually evolved from disking and applying herbicides to implementing highly intensive techniques, such as combining disking and herbicide and fertilizer applications on SI 50 sites and employing regimes using sub-soiling, herbicides, and fertilizers on SI 60 sites. Optimal site preparation practices required applying herbicides and fertilizers regardless of changes in relative biomass prices on SI 70 sites.

Table 4. Changes in financially optimal management strategies (at a 5% real discount rate), land expectation values (LEVs), and available biomass (in green tons) per acre of loblolly pine (*Pinus taeda*) plantations on poorly-drained land as relative price of biomass to sawtimber rises (2006 dollars)

	Relative Price of Biomass to Sawtimber			
Variable	10%	25%	70%	100%
SI 50[1]				
LEV ($/ac)	655	740	1,416	2,230
Site preparation	BDS[2]	BDS	BDS+H[3]	BDS+H
Density (tree/ac)	436	436	727	727
Thinning (yr)	23	23	16	15
Rotation (yr)	35	35	21	20
Biomass (tons/ac/yr)	0.86	1.64	5.07	5.00
SI 60				
LEV ($/ac)	902	1,025	1,828	2,830
Site preparation	BDS	BDS+H+F[4]	BDS+H+F	BDS+H+F
Density (tree/ac)	436	485	727	727
Thinning (yr)	20	18	16	15
Rotation (yr)	33	29	21	20
Biomass (tons/ac/yr)	0.94	2.26	6.04	6.17
SI 70				
LEV ($/ac)	1,251	1,397	2,294	3,436
Site preparation	H+F[5]	H+F	BDS+H	BDS+H+F
Density (tree/ac)	436	436	623	872
Thinning (yr)	16	16	15	15
Rotation (yr)	27	27	21	20
Biomass (tons/ac/yr)	1.16	2.34	6.46	7.19

[1] Base age 25.

[2] 'BDS' indicates site preparation with a three-in-one plowing (i.e., a tractor-mounted setup that deep rips the subsoil, disks, and forms raised beds behind).

[3] 'BDS+H' indicates a combination of site preparation techniques using three-in-one plowing and herbicides.

[4] 'BDS+H+F' indicates a combination of site preparation techniques using three-in-one plowing, herbicides, and fertilization.

[5] 'H+F' indicates a combination of site preparation techniques using herbicides and fertilization.

The optimal initial planting density increased from 436 to 727 trees per acre on sites having a SI 50 and 60, and to 872 trees per acre on SI 70 lands when the relative biomass price rose. On SI 70 poorly-drained lands, the optimal thinning age was the same, year 15. On all other sites the optimal thinning age declined. The optimal rotation length decreased on all sites for maximization of biomass feedstocks in lieu of timber production. Changes in LEVs and biomass availability associated with the rising relative biomass prices are presented in tables 4-5.

Table 5. Changes in financially optimal management strategies (at a 5% real discount rate), land expectation values (LEVs), and available biomass (in green tons) per acre of loblolly pine (*Pinus taeda*) plantations on well-drained land as relative price of biomass to sawtimber rises (2006 dollars)

	Relative Price of Biomass to Sawtimber			
Variable	10%	25%	70%	100%
SI 50[1]				
LEV ($/ac)	613	726	1,387	2,149
Site preparation	D+H[2]	H+F[3]	H+F	D+H+F[4]
Density (tree/ac)	436	545	727	727
Thinning (yr)	17	17	15	15
Rotation (yr)	35	31	20	20
Biomass (tons/ac/yr)	0.86	1.96	4.63	4.81
SI 60				
LEV ($/ac)	887	1,011	1,799	2,780
Site preparation	D+H[5]	H+S+F[6]	H+S+F	H+S+F
Density (tree/ac)	436	485	727	727
Thinning (yr)	18	18	15	15
Rotation (yr)	34	30	20	20
Biomass (tons/ac/yr)	0.99	2.25	5.94	6.01
SI 70				
LEV ($/ac)	1,320	1,471	2,292	3,446
Site preparation	H+F	H+F	H+F	H+F
Density (tree/ac)	436	436	623	872
Thinning (yr)	15	15	15	15
Rotation (yr)	27	27	20	20
Biomass (tons/ac/yr)	1.20	2.38	6.42	7.16

[1] Base age 25.
[2] 'D+H' indicates a combination of site preparation techniques using disking and herbicides.
[3] 'H+F' indicates a combination of site preparation techniques using herbicides and fertilization.
[4] 'D+H+F' indicates a combination of site preparation techniques using disking, herbicides, and fertilization.
[5] 'D+H' indicates a combination of site preparation techniques using disking and herbicides.
[6] 'H+S+F' indicates a combination of site preparation techniques using herbicides, sub-soiling, and fertilization.

5.2. Changes in Discount Rate

As real discount rates increased from 3% to 9%, LEVs decreased dramatically on all sites. Financially optimal management strategies also varied in response to the variation of real discount rates and available biomass changed accordingly (tables 6-7). On poorly-drained lands, optimal site preparation changed from the most intensive combination of three-in-one plowing, herbicide use, and fertilization at a 3% discount rate, to none at a 9% discount rate on SI 50 and 60. On well-drained lands, it changed from the most intensive combination of herbicide use, sub-soiling, and fertilization at a 3% discount rate, to none at a 9% discount rate. Combined herbicide use and fertilization became optimal at all four discount rates regardless of soil drainage. In general, the optimal initial planting density did not respond to variations in discount rates, and the wider spacing of 436 trees per acre (10×10 feet) was generally found optimal.

Annual available biomass for bioenergy production had a slight but irregular change. Resulting stemwood residues as a feedstock for bioenergy production ranged from 0.84 to 0.96 green tons on SI 50, 0.98 to1.32 green tons on SI 60, and 1.12 to 1.24 green tons on SI 70 per acre per year. The sum of stemwood residues and pulpwood ranged from 1.53 to 1.87 green tons on SI 50, 1.89 to 2.18 green tons on SI 60, and 2.31 to 2.43 green tons on SI 70 per acre per year. The optimal rotation length decreased on all sites. The optimal thinning age also declined on SI 60 and 70, but changed irregularly on SI 50.

Table 6. Financially optimal management strategies, land expectation values (LEVs), and available biomass (in green tons) per acre of loblolly pine (*Pinus taeda*) plantation on poorly drained land by discount rate (2006 dollars)

	Discount Rate			
Variable	3%	5%	7%	9%
SI 50[1]				
LEV ($/ac)	1,862	666	277	140
Site preparation	BDS+H+F[2]	BDS[3]	None[4]	None
Density (tree/ac)	436	436	436	436
Thinning (yr)	20	23	21	18
Rotation (yr)	37	35	27	25
Biomass (tons/ac/yr)	*0.87*[5]/1.87[6]	*0.86*/1.64	*0.86*/1.53	*0.92*/1.61
SI 60				
LEV ($/ac)	2,402	914	413	209
Site preparation	BDS+H+F	BDS	BDS	None
Density (tree/ac)	485	436	436	545
Thinning (yr)	21	17	15	15
Rotation (yr)	33	31	26	23
Biomass (tons/ac/yr)	*1.12*/2.16	*0.98*/2.08	*1.01*/2.03	*1.24*/2.10
SI 70				
LEV ($/ac)	2,940	1,268	627	315
Site preparation	H+F[7]	H+F	H+F	H+F
Density (tree/ac)	436	436	436	436
Thinning (yr)	18	16	16	16
Rotation (yr)	30	27	26	23
Biomass (tons/ac/yr)	*1.12*/2.31	*1.16*/2.34	*1.17*/2.38	*1.20*/2.41

[1] Base age 25.

[2] 'BDS+H+F' indicates a combination of site preparation techniques using three-in-one plowing, herbicides, and fertilization.

[3] 'BDS' indicates site preparation with a three-in-one plowing (i.e., a tractor-mounted setup that deep rips the subsoil, disks, and forms raised beds behind).

[4] 'None' indicates that there was no site preparation required for this optimal management regime.

[5] Average annual yield of stemwood residue per acre.

[6] Sum of average annual yield of stemwood residue and pulpwood per acre.
[7] 'H+F' indicates a combination of site preparation techniques using herbicides and fertilization.

Table 7. Financially optimal management strategies, land expectation values (LEVs), and available biomass (in green tons) per acre of loblolly pine (*Pinus taeda*) plantation on well-drained land by discount rate (2006 dollars)

	Discount Rate (%)			
Variable	3	5	7	9
SI 50[1]				
LEV ($/ac)	1,861	644	278	141
Site preparation	H+S+F[2]	D+H[3]	None[4]	None
Density (tree/ac)	436	436	436	436
Thinning (yr)	20	17	18	18
Rotation (yr)	38	35	29	24
Biomass (tons/ac/yr)	*0.92*[5]/1.78[6]	*0.86*/1.60	*0.84*/1.55	*0.96*/1.67
SI 60				
LEV ($/ac)	2,341	900	407	215
Site preparation	H+S+F	D+H	H[7]	None
Density (tree/ac)	436	436	436	545
Thinning (yr)	20	18	16	16
Rotation (yr)	34	34	29	22
Biomass (tons/ac/yr)	*1.01*/1.94	*0.99*/1.94	*1.00*/1.89	*1.32*/2.18
SI 70				
LEV ($/ac)	3,092	1,338	663	339
Site preparation	H+F[8]	H+F	H+F	H+F
Density (tree/ac)	436	436	436	436
Thinning (yr)	17	15	15	15
Rotation (yr)	29	27	26	23
Biomass (tons/ac/yr)	*1.19*/2.32	*1.20*/2.38	*1.21*/2.39	*1.24*/2.43

[1] Base age 25.
[2] 'H+S+F' indicates a combination of site preparation techniques using herbicides, subsoiling, and fertilization.
[3] 'D+H' indicates a combination of site preparations using disking and herbicides.
[4] 'None' indicates that there was no site preparation required for this optimal management regime.
[5] Average annual yield of stemwood residue per acre.
[6] Sum of average annual yield of stemwood residue and pulpwood per acre.
[7] 'H' indicates the site preparation technique of herbicide use only.
[8] 'H+F' indicates a combination of site preparation techniques using herbicides and fertilization.

6. Management Implications for Biomass Production in Mississippi

Our results suggested maximizing MAI of stemwood biomass exclusively for bioenergy production involved more intensive management regimes than maximizing the monetary return. However, these management practices were unprofitable because current low biomass prices and higher operational costs yielded negative LEVs. Therefore, forest landowners may consider managing their lands for the joint production of timber commodities and biomass rather than maximizing sustainable annual output of stemwood biomass exclusively for bioenergy production.

These results also indicated that logging residues from forests may be less expensive and a more feasible option than growing woody biomass from energy-oriented pine plantations. This was consistent with Cook and Beyea (2000) who compared land management for biomass from forests and energy crops. Given current low biomass prices, it was not practical to develop extensive pine plantations exclusively for bioenergy feedstocks. The establishment of a new bio-energy industry should be based on the availability of forest and agriculture residues and other biomass wastes.

Fertilization was not a financially optimal practice on some sites because of increasing prices of petroleum. Factors such as growth and the cost and relative prices of biomass and commercial wood products can have an important impact on determining the optimal site preparation for each site. With rising fertilizer prices, optimal site preparations on high productive lands may change to non-fertilization techniques. Applying fertilizers was optimal on all sites when maximizing the MAI of stemwood biomass, because the biological criterion did not take costs into consideration.

Although incorporating genetics into this chapter goes beyond its basic scope, species selection may achieve the greatest gains when attempting to increase plantation productivity for biomass production (Mead, 2005). Previous studies have indicated that genetic improvements of southern loblolly pine saw additional gains in volume and value, and higher quality genetics realized added benefits from silvicultural treatments (Weir, 1996; Li, McKeand, & Weir, 2000; Martin & Shiver, 2002). Seen as an outstanding short-rotation woody crop, loblolly pine has highly advanced genetic improvements and will be more likely supported by genetic engineering techniques (Li et al., 2000; Dickmann, 2006). In the case of improved genetics, both optimal management strategies (i.e., optimal site preparation techniques, initial tree spacings, thinning activities, rotation length) will change accordingly.

The two general management approaches utilized in this chapter can have very different management implications across the landscape (Bettinger, Boston, Siry, & Grebner, 2009). For instance, maximizing financial benefits required longer rotation length, which has positive impacts on global climate change due to the potential for carbon sequestration (Bjornstad & Skonhoft, 2002). Grebner et al. (2007) discussed how the growth of new carbon sequestration markets may have an impact in afforestation activities in agricultural communities. In the southern U.S., maintaining adequate habitat requirements for some wildlife species is also a major concern for land management (Carley & Grado, 2000). Financially optimal regimes, having wide tree spacings and thinnings, will delay crown closure and stimulate wildlife food production. Also, long rotation plantation management provides mature stands which are favorable habitats for avian species such as red-cockaded woodpecker (*Picoides borealis*) and bald eagle (*Haliaeetus leucocephalus*) (Baker & Langdon, 1990). Although financially optimal regimes may yield better wildlife habitat at wider tree spacing (Yarrow & Yarrow, 1999), narrow tree spacing may still provide some vegetative cover for certain species of wildlife that agricultural fields may not. However, compared with older plantations, biological management regimes with shorter rotations and more intensive site treatments may also have negative ecological impacts such as increased soil erosion and decreased water quality (Richardson & McCarthy, 1994; Tolbert & Schiller, 1996).

With increasing biomass prices, both management strategies will become more profitable. However, managing plantations for the joint production of timber and biomass always yields greater monetary returns than maximizing MAI of stemwood biomass exclusively for bioenergy production, unless biomass prices equal or exceed sawtimber prices. Will this lead to land use changes in Mississippi in the foreseeable future? It is possible in the long-term. In the short-term, rising ethanol prices have shown a shift in agricultural land uses from lands previously producing cotton now producing corn (Grebner et al., 2009). It is more likely that agricultural crops may be gradually displaced with tree plantations. In this case, more environmental benefits and natural ecosystem functions can be provided with increasing pine plantation acreage. The rise of relative biomass prices to sawtimber resulted in more intensive management practices (i.e., more site treatments, closer initial tree spacings, shorter rotation lengths, earlier thinning ages). This was reasonable because management strategies have gradually evolved to maximize total biomass.

Increases in discount rates will reduce the profitability of forest land and intensity of financially optimal management regimes such as fewer site preparation practices and a shortened thinning year and rotation length. Shortened optimal rotation length will result in slightly more residues rather than large diameter wood products. However, changes in the supply of potentially available stemwood residue for bioenergy production are not clear because of the combined effects of variations in optimal site preparations. If the discount rate increases to some extent, land rent from pine plantations will be lower than other land uses. This will lead to land use change and, consequently, affect woody biomass supply.

In recent years, utilization of forest biomass for bioenergy production has been promoted through various governmental programs. Legislation has addressed the challenges of biomass procurement, improvement of ligno-cellulosic conversion technologies and bio-product market development (Guo et al., 2007). Of note, the Energy Policy Act of 2005 included stipulations for forest biomass utilization to prevent hazardous fires, reduce disease and insect infestation, and restore forest health. Grant programs have also been established to improve the commercial value of non-merchantable forest materials or pre-commercial thinnings for bioenergy production. The Improved Biomass Use Grant Program was targeted to develop technologies for biomass utilization, increase job creation, and reduce hazardous forest fuels. These actions might help alleviate the negative influence of global trade on the domestic forest-based marketplace, thus helping to improve forest management on private lands. However, their impacts on the optimal alternative land management still need further investigation.

7. Conclusion

Alternative markets for woody biomass may substantially improve forest management in Mississippi and the southern U.S. Biomass harvested from optimal land management strategies served as the basis for developing a new bioenergy industry. This chapter suggested that around 2.82 million green tons of stemwood logging residues can be produced annually from financially optimal management of loblolly pine plantation in Mississippi. Assuming the recovery rate of 65% for all logging residues, inclusion of recoverable branches and leaves would increase annual biomass availability to approximately 4.07 million green tons.

Though culminating sustainable annual output of stemwood biomass with highly intensive management can yield more than six times of stem residues as feedstock for bioenergy production, this management strategy is unprofitable due to current low biomass prices and high operational costs. Therefore, woody biomass supply will mainly come from financial management regimes, joint production of timber commodities and logging residues, which are more favorable for private forest landowners. However, financially optimal regimes were likely to change with the variation of relative biomass prices and discount rates, consequently affecting woody biomass supply. As the relative prices of biomass to sawtimber increase, management strategies will gradually evolve to maximize total biomass with intensive site treatments and shorter rotation length. It may also cause land use change (i.e., displacing agricultural crops with pine plantation) and, accordingly, more biomass from forests will be available for bioenergy production. Nevertheless, the rise in the discount rate can decrease land rent, resulting in transformation of pine plantations to other land uses and decreases in woody biomass supply.

These two woody biomass management strategies may have different impacts on environments. Maximizing monetary benefits requires longer rotation length than culminating MAI of stemwood biomass. This strategy can help mitigate greenhouse gas emissions because of more carbon sequestration in pine plantation. Therefore, it may have more positive impacts on global climate change. With wide tree spacing and mature stands, this strategy can also provide more favorable wildlife habitats than culminating sustainable annual stemwood biomass availability. However, intensive site treatments for maximum stemwood biomass may result in negative ecological impacts such as soil erosion and water run-off.

These results and implications can be beneficial to both the bio-production industry and private forest landowners in Mississippi and the southern U.S. They are also valuable for government policymakers who can enact and implement biomass incentives at the state level. It is worth noting that the results were dependent on the PTAEDA 3.1 growth and yield model and subject to a setting of merchandizing limits for three wood product commodities, which determines the amount of stem residues left for bioenergy production.

References

Allen, H. L. (2001). Silvicultural treatments to enhance productivity. In J. Evans (Ed.), *The forests handbook,* (pp.129-139). Oxford, UK: Blackwell Science.

Allen, H. L., Dougherty, P. M., & Campbell, R. G. (1990). Manipulation of water and nutrients - practices and opportunities in southern US pine forests. *Forest Ecology and Management, 30*, 437-453.

Baker, J. B., & Langdon, O. G. (1990). Loblolly Pine (*Pinus taeda* L.) In: R. M. Burns & B. H. Honkala (Eds.), *Silvics of North America: Vol. 1. Conifers* (pp. 497–512). Washington, DC: USDA Forest Service.

Bettinger, P., Boston, K., Siry, J., & Grebner, D. L. (2009). *Forest Management and Planning*. London, UK: Academic Press, Elsevier, Inc.

Bjornstad, E., & Skonhoft, A. (2002). Wood fuel or carbon sink? Aspects of forestry in the climate question. *Environmental and Resource Economics, 23*(2), 447-465.

Bullock, B. P., & Burkhart, H. E. (2003). Equations for predicting green weight of loblolly pine trees in the South. *Southern Journal of Applied Forestry, 27*(3), 153-159.

Burkhart, H. E., Amateis, R. L., Westfall, J. A., & Daniels, R. F. (2004). *PTAEDA3.1: simulation of individual tree growth, stand development and economic evaluation in loblolly pine plantations*. Blacksburg, VA: Department of Forestry, Virginia Tech.

Carley, T. J., & Grado, S. C. (2000). Tradeoffs in loblolly pine plantation and white-tailed deer management in the Middle Coastal Plain. In Pelkki, M.H. (ed.). *Hardwoods—an underdeveloped resource?* (pp. 57-62) Lexington, KY: Arkansas Forest Resources Center.

Cason, J. D. (2006). *CO_2 emission mitigation potential of Mississippi forests and impacts to nonindustrial private forest land management.* Unpublished master's thesis, Mississippi State University, Mississippi.

Chang, S. J. (1998). A generalized Faustmann model for the determination of optimal harvest age. *Canadian Journal of Forest Research, 28*(5), 652-659.

Cook, J., & Beyea, J. (2000). Bioenergy in the United States: progress and possibilities. *Biomass and Bioenergy, 18*(6), 441-455.

Daniels, R. A. (1999). *Mississippi timber price report - March/April 1999*. Mississippi State, MS: University Extension Service, Mississippi Agricultural and Forestry Experiment Station.

Dickmann, D. (2006). Silviculture and biology of short-rotation woody crops in temperate regions: then and now. *Biomass and Bioenergy, 30*(8-9), 696-705.

Fisher, R. F., & Binkley, D. (2000). *Ecology and management of forest soils*. New York, NY: John Wiley & Sons, Inc.

Ford, R. C., & Senauer, B. (2007). How biofuels could starve the poor. *Foreign Affairs, 86*(3), 41-53.

Grebner, D. L., Londo, A. J., Sun, C., Grado, S. C., Sumerall, D. C., Dewey, J. C., Nero, B. F., & Maiers, R. P (2007). Potential carbon sequestration opportunities and issues for bottomland hardwood afforestation in the Lower Mississippi Alluvial Valley. *Forum of Public Policy: A Journal of the Oxford Round Table*, *3*(4), 305-312.

Grebner, D. L., Perez-Verdin, G., Sun, C., Munn, I. A., Schultz, E. B., & Matney, T. G. (2009). Woody Biomass Feedstock Availability, Production Costs and Implications for Bioenergy Conversion in Mississippi. In B. Solomon and V. Luzadis (Eds.), *Renewable Energy From Forest Resources in the United States* (pp. 261-280). London, UK and New York: Routledge Taylor & Francis Group.

Guo, Z., Sun, C., & Grebner, D. L. (2007). Utilization of forest derived biomass for energy production in the U.S.A.: status, challenges, and public policies. *International Forestry Review, 9*(3), 748-758.

Henderson, J. E., & Munn, I. A. (2004). The effect of relative product prices on the optimal management of loblolly pine. In S.O. Moffat (ed.) *Bugs, Budgets, Mergers, and Fire: Disturbance Economics* (pp. 188-193). New Orleans, LA: USDA Forest Service, Southern Research Station.

Lauer, D. K., Glover, G. R., & Gjerstad, D. H. (1993). Comparison of duration and method of herbaceous weed control on loblolly pine response through midrotation. *Canadian Journal of Forest Research, 23*, 2116-2125.

Li, B., McKeand, S., & Weir, R. (2000). Impact of forest genetics on sustainable forestry—results from two cycles of loblolly pine breeding in the US. *Journal of Sustainable Forestry, 10*, 79-85.

Martin, S. W., & Shiver, B. D. (2002). Impacts of vegetation control, genetic improvement and their interaction on loblolly pine growth in the southern United States - age 12 results. *Southern Journal of Applied Forestry, 26*(1), 37-42.

Mead, D. J. (2005). Opportunities for improving plantation productivity. How much? How quickly? How realistic? *Biomass and Bioenergy, 28*(4), 249-266.

National Research Council (NRC). (1999). *Review of the research strategy for biomass-derived transportation fuels*. Washington, DC: National Academy Press.

National Council for Air and Stream Improvement, Inc. (NACSI). (2004). *Effects of heavy equipment on physical properties of soils and on long-term productivity: a review of literature and current research* (Technical Bulletin No. 0887). Research Triangle Park, NC: National Council for Air and Stream Improvement, Inc.

Neilson, D., & Flynn, R. (2006). *The international pulpwood resource and trade review* (13th ed.). Rotorua, New Zealand: DANA Publishing.

Oregon Department of Energy. (2003). *Biomass resource assessment and utilization options for three counties in Eastern Oregon.* Lakewood, CO: McNeil Technologies, Inc.

Richardson, C. J., & McCarthy, E. J. (1994). Effect of land development and forest management on hydrologic response in southeastern coastal wetlands: a review. *Wetlands, 14*(1), 56-71.

Schultz, R. P. (1997). *Loblolly Pine: the ecology and culture of loblolly pine*. Washington, DC: USDA Forest Service.

Sharma, M., Burkhart, H. E., & Amateis, R. (2002). Spacing rectangularity effect on the growth of loblolly pine plantations. *Canadian Journal of Forest Research, 32*(8), 1451-1459.

Smidt, M., Dubois, M. R., & Folegatti, B. S. (2005). Costs and cost trends for forestry practices in the South. *Forest Landowner, 64*(2), 25-31.

Tolbert, V. R., & Schiller, A. (1996). Environmental enhancement using short-rotation woody crops and perennial grasses as alternatives to traditional agricultural crops. In W. Lockeretz (ed.) *Environmental Enhancement through Agriculture* (pp. 209-216). Medford, MA: Center for Agriculture, Food and Environment, Tufts University.

Weir, R. J. (1996). The impact of genetics on forest productivity. *Alabama's Treasured Forests*, Spring, 19-21.

Xu, Y., Burger, J. A., & Aust, W. M. (2000). Responses of surface hydrology and early loblolly pine growth to soil disturbance and site preparation in lower coastal plain wetland. *New Zealand Journal of Forestry Science, 30*(1/2), 250-265.

Yarrow, G. K., & Yarrow, D. T. (1999). *Managing Wildlife*. Birmingham, Alabama: Sweet Water Press.

In: Global Change and Forestry
Editors: J. Gan, S. Grado and I. A. Munn
ISBN: 978-1-60876-262-0

Chapter 12

TWO ERAS OF GLOBALIZATION AND HARDWOOD SAWTIMBER

William G. Luppold [*]
and Matthew S. Bumgardner
USDA Forest Service, WV, USA

ABSTRACT

In the early 1970s, the adoption of floating exchange rates resulted in more fluid transfers between international currencies and spurred increased international demand for hardwood lumber produced in the United States. Initially, Germany was the most important European customer for U.S. products while Japan was the most important Asian customer. The consumer cultures in both countries were quality oriented, requiring high-grade hardwood lumber and veneer. However, the major consumer of hardwood lumber remained the domestic furniture industry which required long and wide midgrade boards. This combination of quality-oriented international markets and a large board-oriented domestic market resulted in new technology that obtained maximum value yield from high-quality logs. As a result, the demand for and subsequent value of high-grade hardwood sawtimber surged while prices of mid and lower grade sawtimber stagnated. Since the late 1990s, China and India have become major players in the global economy while the influence of European and Japanese markets has diminished. Furniture production in the United States has decreased as a result of Chinese imports but hardwood demand by the domestic flooring and cabinet industries has increased. In this second global era, factors such as price and service have replaced quality and board size as market drivers. Emphasis on cost has caused individual mills to re-examine production and marketing processes and to reevaluate the use of low- and mid-value sawtimber. These changes have major implications for the future value and management of hardwood timber.

Keywords: Hardwoods; International markets; Lumber.

[*] Email: wluppold@fs.fed.us, phone: (304) 431-2770, USDA Forest Service, Northern Research Station, 241 Mercer Springs Road, Princeton, WV 24740.

1. INTRODUCTION

Hardwood lumber producers face a competitive market, cyclical prices, and occasional structural changes in lumber demand that ultimately influence the valuation of hardwood sawtimber. Such a structural change occurred in the early 1970s with the adoption of floating exchange rates for international currencies for developed western (non-Communist) economies. Floating exchange rates allowed the value of currencies to be determined by market forces rather than by intergovernmental negotiations. These changes initially caused the dollar to be devalued against European and Japanese currencies thus lowering the prices and increasing exports of hardwood lumber produced in the United States. Floating exchange rates also facilitated easier transfer of currencies between trading partners, resulting in an overall increase in international trade between the developed economies.

This increase in European and Japanese export demand caused the price of higher grades of lumber, logs, and stumpage to increase (Luppold, 1996; Luppold & Baumgras, 1995). This new demand encouraged mills to add dry kilns to serve a new customer base that wanted kiln-dried lumber. Still, the largest market for grade lumber continued to be the domestic furniture industry which preferred green lumber of sufficient length to conform to furniture production technologies developed decades earlier.

Another more recent structural change was the rapid decline in lumber demand by the domestic furniture industry and increased furniture imports, primarily from China (Schuler & Buehlmann, 2003). China also became the largest offshore market for exported lumber accounting for one-third of exports to countries other than Canada (USDA For. Ag. Serv., 2006). However, unlike the high-quality, high-value exports to Europe and Japan, exports to mainland China consisted primarily of lower value lumber (USDA For. Ag. Serv., 2006).

We call the major structural changes since the early 1970s the eras of globalization because global economic forces caused these changes. The first era of globalization was associated with the adoption of floating exchange rates between developed western economies. The second was associated with the emergence of mainland China as a major player in the overall market for hardwood product manufacturing. It is difficult to know precisely when these eras began and ended because there were periods of transition and adjustment between them. In both eras, the transition period was marked by several years of declining hardwood lumber production (figure 1). For this chapter we assume that the first era spanned 1975 to 1999 and then transitioned to the second era and continues today. In the remainder of this chapter we examine the first era of globalization and the transition to the second era of globalization, discuss probable trends in domestic and international demand for hardwood lumber in this new era, and explain how domestic lumber production and sawtimber prices might be influenced by these trends.

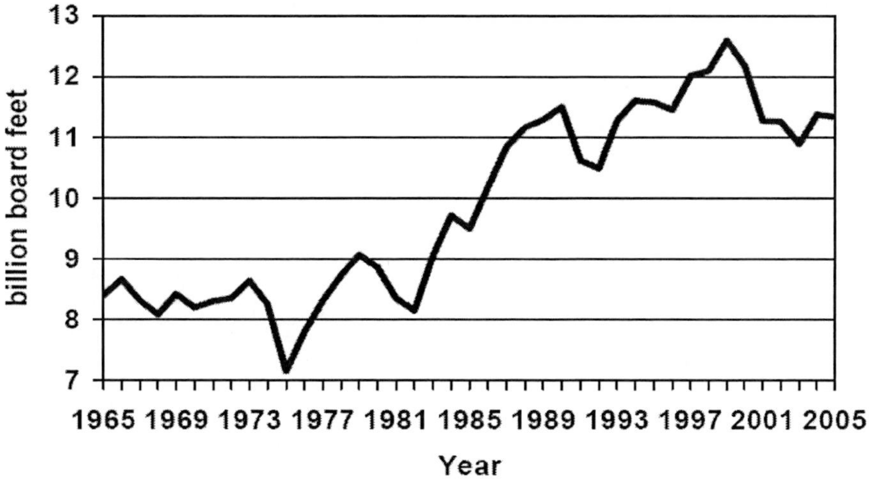

Data source: Luppold and Bumgardner, 2008a.

Figure 1. Eastern hardwood lumber production, 1965 to 2005.

2. The First Era of Globalization: 1975 to 1999

The first apparent change in the hardwood market that resulted from the implementation of floating exchange rates began in 1975 as exports to Europe increased (Luppold & Araman, 1988) and price premiums for shipments of First and Seconds red and white oak lumber were implemented (Hardwood Mark. Rep., 1975). Between 1975 and 1979, lumber exports to Europe increased by 600 percent while overall exports increased by 70 percent. This change in export demand provided the sawmilling industry with a profitable market for high-quality lumber. However, as indicated in table 1, the wood household furniture industry consumed over 700 percent more lumber than was exported in 1972 and 16 percent more lumber than construction appearance uses (millwork, cabinets, and flooring) combined. Unlike the export market, domestic furniture manufacturers primarily purchased green or air-dried lumber and refused to purchase short boards.

High interest rates and stagnant economic growth caused lumber demand and production to decline in 1981; however, exports of hardwood lumber continued to increase as Japan and Taiwan began to import U.S. lumber. Concurrent with increased international demand were increasing demand by the cabinet, millwork, and flooring industries that raised total demand for hardwood lumber to 11 billion board feet in 1987.

The 1990s began with a large decline in production (figure 1) as an economic recession caused furniture production to decline (figure 2). As the decade progressed, lumber demand by the wood furniture industry remained stable but proportional demand declined (table 1). Lumber demand by the pallet industry also declined in the early 1990s, but then rebounded as industrial activity increased in the remainder of the decade. However, the most notable change in the 1990s was the continued growth in exports and increased lumber consumption for flooring, kitchen cabinets, and millwork. By 1997, use of hardwood lumber by producers of construction appearance products were nearly twice that for lumber used by the furniture industry even though furniture shipments were relatively high (table 1).

Table 1. Actual and proportional hardwood lumber consumption by major industry groups 1972, 1977, 1987, 1992, 1997, and 2002

Industry	1972	1977	1987	1992	1997	2002
			million board feet			
Wood furniture	1,926	1,783	1,781	1,546	1,592	1,248
Other furniture	1,078	983	918	1,147	1,065	813
Millwork	614	485	653	644	726	923
Cabinets	293	288	550	898	1,266	1,367
Flooring	706	530	679	755	1,162	1,192
Exports	237	240	688	919	1,213	1,162
Pallets	1,486	1,761	3,349	3,127	4,109	3,666
Railroad ties	850	737	769	578	884	928
Other[a]	1,463	1,309	1,562	1,298	1,279	1,251
Total	8,653	8,116	10,949	10,912	13,296	12,550
			percent[b]			
Wood furniture	22.3	22.0	16.3	14.2	12.0	9.9
Other furniture	12.5	12.1	8.4	10.5	8.0	6.5
Millwork	7.1	5.9	6.0	5.9	5.5	7.4
Cabinets	3.4	3.5	5.0	8.2	9.5	10.9
Flooring	8.2	6.5	6.2	6.9	8.7	9.5
Exports	2.7	3.0	6.3	8.4	9.1	9.3
Pallets	17.2	21.7	30.6	28.7	30.9	29.2
Railroad ties	9.8	9.1	7.0	5.3	6.6	7.4
Other	16.9	16.1	14.3	11.9	9.6	10.0

Data source: Luppold and Bumgardner, 2008b.

[a] Includes other building products, barrels, jewelry chests, toys, musical instruments, and other miscellaneous products.

[b] Columns may not equal 100 percent due to rounding error.

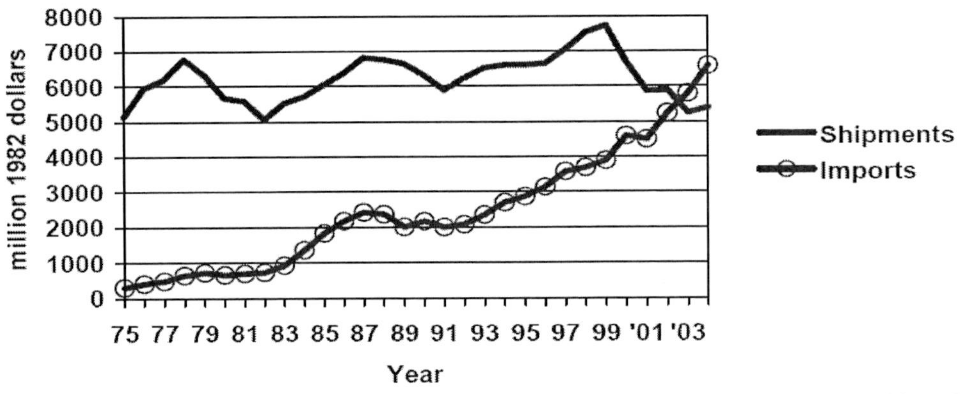

Data source: 1975 to 1988–Nolley 1994; 1989 to 1998- Emanuel and Rhodes 2002; 1999 to 2002-Emanuel and Rhodes 2005; 2003 to 2004 -Akers 2006.

Figure 2. Value of domestic wood household furniture shipments and imports in constant 1982 dollars, 1975-2004.

While supplies of higher grade sawtimber increased between 1975 and 1999, the price of higher grade logs and stumpage also increased because demand for this material exceeded supplies. This increase in demand resulted from a combination of increased international demand for higher quality logs, lumber, and veneer; increased domestic millwork demand for high-grade lumber; and continued demand for long, wide boards by the domestic furniture industry. Higher prices of quality stumpage and logs caused lumber producers to invest in expensive sawing equipment which allowed greater recovery of high-grade lumber from high-grade logs to serve high-value offshore customers and the needs of the domestic furniture industry. However, the competitive nature of the hardwood market caused short-term profits obtained from purchase of this equipment to eventually accrue to the resource, resulting in even higher priced timber.

3. A Period of Transition: 1999 to 2004

Between 1999 and 2004, domestic furniture shipments declined by 30 percent as imports from China and other countries displaced domestic furniture production (figure 2). Pallet producers also reduced lumber consumption because of the continual recycling of pallets and pallet parts. Use of hardwood lumber by the millwork industry also declined slightly. Exports increased slightly but exports to China increased by 150 percent (table 2). Italy and Spain became the most important European markets on a value basis while exports to Germany declined. While the average value per thousand board feet of exported lumber increased for Europe and Japan, the unit value of exports to China decreased by 8 percent (table 2).

Table 2. Volume and imputed price of lumber exported to China, Germany, Italy, Japan, and Spain 1999 and 2004

Country	1999		2004	
	Volume (MMbf)	Price ($/Mbf)	Volume (MMbf)	Price ($/Mbf)
China	32	940	178	849
Germany	38	1,090	21	1,745
Italy	79	1,261	75	1,305
Japan	65	1,261	37	1,299
Spain	67	1,316	73	1,411

Data source: USDA Foreign Agricultural Service, 2006.

The decrease in hardwood lumber demand by the furniture industry since 1999 was considerably larger than that for the value of shipments (figure 2) because of the importation of furniture parts. The two types of furniture that have been least affected by Chinese imports, high-end 18th century reproductions and low-cost, ready-to-assemble furniture constructed primarily of particleboard, have a fairly low ratio of lumber use per unit dollar of sales. Perhaps a better indicator of the plight of the domestic furniture industry is the nearly 40-percent decline in wood household furniture employment between 1999 and the end of 2004 (figure 3). By contrast, employment in the kitchen cabinet industry increased by 17 percent over the same period.

The decline of the furniture industry and the growth in hardwood flooring and kitchen cabinet production demonstrate the increased importance of home construction and remodeling on hardwood lumber demand. Thus, economic indicators such as housing starts and interest rates will become increasingly important to the hardwood industry. This is in sharp contrast to the late 1970s when furniture consumed 65 percent more lumber than all other appearance-based lumber uses combined. In addition, the offshore export market has shifted from being influenced by users of high-priced lumber, e.g. Germany and Japan, to users of lower priced lumber, primarily China (table 2). This is likely to result in increased global competition in export markets.

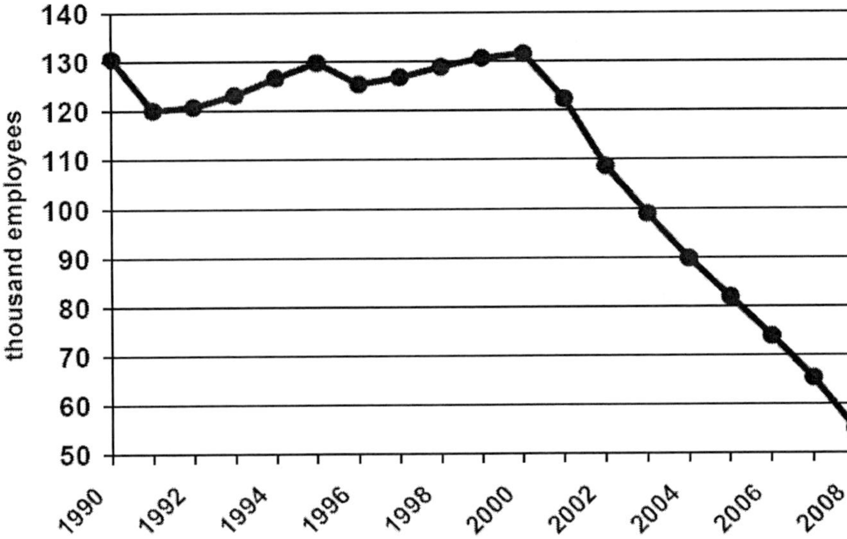

Data source: USDL Bureau of Labor Statistics, 2008. Estimate for 2008 based on employment in August 2008.

Figure 3. Employment in wood household furniture industry, 1990 to 2008.

4. The Second Era of Globalization: 2005 and Beyond

Since the emergence of the Chinese furniture industry, there has been increased emphasis on pricing and costing of U.S. hardwood products at every market level. Chinese plants have been built to produce high volumes of furniture at a low cost. This price leadership business model extends to reducing the cost of raw material, which ultimately has caused China to seek out the lowest cost material. This has led to the development of new supplies of hardwood lumber and logs from central Asia and eastern Europe.

In the future it will be crucial to identify the extent of alternative timber sources available to Asian manufacturers, and monitor their sustainability, to assess the potential impact of alternative lumber species. It also will be necessary to understand the level of acceptance by U.S. consumers of products made from this timber. For example, rubberwood increased from 1 percent of the bedroom and dining room showings at the High Point, NC Furniture Market in 2000 to 6 percent of the showings in 2005 (Appalachian Hardwood Manuf., 2002, 2005).

This suggests that lower priced imported species can be introduced successfully to the U.S. market. Promotion of U.S. species will continue to be important both in pushing the advantages of domestic species to offshore manufacturers and developing pull demand from U.S. consumers.

The high cost of timber and timber processing in the United States makes it difficult for domestic hardwood sawmills to compete in this new global era. Providing additional customer services could increase profitability for both hardwood lumber producers and customers. At the producer level, value will be inherently linked to the separation of lumber in a manner that will reduce the cost of production and/or increase the profit margin of wood purchasers. This may include better color sorts that allow secondary processors to use less costly finishing systems to length and width separations that result in a higher yield of lumber into dimension. However, the probability that markets will continue to evolve means that hardwood lumber producers must be sufficiently flexible to react to continual change with a continual emphasis on reducing cost.

5. Conclusion

The structural changes that have occurred in the domestic hardwood lumber market since 1999 have caused demand to decline and the hardwood lumber industry to contract, resulting in a more competitive environment for surviving producers of hardwood lumber. Although the efforts of China to seek out new supplies of hardwood timber may have a dampening effect on demand and price of U.S. timber in the next decade, the decentralized nature of the hardwood lumber industry allows for independent ideas to evolve and solutions to develop. Implementation of these solutions will temporarily increase the profitability of sawmills. However, the competitive nature of the hardwood market also ensures that short-term increases in sawmill profitability will eventually transfer to stumpage price.

References

Akers, M. (2006). Bulletin of hardwood markets statistics : first half 2005. Research Note NE-386. Newtown Square, PA: USDA Forest Service, Northeastern Research Station.

Appalachian Hardwood Manufacturers, Inc. (2002). Furniture manufacturers offer fresh look at Spring '02 Market. http://www.appalachianwood.org.

Appalachian Hardwood Manufacturers, Inc. (2005). Wood furniture maintains numbers at Spring Market. http://www.appalachianwood.org.

Emanuel, D., & Rhodes, C (2002). Bulletin of hardwood markets statistics : 1989-2000. Research Note NE-375. Newtown Square, PA: USDA Forest Service, Northeastern Research Station.

Emanuel, D., & Rhodes, C. (2005). Bulletin of hardwood markets statistics : 2004. Research Note NE-38. Newtown Square, PA: USDA Forest Service, Northeastern Research Station.

Hardwood Market Report. (1975). Memphis, TN. *Hardwood Market Report. 53(30),* 2-5.

Luppold, W.G. (1996). An explanation of hardwood lumber pricing. *Forest Products Journal. 46(5)*, 52-55.

Luppold, W. G., & Araman, P.A. (1988). Hardwood trade trends: U.S. exports. Research Paper NE-611. Broomall, PA: USDA Forest Service, Northeastern Forest Experiment Station.

Luppold, W.G., & Baumgras, J.E. (1995). Price trends and relationships for red oak and yellow-poplar stumpage, sawlogs, and lumber in Ohio: 1975-1993. *Northern Journal of Applied Forestry. 12(4)*, 168-173.

Luppold, W., & Bumgardner, M. (2008a). Regional analysis of hardwood lumber production: 1963 - 2005. *Northern Journal of Applied Forestry. 25(3),* 146-150.

Luppold, W., & Bumgardner, M. (2008b). Forty years of hardwood lumber consumption: 1963 to 2002. *Forest Products Journal. 58(5),* 6-12.

Nolley, J. (1994). Bulletin of hardwood markets statistics : winter 1993. Research Note NE-358. Radnor, PA: USDA Forest Service, Northeastern Research Station.

Schuler, A., & Buehlmann, U. (2003). Identifying future competitive business strategies for the U.S. furniture industry: Benchmarking and paradigm shifts. General Technical Report NE-304. Newtown Square, PA: USDA Forest Service Northeastern Research Station. 15 p.

USDA Foreign Agricultural Service. 2006. http://www.fas.usda.gov/ustrade/ USTEx FAS.asp?QI=.

USDL Bureau of Labor Statistics. 2006. http://data.bls.gov/cgi-bin/srgate. Series ID: ceu3133712201.

In: Global Change and Forestry
Editors: J. Gan, S. Grado and I. A. Munn
ISBN: 978-1-60876-262-0

Chapter 13

An Introduction to the Southern United States Wood Supply System: A Value Chain Approach

Clayton B. Altizer, Laura A. Grace* and William B. Stuart
Department of Forestry, Mississippi State University, MS, USA

Abstract

Supply chains typically depict a process which accounts for the acquisition of raw materials, conversion of these materials into a merchantable product, and the delivery of these products or goods to the consumer. Value chains are the "other side" of a supply chain, responsible for distributing elements of value back to the suppliers of those goods and services. A well-constructed value chain is much more complex than a supply chain, the economic and social interactions more difficult to trace. Traditional emphasis has been on only two segments of the value chain, timber growers and converting firms. The intermediate enterprises and participants have been considered mere producers of service, therefore largely "outside" the wood supply system (WSS). Increasing logging operating expenses coupled with decreasing prices paid for services rendered affected the entire WSS and the ramifications and repercussions could be severe and long-term in nature. The economic activity generated by timber harvesting contributed heavily to rural economies; the subsequent currency turnover of harvesting generated revenue in such communities was high. The value chain for forestry extended completely across the physical and political landscape, and the socio-economic impacts benefitted locations ranging from urban centers to rural communities and land holdings.

Keywords: Wood; Harvesting; Supply system; Value chain; Southern U.S.

* Email: lgrace@cfr.msstate.edu, Department of Forestry, Mississippi State University, Box 9681; MS State, MS 39762.

1. Introduction

Production forestry plays a vital role in the economic development of the southern United States. Henderson et al. (2008) stated that over $1 billion worth of forest products were harvested from Mississippi's forest lands annually, with timber harvesting currently generating over 15,000 jobs. This influx of revenue was assumed to be divided among the key players in the wood supply system (WSS) (i.e., forest landowners, logging contractors, and forest products mills). Much of the economic benefits stay in local coffers recognizing that some of this revenue leaves local, state, and regional jurisdictions.

A well-constructed value chain model, documenting the distribution of elements of value back to integral WSS suppliers, is useful for depicting these interactions and demonstrating how important production forestry is for the livelihood of rural southern communities. In many instances, a mill closure and the subsequent job loss for such communities has devastating effects. The town of Electric Mills, MS was literally abolished after the demise of Sumter County Lumber Company in the 1940's (Burkhardt 2004). Social and economic impacts of these unfortunate events are widespread and long-term in nature. A better understanding and appreciation of the entire process is paramount for all stakeholders involved in the WSS and those responsible for managing it.

2. Current Wood Supply System

The WSS in the southern U.S. is comprised of three primary stakeholders: landowners, logging contractors (logger), and forest products mills. One of the social and economic dilemmas facing production forestry today is that the relationships between key players in the WSS are poorly understood and not well documented. Many community leaders during the late 1960's considered pulpwood harvesting as a profession requiring little formal education and menial in nature (Ronan et al. 1970). Research has traditionally focused on growth and increasing yields of individual trees or stands, or on finished products once raw materials arrive at the manufacturing plant. The broad area consisting of harvesting timber and transporting this raw material to the consuming mills has generally been taken for granted and often discounted. Most federal research funding has been directed at other elements, primarily soil and water, affecting the WSS. The number of faculty concentrating on the socio-economic contributions of the WSS in eastern U.S. forestry schools and the amount of attention focused on the survival of wood supply firms in the wake of Hurricane Katrina attest to this deficiency. This WSS is critical to the field of forestry and forest products, and has a myriad of social and economic impacts for society as a whole.

Social accounting software, such as Impact Analysis for Planning (IMPLAN), is commonly used to trace the socio-economic interactions among stakeholders for select industrial and commercial sectors. Contributions of some participants in the WSS are often undervalued in this system, primarily a result of misclassifications or omissions inherent to the underlying data used to drive the software. This was precisely the reason that the development of a value chain model was vital to accurately document the socio-economic contributions of the WSS.

Researchers at the Forest and Wildlife Research Center (FWRC) at Mississippi State University have acknowledged the importance of maintaining the structural integrity of the WSS and have responded by placing a greater emphasis on examining its timber harvesting component. This collaboration has produced a series of reports documenting some alarming trends afflicting the current health and status of the logging sector. Most notably, the long-term logging cost index developed through this research effort documented a 42% increase in the overall total cost per delivered ton of wood fiber from 1995 to 2005 (Stuart et al. 2006).

When compared to the consumer price index (CPI) and the producer price index for logging services (PPIL), both indices produced by the U.S. Department of Labor, – Bureau of Labor Statistics, an even more troubling trend is apparent. For 2005, the logging cost index was 12% higher than the CPI; the largest disparity between the two figures for the entire 10-year study period (Stuart et al. 2006). The CPI is the price a consumer pays for a myriad of essential goods and services and commonly used as an indicator of the cost of living (i.e., inflation). For 2005, Stuart et al. (2006) further contended that a 52 point divergence was apparent between the logging cost index and the PPIL for logging services (i.e., the price reported as paid). In fact, the price paid for logging services reported by the PPIL had decreased 10% since the base year of 1995.

3. Supply Chains vs. Value Chains

The focus on supply chain relationships and management in the forest products industry and WSS has been economic in nature, and generally centered on the wood consumer. Meeting production demands, keeping raw material costs at the mill gate competitive, and ensuring that forest certification guidelines are met are common supply chain concerns. Traditional emphasis has been on only two segments of the system: the timber growers and the converting firms. The intermediate enterprises and participants have been considered mere service providers, therefore, largely "outside" the WSS. Much attention has been paid to make certain that the WSS was efficient but little emphasis has been placed on examining social and policy relationships; and responsibilities associated with managing the process of timber conversion from the stump, through the mill gates, to the finished product.

According to Beamon (1998), "a supply chain may be defined as an integrated process wherein a number of various business entities (i.e., suppliers, manufacturers, distributors, and retailers) work together in an effort to: (1) acquire raw materials, (2) convert these raw materials into specified final products, and (3) deliver these final products to retailers." This chain is typically depicted as a forward flow of materials and a backward flow of finances and information. The backward flow is the framework for a value chain. Value chains are the "other side" of the supply chain, responsible for distributing elements of value back to the suppliers of goods and services.

Supply chains have been used more frequently than value chains even though the terms are often used interchangeably. The use in describing the WSS was relatively recent. Both value and supply chains were more common in the field of agricultural economics, specifically in relation to topics directly involved in the food chain. Salin (2000) examined the cattle-beef market and focused on the importance of responsiveness to consumer needs and efficient delivery of goods to consumers. Ward and Stevens (2000) stated that for many agricultural products, mainly beef and dairy products, the identity of the initial product remains virtually clear along the distribution chain. For other products, including those manufactured from wood fiber, identity of the initial product can be "lost" as it is transformed into the finished product.

Development of an integrated supply chain required the management of material and information flows at three levels: strategic, tactical, and operational (Mason-Jones and Towill 1999). These approaches were commonly found in business system engineering designs and information systems. Consumers were becoming more diverse in their demands which have created increased pressures on service industries to provide high quality diversified products at a low cost (Talluri et al. 1999). Modern society has added a fourth dimension, that of protecting the environment.

Davis (1993) stressed three distinct sources of uncertainty confronting supply chains: suppliers, manufacturing, and customers. Late deliveries from suppliers, machine breakdowns in the manufacturing process, and changing consumer preferences were a few, among many, scenarios that can disrupt the smooth flow of an ideal supply chain. The WSS chain is complex, beginning with the grower (i.e., landowner), then the producer (i.e., logger), and finally the supplier (i.e., dealer or broker).

Chains differed from traditional marketing channels in the degree of cooperation among firms involved in the process (Salin 2000). With a supply chain, it was much easier to identify stakeholders and attach appropriate responsibilities and degrees of separation. A well-constructed value chain was much more complex, and economic and social interactions were more difficult to trace. For the purposes of simplification, supply chains illustrated the relationships necessary in converting a raw material into a finished merchantable product through a specific mill. These chains were essentially concentration systems, collecting physical materials from many sources and concentrating them on a specific mill or product. On the other hand, value chains depicted the entire process in a more complex 3-dimensional model which encompassed all that either contributed or benefited from the process being analyzed. Value chains are distribution systems passing an abstract entity, "value," back to the suppliers of goods, services, labor, and capital necessary for the physical commodity production.

The WSS differs from other raw material supply systems. Most generate only one product at the time of harvest or exploration and the movement of that product to a single, specific converter. The harvest of even a tract of intensely managed forest land usually results in the production of multiple products that, in turn, are moved to multiple markets or mills. The effectiveness of the production system, therefore, depends on accommodating multiple markets (i.e., sawtimber, chip-n-saw, and pulp furnish).

4. Methods

This chapter presents a conceptual model depicting the southern U.S. WSS value chain. This system developed differently from those in New England, the Lake States, and the Pacific Northwest. Timber dependent communities in the southern U.S. evolved around previously established communities, eliminating the need to import a workforce; a common practice used in other parts of the country (Stewart 2004). Management practices tailored for southern U.S. timber species have differed from recommendations for species commonly grown in other regions of the U.S. Furthermore; public, private, and industrial land ownership patterns have differed regionally and the contractual relationship between the logging force and the mills have differed in other parts of the country as well.

All stakeholders in the WSS were documented and their sphere of influence was traced throughout the WSS process. This model should serve as a foundation for a more complex depiction of a WSS value chain. It should be noted that, in reality, stakeholders comprising the model function and interact in a 3-dimensional structure. However, to better describe the model and facilitate comprehension, it was unraveled and presented in a 2-dimensional format.

Landowners, logging and transportation firms, and forest products mills have a direct or vested interest in the process (figure 1). It was important that these entities, while sometimes of lower profile, were included in the stakeholder process. Failure to include all results in an incomplete value chain and leads to flawed decision-making. The model constructed quickly evolved into a considerable network, requiring additional expansion. The structure was most easily illustrated by dividing the model into several different sections for discussion, and following one sample branch of the network. Other network branches can be expanded and explored in the same manner. The model accounts for both direct and indirect relationships at several levels and these relationships extend in both directions.

Primary relationships of any entity in the WSS chain are those with a direct business relationship, such as between the mill and a chemical or energy supplier or between the logger and a wood dealer. A secondary relationship for the mill is that between a primary supplier or customer and their customer or supplier. A mill, relying on a dealer system for procurement, has a primary relationship with the wood procurement division of the plant, a secondary relationship with the wood dealer, and a tertiary relationship with the logging contractor supplying wood through that dealer and with the landowner from whom the dealer acquires timber. A fourth level or quaternary relationship is one step further removed. The mill has a quaternary relationship with the labor force working for the logging contractor, the fuel supplier to that operation, and the equipment dealership supplying logging machinery. The grocery store where that worker's family buys food and the bank that holds their mortgage and savings has a fifth level or quinary relationship with the mill.

An "arms length relationship" is a term used to explain legal relationships between separate entities and commonly used in the WSS (Black 1990). Primary, secondary, tertiary, quaternary, and quinary simply describe how many "arm lengths" separate the entities. They form the value chain and the action of any one of these entities has an effect on the others. That effect may be diminished or amplified by the distance between the two parties. Some actions are diffused as they move down the supply chain while others are amplified. All direct suppliers and customers are stakeholders in the WSS chain for they have made an investment of money, skill, energy, and time in the functioning of the process. Process changes and variability will affect those investments.

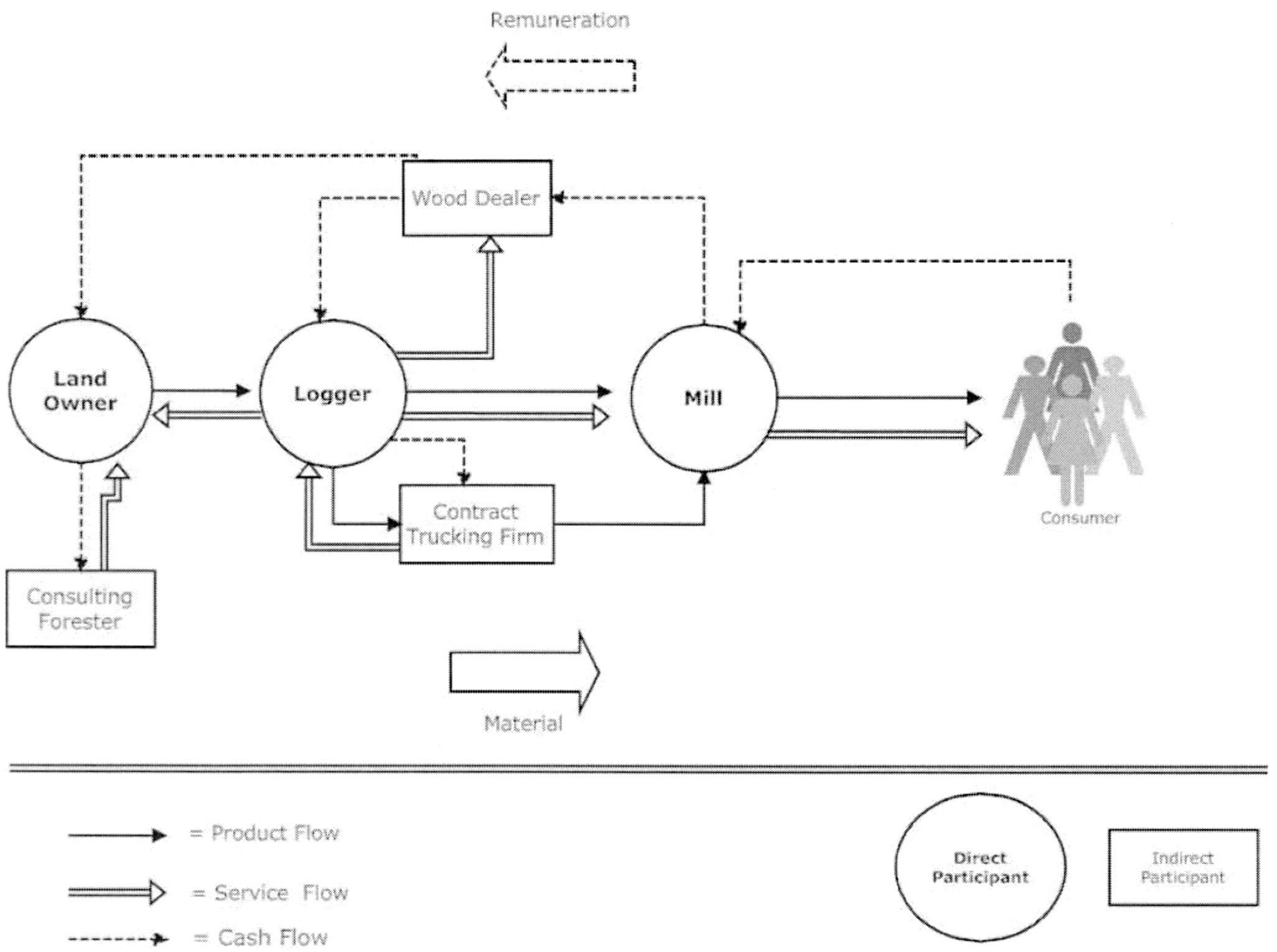

Figure 1. Value chain model for the southern United States wood supply system.

The model can be best understood by selecting a starting point and tracing multiple direct and indirect relationships through the entire system. The mill was chosen to be the center of the model for this discussion (figure 2). The choice of the mill as the center was not intended to reflect a placement of importance, but simply an arbitrary, but important, starting point. Solid connector lines between boxes indicated direct or "primary" relationships. Direct suppliers were shown at the top of the chart; the direct customers were to the right. These were the stakeholders who have invested in the mill's WSS chain in expectation of a monetary return. Indirect evaluators were located below the mill in the diagram and were joined by dashed lines. They were not separated as suppliers and customers because in many instances, such as the political structure of the community surrounding the mill, they serve both roles. These suppliers and customers seldom have a direct investment in the mill or supply chain, but may have a financial or emotional investment affected by actions of the supply chain. Forest certification systems have been developed and do have an impact on the WSS, however, most of these programs are timber, soil, and water oriented. "Chain of custody" essentially has to do with forest management conditions where the timber was grown and has very little emphasis on the "sustainability" of businesses involved in the WSS. Suppliers were grouped into broad categories by goods or services offered, although this list was truncated for simplification. The order of listing does not imply importance or the amount of money spent.

Figure 3 illustrated the value chain model for the wood procurement division of the mill in the previous figure. Boxes at the top were again direct suppliers, those at the right were direct customers, and those at the bottom were indirect customers and suppliers. The solid line between the procurement box and the mill box denoted a primary relationship between fiber procurement and the mill. Similarly, boxes connected to the procurement box by solid line enjoy a primary relationship with each other and with mill procurement. For example, wood dealers under contract with fiber procurement may be serving the same role for a sawmill, which in turn is a direct supplier of residue chips to the procurement organization of the mill. Those relationships that are primary to procurement are secondary to the mill, as they must flow through procurement before reaching the mill.

The indirect customers and suppliers for fiber procurement fall into the same broad category as for the mill, but were likely quite different groups or segments of larger groups with different or specific interests. Where environmental groups with concerns over air quality and point source water pollution were attracted to the mill, those associated with procurement were more likely to have concerns about endangered species, clearcutting, and non-point source pollution. Figure 4 delved one level deeper in the value chain and showed functional relationships for wood dealers, an example of secondary stakeholders of the mill. The position of boxes relative to the central wood dealer box was the same as in the previous examples. Independent and contract loggers supplying wood through this dealership have a primary relationship with it, a secondary relationship with procurement, and a tertiary relationship with the mill.

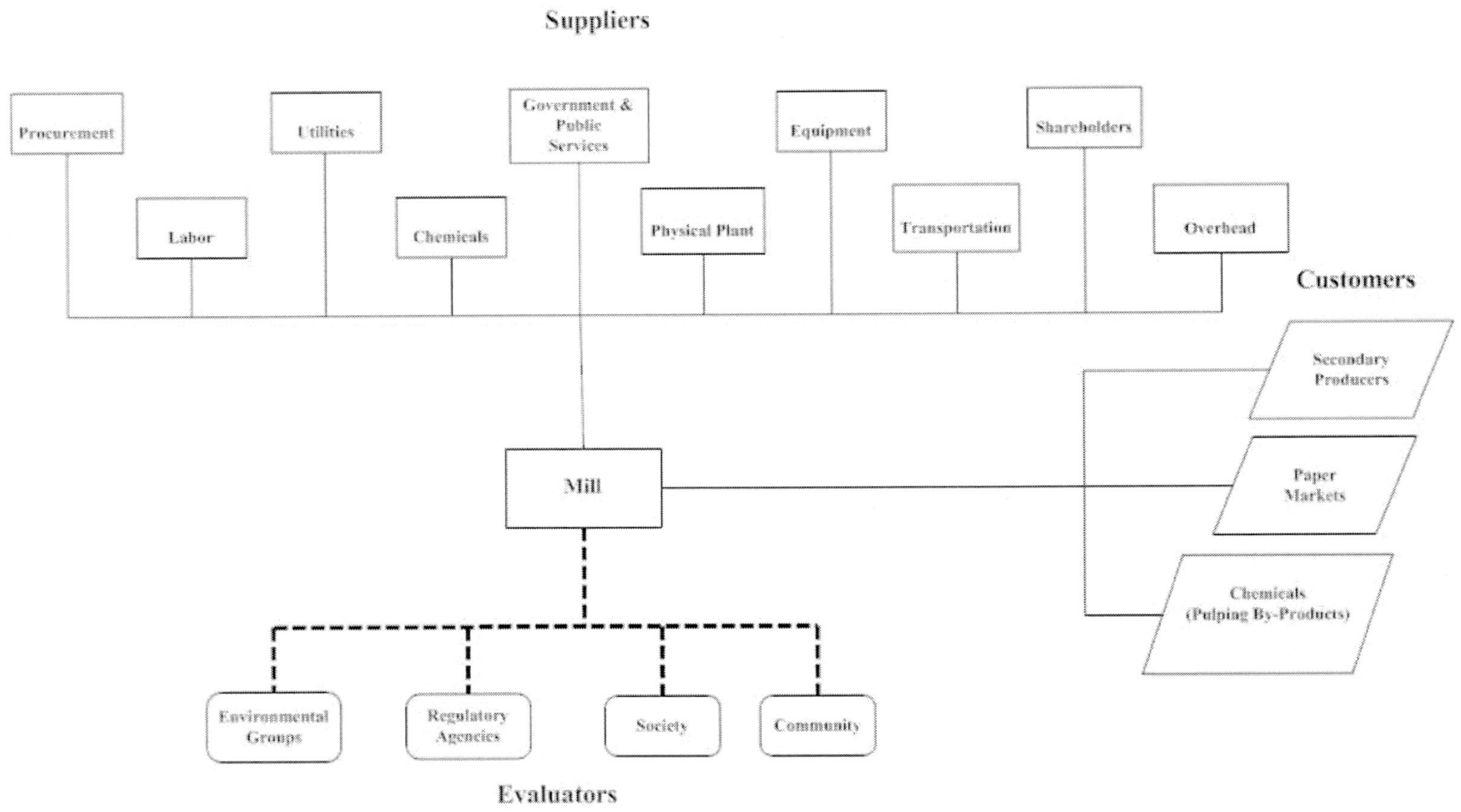

Figure 2. Forest products mill's value chain relationships with southern United States wood supply system stakeholders.

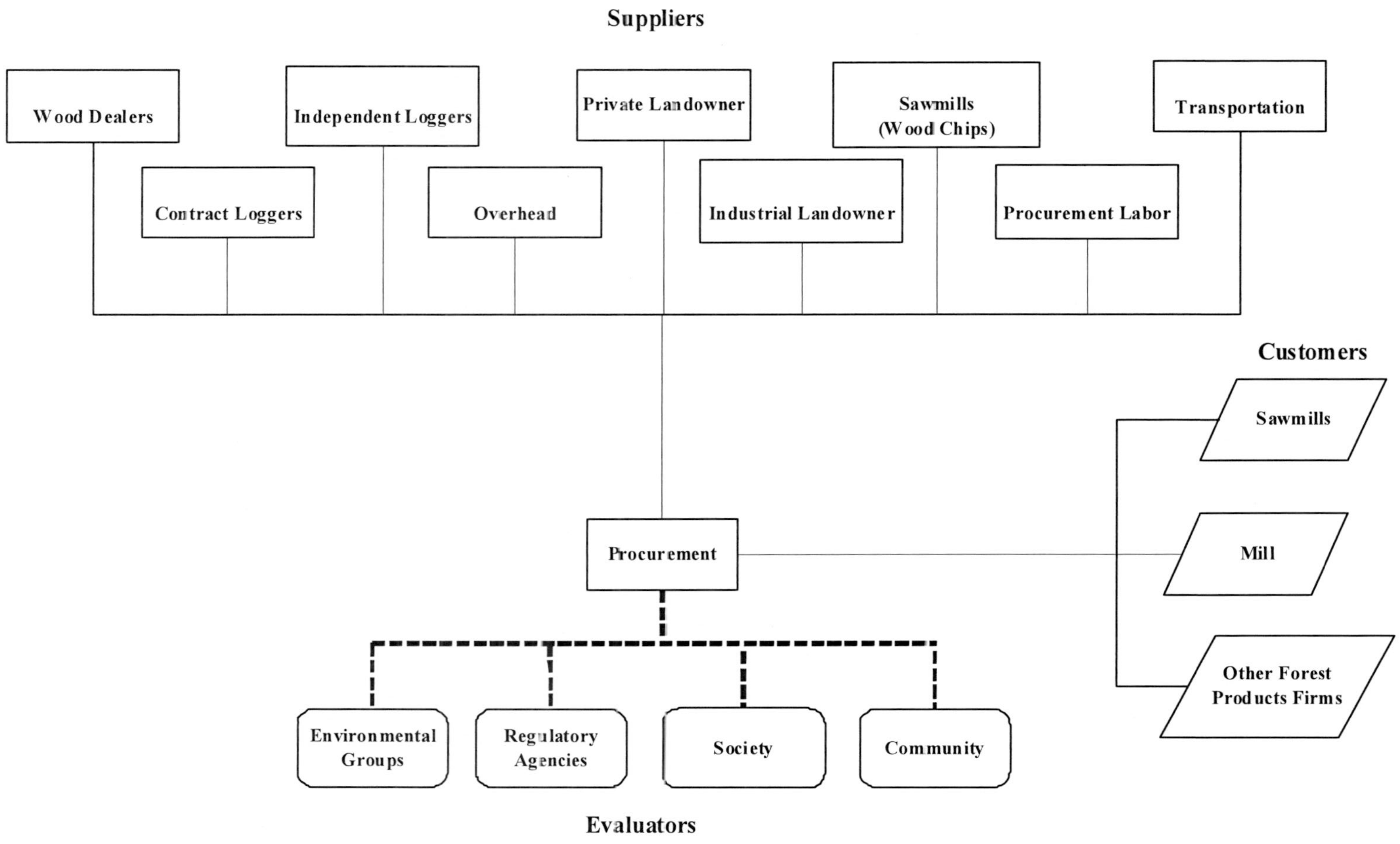

Figure 3. Wood procurement division's value chain relationships with southern United States wood supply system stakeholders.

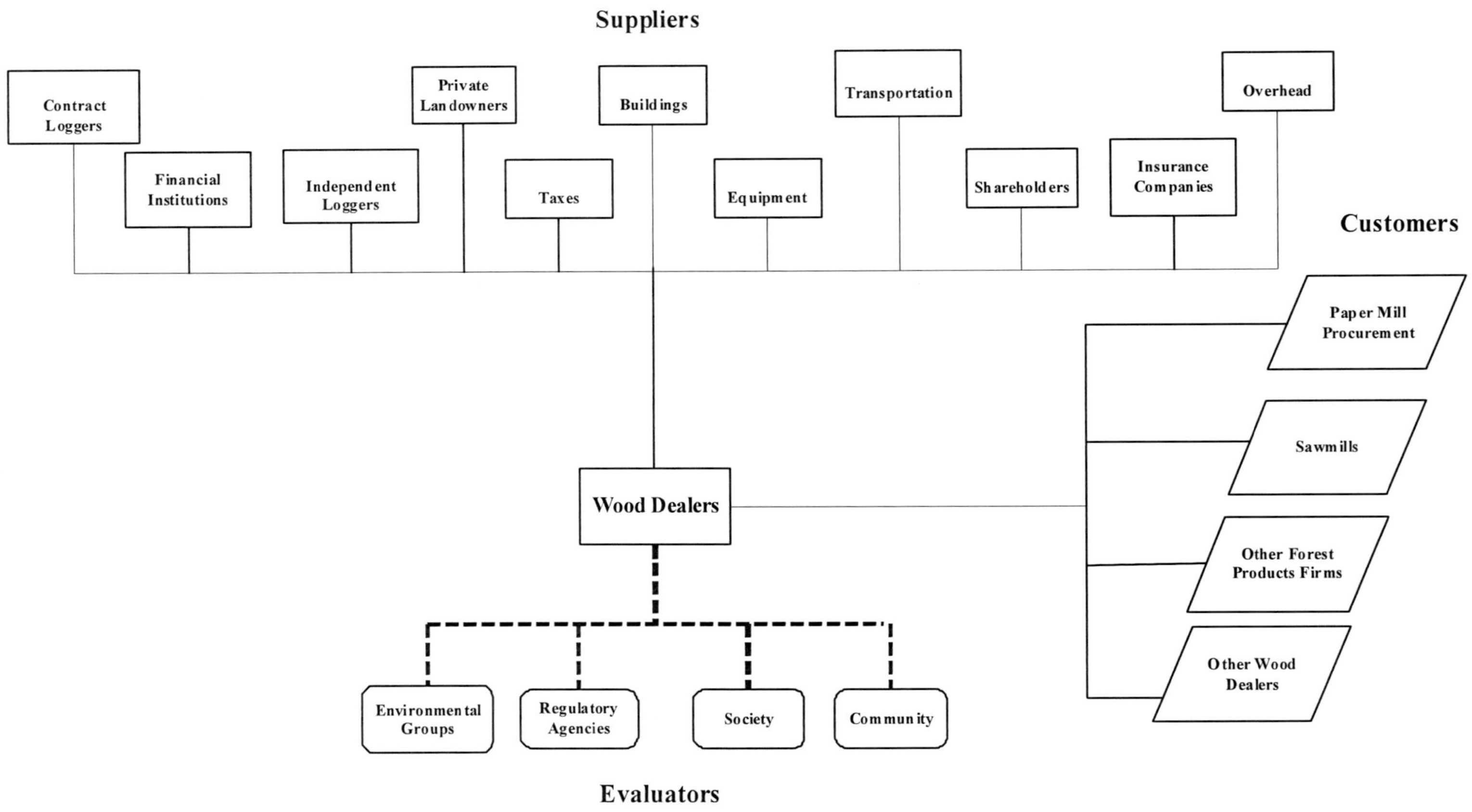

Figure 4. Wood dealer's value chain relationships with southern United States wood supply system stakeholders.

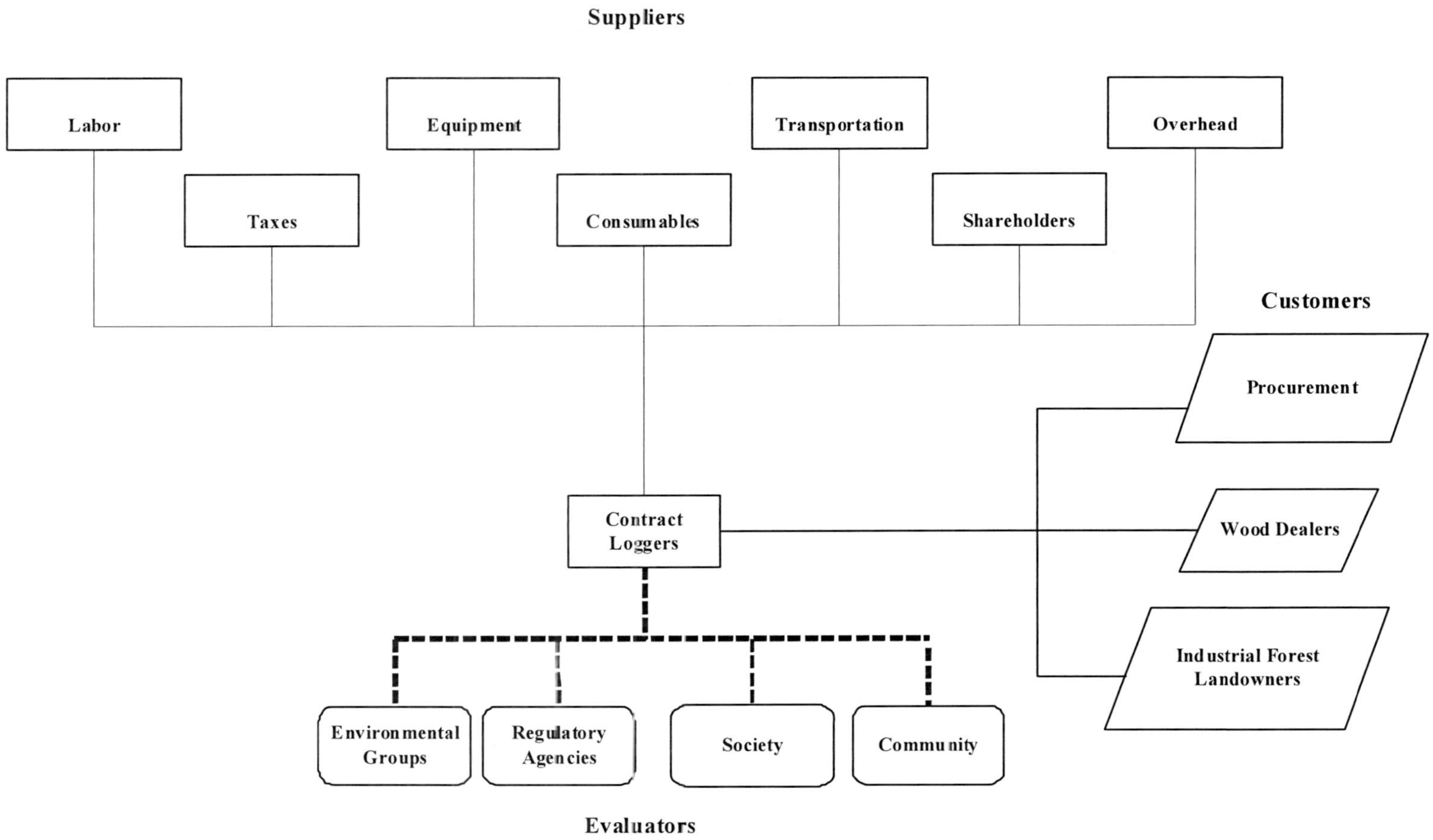

Figure 5. Contract logger's value chain relationships with southern United States wood supply system stakeholders.

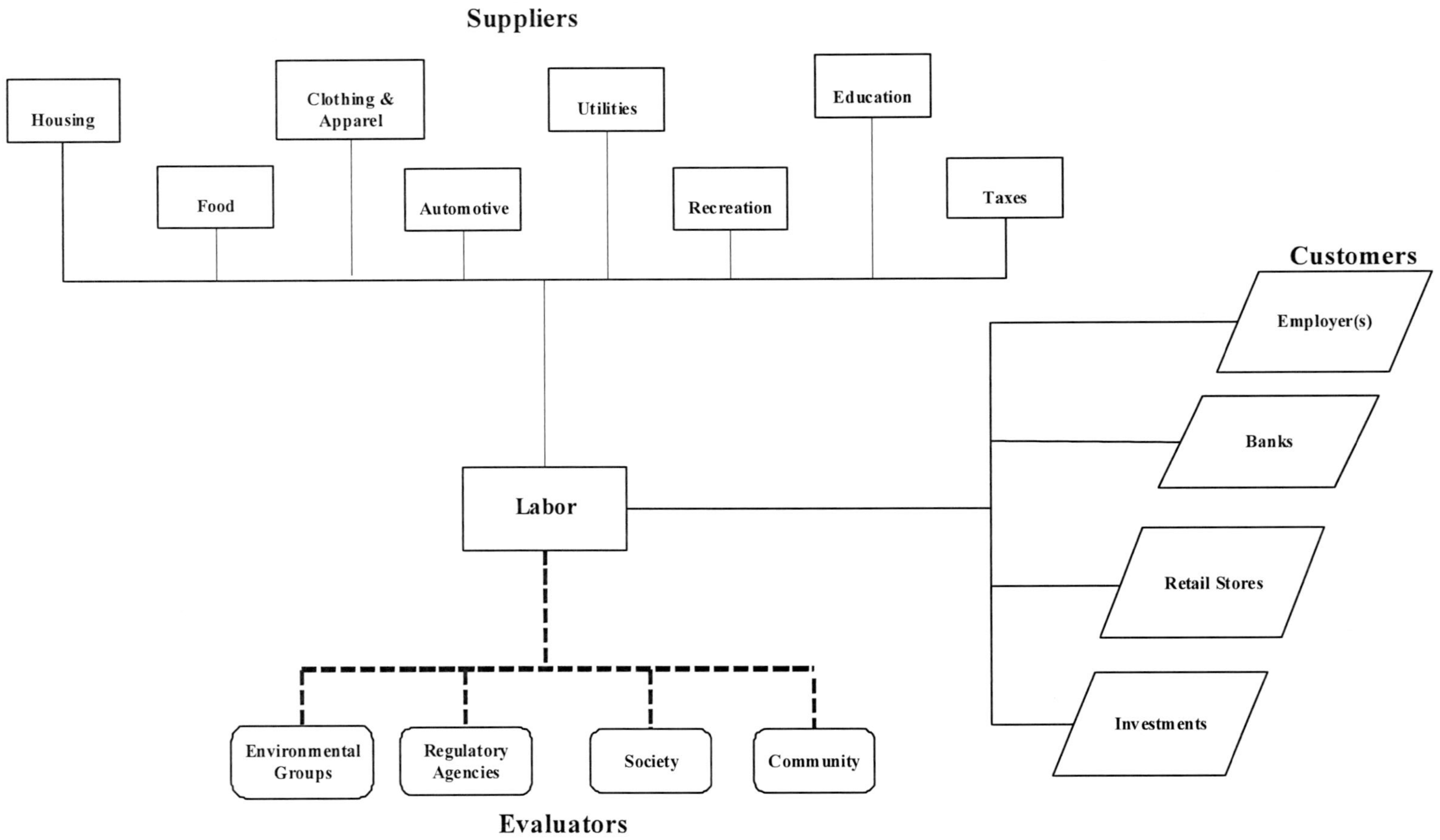

Figure 6. Logging employee's value chain relationships with southern United States wood supply system stakeholders.

Indirect customers and suppliers again fall into the same broad categories, but operate on a different scale. As the scope of operation becomes more localized, these entities do as well. At this level, local chapters of larger environmental groups were important, as were relationships with regional planning groups, county government, local newspapers, and civic groups. The cost of meeting concerns of these groups is more easily identified. Expenditures required to meet concerns of these groups were a part of normal business practice and supported the firm's position in the local business and forestry communities. Potentially larger costs and less predictable costs were those of lost business opportunities and possible legal proceedings.

Figure 5 illustrated the movement one more link down the chain to the logging contractor who has a tertiary relationship with the mill. The nature of both direct and indirect customers and suppliers change again. Many direct relationships at this level were community based and scattered throughout the procurement region and included labor drawn from the local pool, fuel purchased from a local supplier, and equipment purchased from a local dealer. The nature of the indirect customers and suppliers was also more localized. Public relations may focus on the adjacent landowner, government relations on the county road engineer, and environmental concerns on a specific stream segment or wildlife species. Again, costs become more concrete. Business opportunities were tied closely with local business reputation. Additional costs were quite often in the form of good will, fines for regulatory violations, performance bonds, permits, and insurance costs.

Labor, working for the logging contractor, has a quaternary relationship with the mill, and also has a very personal interest in the paycheck that originates there. The direct suppliers to the labor working for the logging contractor (figure 6) were largely commercial or mercantile and represented a quinary relationship with the mill. The nature of the indirect suppliers and customers also become very localized. Indirect suppliers at this level can make their presence felt through the local business community, through the political process, through the financial system, or simply by withdrawing their services.

5. CONCLUSION

Today's harvesting sector must deal with rapidly increasing operating costs, unpredictable cash flow, and a general uncertainty of future production capacities. Corporate mergers and acquisitions, as well as constantly changing mill and wood procurement management, have created a cloud of uncertainty regarding the future role of independent contractors in the WSS. In this sense, the struggle of farmers trying to provide our country with a steady food supply and independent loggers trying to produce enough raw material to meet our Nation's demand for wood fiber were essentially the same. Likewise, many problems facing agriculture and production forestry are identical, and both sets of problems have the potential to adversely affect the future roles of each profession.

Effective management of the value chain was critical to WSS chain management, for if the suppliers of goods and services were not satisfied that they were being properly remunerated, they would withdraw their services. Increasing operating expenses coupled with decreasing prices paid for services rendered affected the entire WSS and the ramifications and repercussions could be severe and long-term in nature. The value chains for many industries are relatively short, extending through the town or city where the plant is located. The value chain for forestry extended completely across the physical and political landscape, and included numerous socio-economic impacts for both urban and rural locations. Such complexity warrants further exploration and expansion of this model to accurately depict the WSS in detail and subsequent socio-economic effects on local and regional communities.

REFERENCES

Beamon, B. M. (1998). Supply chain design and analysis: Models and methods. *International Journal of Production Economics, 55*, (3), 281-294.

Black, H. C. (1990). *Black's Law Dictionary* (Sixth Edition). St. Paul, MN: West Publishing Company.

Burkhardt, C. E. (2004). Electric mills. *The Mississippi Logger, 14* (7), 1-2.

Davis, T. (1993). Effective supply chain management. *Sloan Management Review, 34* (4), 35-45.

Henderson, J. E., I. A. Munn, G. Perez-Verdin, D. L. Grebner (2008). Forestry in Mississippi: The impact of the forest products industry on the post-Katrina Mississippi economy-An input-output analysis. *Forestry and Wildlife Research Center, Bulletin FO374, Mississippi State University*. 31pp.

Mason-Jones, R. & D. R. Towill. (1999). Total cycle time compression and the agile supply chain. *International Journal of Production Economics, 62*, (1-2), 61-73.

Ronan, W. W., C. R. Tatro, S. Aral & G. P. Latham. (1970). *Psychological, sociological, and demographic variables affecting the pulpwood harvesting labor force in a southeastern United States community*. Atlanta, GA: American Pulpwood Association-Harvesting Research Project.

Salin, V. (2000). Information technology and cattle-beef supply chains. *American Journal of Agricultural Economics, 82* (5), 1105-1111.

Stewart, L. D. (2004). *Early development of the Mississippi wood supply system: 1699-1930.* M.S. Thesis. Department of Forestry, Mississippi State University. 143pp.

Stuart, W. B., L. A. Grace, C. B. Altizer, & J.J. Smith. (2006). 2005 preliminary indices. *Wood Supply Research Institute*. 31pp.

Talluri, S., R. C. Baker, & J. Sarkis. (1999). A framework for designing efficient value chain networks. *International Journal of Production Economics, 62*, (1-2), 133-144.

Ward, R. W. & T. Stevens. (2000). Pricing linkages in the supply chain: The case for structural adjustments in the beef industry. *American Journal of Agricultural Economics, 82* (5), 1112-1122.

In: Global Change and Forestry
Editors: J. Gan, S. Grado and I. A. Munn

ISBN: 978-1-60876-262-0

Chapter 14

A Comparative Analysis of Softwood Production from Industrial and Nonindustrial Forest Ownerships in the U.S. South[1]

Xianchun Liao[1] and **Yaoqi Zhang**[2]*
[1] Department of Forestry, Mississippi State University, MS, USA
[2] School of Forestry and Wildlife Sciences, Auburn University, AL, USA

Abstract

Inadequate long-term timber supply from nonindustrial private forests (NIPF) has long been concerned in forest literature. However, there was a gap to understand stumpage production behavior by forest industry (FI) and NIPF ownerships and by sawtimber and pulpwood products at the same time. This chapter investigated determinants of softwood production in the U.S. South by comparing the two ownerships. An econometric model was derived under the framework of profit maximization. A two-stage least squares (2SLS) technique with time series data from 1953 to 2002 was used in this study. The estimated results revealed that supply price elasticities of 0.70 for sawtimber and 0.90 for pulpwood for FI owners were larger than those of 0.29 for sawtimber and 0.32 for pulpwood for NIPF owners, which were within the price elasticity range from previous studies in general.

Keywords: Stumpage market; Profit maximization model; Two-stage least squares; Price elasticity.

---

[1] This chapter was revised based on the paper entitled "An economic analysis of softwood production in the U.S. South: A comparison of industrial and nonindustrial forest ownerships" originally published in the *Forest Products Journal* 59(11): 69-74. Witten permission was granted by the Journal.

* E-mail: xliao@cfr.msstate.edu, Department of Forestry, Mississippi State University, MS 39762.

1. INTRODUCTION

Inadequate long-term timber supply from nonindustrial private forests (NIPF) has long been concerned in forest sector. For example, in the U.S. South, a significant amount (34%) of the regional softwood production came from the relatively small share (17%) of forested area owned by forest industry (FI), while 62% of the production came from a much larger share of forestlands (71%) held by NIPF landowners in 2002 (Smith et al., 2004).

To stimulate the timber supply from NIPF, various forestry incentive programs have been implemented to facilitate timber harvesting from NIPF (Kluender et al., 1999). These programs included direct incentive programs, such as the Forest Incentives Programs (FIP), the Stewardship Incentives Program (SIP), and the Conservation Reserve Program (CRP), and indirect incentive funds, such as research, extension, and technical assistance.

However, NIPF owners' response to these programs has long been week. According to Kluender et al. (1999), these government programs designed to induce future timber production have increased reforestation, while there has been a much smaller corresponding increase in timber harvest. Beach et al. (2005) also pointed out that NIPF landowners have been less responsive to market signals than many researchers might expect using a meta-analysis. The study revealed that the influence of timber price was positive and significant in only 58% of studies with one study finding a negative effect and the rest finding no significant effect. Therefore, understanding the difference in production behavior between FI and NIPF has been a concern in the forestry literature and an important aspect in public policy and management planning attempting to improve NIPF output (Boyd & Hyde, 1989; Hardie & Parks, 1996).

Three group models were extensively used to assess forest policy effects. They included spatial equilibrium model, profit-maximization model, and utility-maximization model. Timber assessment market model (TAMM) and North American pulp and paper model (NAPAP) have long been widely used as spatial equilibrium models (Adams & Haynes, 1980, 1996; Ince, 1999; Hayes, 2003). Both models optimized consumer and producer surplus to compute annual market equilibria, following Samuelson's theoretical framework (Samuelson, 1952). These studies provided accurate prediction, however, they were based on various assumptions about the characteristics of the stumpage market, particularly the participant's responses to internal as well as external market changes (Haynes, 2003, p 47; Newman, 1987). Among profit-maximization models, a structural econometric model has been used to quantify both demand and supply factors at aggregate level. For example, Newman (1987) presented an econometric analysis of the Southern softwood stumpage market in the US, while Carter (1992) investigated Texas pine pulpwood stumpage market using time series data from 1964-1986. These studies estimated market behavior in full detail, however, they did not consider different ownerships. One exception was that Newman and Wear (1993) compared the production behavior of industrial and nonindustrial private forestland in the U.S. Southeast using profit-maximization model. However, the study used cross-sectional data, which might not pick up all the past dynamical variability between stumpage supply and price. Moreover, the study treated stumpage prices as exogenous variables, which might not have economic justification because stumpage prices would be endogenous at the forest sector in general. Considering the heterogeneity of NIPF landowners' objectives, management activities, and preference, the majority of previous studies on NIPF in the U.S. have tended to view landowners as individual decision makers using utility-maximization model at landowner level (Binkley, 1981; Dennis, 1989; Kuuluvainen et al.. 1996; Prestemon & Wear, 2000; Pattanayak, 2002). For an excellent review of literature on family forests see Beach et al. (2005). These studies revealed that NIPF landowners have been less responsive to market signals. However, these studies were based on an assumption, in which demand and supply was not separatable. This was far from true for many landowners holding large tract for timber production purpose. Overall, the review revealed that there was a great need to understand stumpage production behavior by industry and NIPF ownerships and by sawtimber and pulpwood products at the same time.

This study was motivated to estimate stumpage supply in the U.S. South by comparing the production behavior between FI and NIPF. This study hypothesized that NIPF owners' behaviors were the same as industrial owners who pursue profit maximization. A two stage least squares (2SLS) method with time series data from 1953 to 2002 was applied to estimate the model. The paper was organized as follows: First, the theoretical model of stumpage supply was presented. Then, the data sources were presented and the empirical estimation using 2SLS followed. Next, the regression results were interpreted. This study ended with a summary and conclusion.

2. Data Sources

Data sources were summarized in tables 1 and 2. Softwood stumpage was the total quantity of softwood timber from the U.S. South. Data for softwood harvest on FI and NIPF for 1953-2002 were from Adams et al. (2006), but the *data combined pulpwood/sawtimber. P*ine pulpwood production for 13 southern states was from the Southeastern and Southern Forest Experiment Stations of the USDA Forest Service for 1950-2002, but the data were not classified by two ownerships (FI and NIPF). A *ratio of sawtimber to pulpwood for 1950-1980 was from Adams et al. (1988) and for 1981-2002 from Smith et al. (2006) was estimated according to the source from the* Southeastern and Southern Forest Experiment Stations. *Then production was allocated for sawtimber and pulpwood by the two ownerships, respectively.* The average volume-weighted stumpage price of softwood timber for 1950-1976 was from Ulrich (1989) and for 1977-2002 from Howard (2003). Lumber price of Southern pine for 1953 to 1976 was from Adams et al. (1988) and for 1977 to 2002 from Random Lengths (1977-2002). Net volume of softwood sawtimber for 1950-2002 and total volume were from Smith et al. (2001) by two ownerships (FI and NIPF). Because the USDA Forest Service reports on inventory data were conducted at approximate ten year intervals, t*he missing data were found based on the formula from Newman (1987). The formula was specified as the following:* $v_t = v_{t-1} + [G^* - (S_t - S^*)$ *where* G^* *was the average annual net growth between survey years and* S^* *was the average stumpage production between survey years.* The producer price index of the paper and allied products was from the Bureau of Labor and Statistics (BLS) (1982=100). The deflator was the Producer Price Index used for all prices from the BLS (1982=100). *All data were annual and the time series covers the period from 1953 to 2002 (50 observations).*

Table 1. Data description and sources

Data	Abbreviation	Measurement	Source
Softwood supply (i=1 for sawtimber, i=2 for pulpwood; o=1 for FI and 2 for NIPF)	Q_o^i	MCF	Adams et al. (2006); Smith et al. (2004)
Stumpage price of softwood	p_o^i	U.S.$/MBF for sawtimber, U.S.$/cord for pulpwood	For 1977-2002 from Howard (2003), for 1950-1976 from Ulrich (1989).
Lumber prices of southern pine	LP	U.S.$/MBF	For 1977-2002 from Random Lengths, for 1953-1976 from Adams et al. (1988)
Inventory	v_o^i	MCF	Adams et al. (2006); Smith et al. (2004)
Paper and allied products	FP	Index (1982=100)	U.S. Bureau of Labor Statistics
U.S. Producer Price Index	PPI	1982=100	U.S. Bureau of Labor Statistics

Note: MCF indicates million cubic feet and MBF means thousand board feet.

Table 2. Data summary

Variables	Mean	Std	Min	Max
Softwood supply for sawtimber for FI	668.61	308.36	226.38	1096.09
Softwood supply for pulpwood for FI	673.27	225.31	341.52	1047.93
Softwood supply for sawtimber for NIPF	1292.82	383.13	736.15	1975.57
Softwood supply for pulpwood for NIPF	1337.01	199.12	926.86	1723.82
Sawtimber stumpage price	168.49	72.03	85.44	326.98
Pulpwood stumpage price	14.47	3.12	10.62	23.50
Lumber prices of southern pine	241.01	47.61	183.86	353.32
Sawtimber inventory for FI	9901.27	888.47	7973.02	10923.05
Pulpwood inventory for FI	12814.45	1786.28	8567.98	14952.19
Sawtimber inventory for NIPF	24629.09	6113.82	13742.42	31796.53
Pulpwood inventory for NIPF	32938.42	5041.95	22617.58	38324.38
Paper and allied products	79.76	45.87	28.00	159.00

3. Method

Profit maximization model was widely used for forest industry (FI) (e.g., Adams and Haynes, 1980; Brännlund et al., 1985; Newman, 1987). For nonindustrial private forest (NIPF), there was some evidence to support profit maximization model (e.g., Adams and Haynes, 1980; Brännlund et al., 1985; Newman, 1987). At the aggregated level, profit maximization was appropriate because many large-scale NIPF owners who mainly pursued profit from timber production made significant contribution to stumpage supply, although the number of these large-scale family forest owners was less than that of small-scale family forest owners in the U.S. South. The further fact of separation of timber production and consumption in the U.S. made it still possible to use profit-based model for NIPF owners. Therefore, in this study, an aggregate stumpage supply was derived from a profit maximization model, following the early authors' framework of Johansson & Löfgren (1985), & Brännlund et al. (1985). The present profit function could be defined as:

$$\pi^i_o(p^i_o, w^i_o, v^i_o) = \underset{Q^i, L^i}{Max}(p^i_o Q^i_o - w^i_o L^i_o \,|\, v^i_o) \tag{1}$$

where i = 1 for sawtimber and 2 for pulpwood, o = 1 for FI and 2 for NIPF, Q was the set of feasible cutting possibilities, *p* was the stumpage price, *w* was per unit harvesting labor cost, *L* was labor input, *v* was the inventory. Timber production was constrained by inventory. Assuming that the present profit function was convex in *p* and *w* and applying Hotelling's lemma, the firm's supply of the stumpage in period *t* was a function of market price and the prices of all inputs in production. The supply function was found by taking the first derivative of the profit function. Thus,

$$\frac{\partial \pi^i_o}{\partial p^i_o} = Q^i_o(p^i_o, w^i_o, v^i_o) \tag{2}$$

$$\frac{\partial \pi^i_o}{\partial w^i_o} = -L^i_o(p^i_o, w^i_o, v^i_o) \tag{3}$$

Because the data about harvesting cost were not available, the amount of growing stock served as an inverse proxy for it suggested by Newman (1987). The reasoning was that growing stock was viewed as a measure of accumulated forestry capital adjusted through time by forest regeneration costs, forest growth, and timber cutting (Newman & Wear, 1993). Moreover, numerous factors influenced the individual output of roundwood such as multiple potential outputs (sawlog, pulp and paper log, poles). For example, pine sawtimber stumpage might influence the output of pine pulpwood suggested by Newman (1987) and vice versa. Thus, the supply specification for individual landowner (*n*) was as the following:

$$\mathrm{Q}^i_{\mathrm{ont}} = Q^i_{ont}(\underset{+}{p^i_{ont}}, \underset{+}{v^i_{ont}}\ \underset{\pm}{p^j_{ont}}) \tag{4}$$

where i=1 for sawtimber, i=2 for pulpwood; j was reverse and was not equal to i. By definition, own price elasticity referred to changes in quantities of a good (e.g. pulpwood) supplied due to changes in price of this good (pulpwood). In this study, the own price for the pulpwood had a positive effect on supply. Timber inventory had a positive effect on the output because the marginal harvesting costs decreased as inventory increased. In contrast, cross-price elasticity measured the effect of changes in other goods' prices (e.g. sawtimber) on a given good (pulpwood). The sign for the cross price was uncertain because the complementary production effect might be larger than the substitution effect. If all the forest owners in the region maintained the same production, the regional stumpage supply specification could be found by aggregating the N individual forest owner's production functions. Thus,

$$Q^{i}_{ot}(p^{i}_{ot}, v^{i}_{ot}, p^{j}_{ot}) = \sum_{n=1}^{N} Q^{i}_{ojt}(p^{i}_{ont}, v^{i}_{ont}, p^{j}_{ont}) \tag{5}$$

This equation served as a theoretic model for this analysis and showed that the stumpage supply depended on own price and inventory. Keep in mind, transportation costs were assumed a relatively constant fraction of the stumpage price and did not affect the short-run supply in the region as well.

4. EMPIRICAL ANALYSIS

Equation 5 was further adapted in the following way. A two-stage least squares (2SLS) procedure was used to correct for endogenous bias in the stumpage supply model because market price and output quantity might be determined jointly. The 2SLS approach included a first stage regression, estimating how market price changes were influenced by economic variables. Then, predicted price values from this first stage regression were used in place of output price in the second stage (Brown & Zhang, 2005). The two-stage empirical model was as follows:

Stage 1

$$\mathrm{p}^{i}_{\mathrm{ot}} = \alpha^{i}_{o} + \beta^{i}_{o1} v^{i}_{ot} + \beta^{i}_{02} p^{j}_{ot} + \varphi Z^{i}_{ot} + \varepsilon^{i}_{ot} \tag{6}$$

Stage 2

$$\mathrm{Q}^{i}_{\mathrm{ot}} = \alpha^{i}_{o} + \beta^{i}_{o1} v^{i}_{ot} + \beta^{i}_{02} p^{j}_{ot} + \varphi \hat{p}^{i}_{ot} + \varepsilon^{i}_{ot} \tag{7}$$

where Z^{i}_{ot} was an instrumental variable for stumpage price and $\hat{p}^{i}_{ot}$ was predicted values for stumpage price at year t from the first stage regression. Instrument choices were most reasonable in this study. An instrument should be (a) correlated with the endogenous explanatory variable and (b) uncorrelated with the error term in the equation. We regressed pulpwood stumpage price on paper and allied products price index (instrument variable) and other independent variables using a reduced-form. The results showed that the instrument variable (IV) was correlated (coefficient = 0.09) with pulpwood price at the 5% significant level (p-value = 0.011) and R^2 was 0.73. Unfortunately, we could not test (b) using the data because it was impossible to check the correlation between IV and the error term, which was not observable (Wooldridge, 2000, p 463). Based on economic theory, final paper price did not enter into pulpwood production, which implicitly assumed that there was no correlation between the IV and the error term. Likewise, we regressed sawtimber stumpage price on lumber price (instrument variable) and other independent variables using a reduced-form. The results indicated that the instrument variable was correlated (coefficient = 0.64) with pulpwood price at the 1% significant level (p-value = 0.0001) and R^2 was 0.82.

5. Empirical Results

Linear and log-linear forms have been estimated by two stage least squares (2SLS). The results with linear form were presented here because linear form performed better than log-linear form in terms of coefficient significance. Table 3 presented coefficients of the estimated profit maximization function for FI and NIPF. Overall, the R^2 values for all equations were high, which meant the explanatory variables significantly explained the dependent variables. The coefficients had the expected sign and all of them were significant at the 5% or 10% levels for both FI and NIPF ownerships.

Table 3. Estimates of coefficients for both FI and NIPF using profit maximization model

Variable	FI		NIPF	
	Sawtimber	Pulpwood	Sawtimber	Pulpwood
Constant	-745.02** (334.60)	-853.81** (118.98)	98.33 (108.56)	203.23* (111.07)
Inventory	0.05* (0.029)	0.07** (0.02)	0.03** (0.01)	0.02** (0.003)
Pulpwood price	28.60** (12.45)	42.03** (13.01)	a	29.70** (6.72)
Sawtimber price	2.80** (0.66)	a	2.23** (0.87)	0.66** (0.27)
Obs.	50	50	50	50
R-squared	0.70	0.74	0.79	0.85

Note:
[1] ** and * indicate significances at 5% and 10% levels, respectively.
[2] Numbers in parentheses denote standard error.
[3] All variables are in level form.
[4] a means the variable is not significant and dropped off from the model.

The results showed that all own prices were significantly positive at the 5% level, which was consistent with the literature in that an increased own price of sawtimber or pulpwood increased the supply of the assortment (e.g. Brännlund et al., 1985). Timber inventory variables were significantly positive at the 5% or 10% level, which was also consistent with the claims in the literature that the marginal harvesting costs decreased as inventory increased (e.g., Newman, 1987). The positive cross-price effects between sawtimber and pulpwood for FI and between pulpwood and sawtimber for NIPF indicated that they were complementary in the short run. However, the effects between pulpwood and sawtimber for FI and between sawtimber and pulpwood for NIPF were insignificant and excluded from the equations. A possible explanation was that the complemetary production effect was greater than the substitution effect in this study. The complementary production effect indicated that an increase in sawtimber production would augment pulpwood supply. This was possible because firms often sold pulpwood as a byproduct of a solidwood harvest (Brännlund et al., 1985; Newman, 1987).

To measure the impacts of the explanatory variables on stumpage supply, the elasticities were calculated at the mean of the variables (see table 4). The own price elasticities were generally high, 0.70 for sawtimber and 0.90 for pulpwood, for FI owners, while they were low, 0.29 for sawtimber and 0.32 for pulpwood, for NIPF owners. The respective elasticities were significantly different between the two ownerships; however, the result was consistent with those reported for the U.S. South (Newman and Wear 1993). The possible explanation was that FI owners responded to market signals, such as price change, more strongly than NIPF owners. Pulpwood supply showed relatively more elastic responses to own price than sawtimber for both ownerships. The possible explanation was that pulpwood could be produced from growing stocks at almost any age whereas sawtimber could only be produced from larger trees at an older stage (Newman &Wear, 1993). Inventory elasticities for FI were higher than those for NIPF, which was consistent with the literature in that NIPF owners obtained nontimber benefits from the growing stock remaining in place while FI owners received financial profits from the timber (Newman & Wear, 1993).

The estimated elasticities in this study could only be partially compared with existing values in the literature because of difference in methodology, data sources and regional focus. Table 5 compared price and inventory elasticities from this study and other studies for the U.S. South (Adams & Haynes, 1980; Haynes & Adams, 1985; Newman & Wear, 1993). For example, Newman & Wear (1993) estimated supply price elasticities for sawtimber (0.27 for FI and 0.22 for NIPF) and pulpwood (0.58 for FI and 0.33 for NIPF) in the Southeast separately. Few studies on supply price elasticities in the U.S. South existed in the literature in terms of the two ownerships and the two timber categories.

Table 4. Elasticities of the stumpage market for the U.S. South, from this study and other studies

Source	Region and timber type	Supply	Inventory
This study	Forest industry sawtimber (S)	0.70	0.79
	Forest industry pulpwood (S)	0.90	1.36
	NIPF sawtimber (S)	0.29	0.63
	NIPF pulpwood (S)	0.32	0.44
Adams and Haynes 1980	Forest industry stumpage (SC)	0.47	0.41
	Private stumpage (SC)	0.39	0.66
	Forest industry stumpage (SE)	0.47	0.49
	Private stumpage (SE)	0.30	0.72
Haynes and Adams 1985	Forest industry stumpage (SC)	0.63	1.00
	Private stumpage (SC)	0.17	1.00
	Forest industry stumpage (SE)	1.20	1.01
	Private stumpage (SE)	0.17	1.00
Newman and Wear 1993	Industry sawtimber in SE	0.27	
	Industry pulpwood in SE	0.58	
	NIPF sawtimber in SE	0.22	
	NIPF pulpwood in SE	0.33	

Note: S, the Southern United States; SC, South Central United States; SE, Southeast United States.

6. CONCLUSION

Using a profit maximization model with time series data from 1953 to 2002, this study estimated stumpage supply for both FI and NIPF owners in the U.S. South. The results indicated that a profit maximization model was appropriate for NIPF owners at the aggregate level because the coefficients had the expected sign and all of them were significant at the 5% or 10% levels for NIPF ownership. An intuitive explanation was that small parcels were less likely to be harvested than larger parcels and large portion of stumpage production came from large tract owners for timber production purpose. This result was consistent with the results in literature (Newman & Wear, 1993). This also implied that the conclusion of NIPF landowners' less responsiveness to market signals based on utility-maximization model at individual level should be modified, because NIPF owners holding tracts over 500 acres, representing less than one percent of all private owners, controlled almost 47 percent of the South's private timberlands (Wicker, 2001).

The results of profit-maximization also revealed that supply price elasticities of 0.70 for sawtimber and 0.90 for pulpwood for FI owners were larger than those of 0.29 for sawtimber and 0.32 for pulpwood for NIPF owners, which in general were relatively larger than previous studies (Adams & Haynes, 1980; Newman & Wear, 1993). The finding suggested that they did respond to market signals, although they were not willing to respond to changing market conditions as strongly as FI owners. Coupled with literature review, the implication for policymakers is that any incentive program intended to promote timber supply from NIPF should be adapted to NIPF owners' objectives, management activities, and preferences. For instance, cost share program should be adjusted to increase ratio to landowners holding large tracts for timber production purpose, while developing other subside programs for NIPF owners who provide nontimber services for the society.

Furthermore, the results of profit-maximization indicated that pulpwood supply had relatively more elastic responses to own price than sawtimber regardless of ownership. The possible explanation was that pulpwood could be produced from growing stocks of nearly any age, while sawtimber was produced only from larger and older trees (Newman & Wear, 1993). The policy implication is limited for this finding.

Finally, the results revealed the relatively low estimated inventory elasticity for stumpage supply by NIPF compared to FI. One possible explanation for the difference was the joint production of nontimber goods and services on NIPF lands. This finding also conformed to the claims in the literature that government programs intended to promote future timber production have increased reforestation or inventory, but a much smaller corresponding increase in timber harvest (Kluender et al. 1999). The implication for policymakers is similar to price variable.

Overall, these results of profit-maximization were consistent with results in literature. They are helpful for us to understand private landowners' response to market drivers, and thus, promote supply of timber from NIPF. Nevertheless, caution should be taken in reaching any definite conclusions from our findings due to data availability and technical constraints. We have witnessed the dramatic shift of timberland ownership from traditional forest industry to timberland investors such as timber investment management organizations (TIMOs) or real estate investment trusts (REITs). Production behavior of this ownership is unknown so far.

Therefore, further research is needed to examine the production behavior of TIMOs and REITs.

REFERENCES

Adams, D.M., Jackson, K.C., & Haynes, R.W. (1988). Production, consumption and prices of softwood products in North America: regional time series data, 1950–1985. Resource Bull. PNW-RB-151, USDA For. Serv. Pacific Northwest Res. Stn., Portland, OR. 49 pp.

Adams, D. M., & Haynes, R.W. (1996). The 1993 timber assessment market model: structure, projections, and policy simulations. Gen. Tech. Rep. PNW-GTR-368. Portland, OR: U.S. Department of Agriculture, Forest Service, Pacific Northwest Research Station. 58 pp.

Adams, D.M., & Haynes, R.W. (1980). The 1980 softwood timber market assessment model: Structure, projections, and policy simulations. *Forest Science, Monograph. 22*, 64 pp.

Adams, D.M., Haynes,R.W., & Daigneault, A.J. (2006). Estimated timber harvest by U.S. region and ownership, 1950-2002. Resource Bull. PNW-GTR-659, USDA For. Serv. Pacific Northwest Res. Stn., Portland, OR. 64pp.

Beach, R.H., Pattanayak, S.K., Yang, J.C., Murray, B.C., & Abt, R.C. (2005). Econometric studies of nonindustrial private forest management: a review and synthesis. *Forest Policy and Econ. 7(2005),* 261-281.

Binkley, C. (1981). Timber supply from non-industrial forests: a microeconomic analysis of landowner behavior. Bull. No. 92. Yale University Press, New Haven, CT.

Boyd, R., & Hyde, W.F. (1989). *Forestry sector intervention*. Iowa State Univ. Press, Ames, IA. 295pp.

Brännlund, R., Johansson, P.O., & Löfgren, K.G. (1985). An economic analysis of aggregate sawtimber and pulpwood supply in Sweden. *Forest Science, 31,* 595-606.

Brown, R., & Zhang, D. (2005). Estimating supply elasticity for disaggregated paper products: a primal approach. *Forest Science, 51(6),* 570-577.

Bureau of Labor Statistics. The Producer Price Index (PPI). January 2007 accessed (http://www.bea.doc.gov).

Carter, D.R. (1992). Effects of supply and demand determinants on pulpwood stumpage quantity and price in Texas. *Forest Science. 38(3),* 652-660.

Dennis, D. (1989). An economic analysis of harvest behavior: integrating forest and ownership characteristics. *Forest Science. 35(4),* 1088– 1104.

Hardie, I.W., & Parks, P.J. (1996). Program enrollment and acreage response to reforestation cost-sharing programs. *Land Economics. 72(2),* 248 –260.

Haynes R.W. (2003). An analysis of the timber situation in the United States: 1952 to 2050. Gen. Tech. Rep. PNW-GTR-560. U.S. Department of Agriculture, Forest Service, Pacific Northwest Research Station, Portland, OR. 220 pp.

Haynes, R.W., & Adams, D.M. (1985). Simulations of the Effects of Alternative Assumptions on Demand-Supply Determinants on the Timber Situation in the United States. USDA Forest Service, Forest Resource and Economics Research, Washington, D.C.

Howard, J.L. (2003). U.S. Timber Production, Trade, Consumption, and Price Statistics 1965-2002. Gen. Tech. Rep. FPL-RP-615. USDA Forest Service, Madison, WI. .6pp.

Ince, P.J. (1999). Long-range outlook for U.S. paper and paperboard demand, technology, and fiber supply-demand equilibria. *In*: Procceeding of the Society of American Foresters 1998 national convention. Bethesda, MD: Society of American Foresters: 330–343.

Johansson, P.O., & Löfgren, K.G. (1985). *The Economics of Forestry and Natural Resources.* Basil Blackwell, Oxford. 300p.

Kluender, R.A., Walkingstick, T.L., & Pickett, J.C. (1999). The use of forestry incentives by non-industrial forest landowner groups: is it time for a reassessment of where we spend our tax dollars? *Natural Resources Journal. 39,* 799–818.

Kuuluvainen, J., Karppinen, H., & Ovaskainen, V. (1996). Landowner objectives and non-industrial private timber supply. *Forest Science. 42(Aug),* 300– 309.

Newman, D.H. (1987). An econometric analysis of the southern softwood stumpage market: 1950-1980. *Forest Science. 33(4),* 932-945.

Newman, D.H., & Wear, D.N. (1993). Production economics of private forestry: a comparison of industrial and nonindustrial forest owners. *American Journal of Agricultural Economics. 75,* 674-684.

Pattanayak, S., Murray, B., & Abt, R. (2002). How joint is joint forest production? An econometric analysis of timber supply conditional on endogenous amenity values. *Forest Science. 47 (3),* 479– 491.

Prestemon, J., & Wear, D. (2000). Linking harvest choices to timber supply. *Forest Science, 46 (3),* 377–389.

Random Lengths. (1974-2001). *Forest Porduct Market Prices and Statisitcs Yearbook.* Annual Issue, Random Lengths Publications, Inc., Eugene, OR.

Samuelson, P.A. (1952). Spatial price equilibrium and linear programming. *The American Econ. Review. 42,* 282-303.

Smith, W. B., Miles, P. D., Vissage, J. S., & Pugh, S. A. (2004). Forest Resources of the United States, 2002. Gen. Tech. Rep. NC-241. U.S. Department of Agriculture, Forest Service, North Central Research Station, St. Paul, MN. 73pp.

Smith, W. B., Vissage, J. S., Darr, D. R., & Sheffield, R.M. (2001). Forest Resources of the United States, 1997. U.S. Department of Agriculture, Forest Service, North Central Research Station, St. Paul, MN. January 2007 accessed (http://www.ncrs.fed.us).

Ulrich, A. (1989). U.S. timber production, trade, consumption and price statistics, 1950–1987. Misc. Publ. 1471. U.S. Department of Agriculture, Forest Service, Washington, DC. 77 pp.

Wicker, G. (2001). Motivation for private forest land owners. U.S. Department of Agriculture, Forest Service, Southern Region, Atlanta, GA. May 2008 accessed (http:/www.srs.fs.fed.us/sustain/draft/socio4).

Wooldridge, J.M. (2000). *Introductory Econometrics: A Modern Approach.* South-Western College Publishing, Boston, MA. 824pp.

Part III. Forest Resource Management and Multiple Benefits

In: Global Change and Forestry
Editors: J. Gan, S. Grado and I. A. Munn
ISBN: 978-1-60876-262-0

Chapter 15

THE CHANGING SMALL-SCALE FAMILY FORESTRY IN THE UNITED STATES: 1993-2003[1]

***Yaoqi Zhang*[*1], *Xianchun Liao*[2], *Brett. J. Butler*[3] and *John Schelhas*[4]**

[1] School of Forestry and Wildlife Sciences, Auburn University, AL, USA
[2] Department of Forestry, Mississippi State University, MS, USA
[3] USDA Forest Service, Northern Research Station, PA, USA
[4] USDA Forest Service, Southern Research Station, NC, USA

ABSTRACT

The forestland ownership has been changing rapidly in the past two decades in the United States. One of the changes is growing number of *small-scale family forestry. This chapter examines the structural variation of forest landholding size among the states using data from the United States from 1993 to 2003. The results from* regressions models analyzing the factors influencing the mean size and structural variation among states and between the two periods indicate population density, the age structure, the share of rural/urban population, income, equity have influence on the structure of landholding size. *It argues that the economic efficiency of small-scale forestry has resulted primarily from increasing demand for non-timber amenities of the forests as society is changing from industrial to post-industrial era.*

Keywords: Non-industrial private forests; Seemingly unrelated regression; Ordinary least squares regression; National Woodland Owner Survey; Parcelization.

---

[1] This chapter was revised based on paper "The Increasing Importance of Small-Scale Forestry: Evidence from Family Forest Ownership Patterns in the United States" published in Small-Scale Forestry. Permission was granted by Spinger.

[*] E-mail: yaoqi.zhang@auburn.edu, School of Forestry and Wildlife Sciences, Auburn University, AL 36849.

1. Introduction

Currently, there are an estimated 248 M ha of forestland in the contiguous United States (Smith et al. 2004). Nearly two-thirds, or 157 M ha, are privately owned and two-thirds of this land, or 105 M ha, is owned by 10.3 M families and individuals (Butler and Leatherberry 2004). The number and relative importance of family forest owners varies considerably across the country. The north has 46% of the family forest owners in the USA, the south has 42%, and the west has 12%.

The distribution of size of family forest landholdings has been undergoing dramatic change. The number of family forest owners in the contiguous United States increased from 9.3 M in 1993 (Birch 1996) to 10.3 M in 2003 (Butler and Leatherberry 2004). Research also suggests that both the share and total acreage in small parcels (less than 20 ha) have increased in the last 10 years (Butler and Leatherberry 2004). DeCoster (1998) noted that if this trend continues, nearly 95% of the nation's private forestlands will be owned by individuals with less than 40 ha by 2010. *It is widely believed that the average size of family forest holdings in the USA is shrinking through a process called parcelization. However, most of the evidence has been anecdotal, rather than from empirical research.*

Forest parcelization is the process of taking a contiguous parcel of land owned by a single owner and dividing it among two or more owners. This process should not be confused with forest fragmentation, which is the process of division of the *physical* forest resource into smaller parcels (rather than the ownership of the forest). To assess and analyze parcelization, previous research (e.g. Mehmood and Zhang 2001) has examined mean landholding size. If the structure of the landholdings has a statistically normal distribution and the shape of the distribution does not change, the mean is a useful indicator; otherwise, the mean can be misleading. For example, it is possible for parcelization and consolidation to co-exist without change in the mean (Ripatti 1996, Zhang et al. 2005).

The study reported here has similarities with that of Mehmood and Zhang (2001). However, the 1978 and 1994 data used by Mehmood and Zhang are not fully comparable because the 1978 data included forest industry land and the 1993 and 2003 data are limited to family forest owners. It is therefore argued that the data used here are more appropriate for analyzing parcelization. Although the analysis may provide some insights about change across time, as Mehmood and Zhang did, the primary focus is on variation across states. The main contribution of this study is the investigation of changes in the structure of landholding sizes (i.e. the share of family owners by size of forest landholdings), rather than only average landholding sizes.

Analyzing parcelization for family forest owners in the USA is highly challenging because of the lack of transition probability matrices of transfers from and to other ownership types (e.g. forest industry and government agencies) and gains and losses of forest land (e.g. from agriculture and to development). As an alternative, this study examines variation in the structure of family forest holdings among states. Even though spatial variation could differ greatly from time trends, the findings do explain variations between states and shed light on causal, or at least correlated, factors. If states with higher population densities and higher incomes have smaller holding sizes, it can be surmised that these factors are correlated with parcelization, and other states will likely face similar trends if populations grow and incomes increase in the future.

Previous studies relating to the size of landholdings have often centred on community well-being (e.g. Gilles and Dalecki 1988; Swanson 1988; Labao and Schulman 1991). Most studies conclude that the persistence of small-scale farms is positively related to social well-being (e.g. Goldschmidt 1978). Only a handful of studies have explicitly examined the detrimental aspects of declining farm size. Schmitt (1991) and Allen and Lueck (1998) used transaction cost theory to explain the proliferation of family farms and the resistance to large-scale farms in Europe and North America. Allen and Lueck (1998) showed that seasonality and random shocks are the main features that distinguish family farms from 'industrial' forms of landownership. When farms are successful in mitigating the effects of seasonality and random shocks to output, farm organization gravitates towards factory-type processes and large-scale farms.

In the case of forestry, there have been a number of studies of relationships between strength of local forest industry and community well-being. These studies have found that concentration of forestland ownership is negatively related to community well-being (Fortmann et al. 1991; Sisock 1998; Bliss et al. 1998; Joshi et al. 1998). A key difference, however, is that, unlike farm owners, forestland owners are increasing in number. Thus research on forest land ownership change is critical. Research in Finland by Ripatti (1996) was the first study, and is perhaps still the most comprehensive study, of the causes of forest parcelization. The study by Mehmood and Zhang (2001) was one of the first empirical examinations of the causes of forest parcelization in the USA. These authors found that death rate, urbanization, income, regulatory uncertainty and financial assistance for landowners all have significant impacts on the change in average holding size in the USA. But this study only examined change in mean landholding size and did not examine change in the overall structure of forest landholdings.

Gobster and Rickenbach (2003) suggested that economics, demographics, values and motivations, globalization and new technology, natural capital and forest policies are the primary drivers of forestland parcelization. DeCoster (2000) suggested that parcelization is driven by social, political and economic forces that consistently under-tax and over-serve developed areas, while over-taxing and under-serving rural land owners, including family forest owners.

Using transaction cost theory, Zhang et al. (2005) proposed an explanation for the increasing number of small non-industrial private forest ownerships in the USA. The major argument was that the higher transaction costs of non-timber amenities from forests, along with increasing demand for these services, increases the number of people who own smaller forest holdings.[2] As non-timber values increase, forestland is more likely to be owned by individuals to provide for their personal consumption of non-market amenity goods and services. This argument was supported with some historical evidence. The current paper approaches this question from a different perspective, using cross-regional evidence and statistical analysis to demonstrate the increasing importance of small-scale forestry and suggest some factors that might be behind changes in landholding size.

The paper first formulate some economic rationales for holding size is presented, followed by the specification of a landholding size econometric regression model and data

---

[2] The logic of this argument is that people who use timberlands frequently (e.g. as hunting lease, bird watching, or simply enjoying the peace and or amenities) would be better off (in terms of saving transaction costs) by owning the timberland purchasing the services owned by someone else.

sources used to test the model. Empirical results are then reported. Finally, the results of the analysis and their implications for family forest owners in the USA and other countries are presented.

2. Why Small-Scale Forestry?

Small-scale forestry is affected by substantial transaction costs and poorly developed markets for commodities and services. In developed countries, especially post-industrial countries, there is an increasing demand for consumption of non-timber goods and services from forests (amenity, recreation and wildlife) by growing urban populations, as opposed to the production value of forests for timber. Mather (2001) reviewed how forestry practices and policies in Europe and North America have changed as the importance of forest amenities has increased relative to timber production.

Forestland tenure and holding size change largely in response to changing transaction costs (Zhang 2001; Zhang et al 2005). For example, from the 1950s to the 1980s the forest industry purchased large areas of timberland, but the industry began to sell timberland in the 1990s in part because of a change in transaction costs (especially the reduction of transaction costs of wood markets) along with disadvantageous taxation and capital costs. Consequently, the wood processing industry is changing from in-house supplies of wood from industry-owned timberland to acquisitions from other timber owners through markets. In a similar manner, in cases of forest industry-owned timberland, silviculture can be either conducted in-house or through outside contractors (Wang and Van Kooten 2000). Using in-house silviculture or harvesting saves the transaction costs of hiring outside contractors but adds to internal administrative costs, while contacting-out the activities saves the administrative costs but increases market transaction costs. Economic analysis finds that optimal holding size occurs when net marginal utility is equal to marginal cost (the market price of forestland plus holding costs, including taxes, management costs and risk). Over time, owners will adjust their holding sizes in response to changes in input and output prices (Zhang et al 2005).

To further explain, figure 1 represents the behavior of a family forest owner whose primary objectives are not related to timber production. In figure 1, marginal cost refers to forestland price and holding costs. The forestland price is fundamentally the present value of all future value generated from timber production and non-timber products and services. Forestland price is determined by joint market forces. Buyers are price takers, or must make their decision based on the price offered to them. An increase in price, for example from P_0 to P_1, will decrease the average holding size and increase the number of owners.

The marginal value of timberland is the utility or benefits owners receive. The utility curve shifts according to owners' preferences. The two utility curves represent either one owner's preferences at two different times or two owners' preferences at the same time. Assuming U_0 and U_1 represent current and future utility respectively, the different shapes imply that the family forest owner represented by U_1 place greater weight on non-timber values than the owner represented by U_0. A shift from U_0 to U_1 can be caused by changes in owners' incomes or related changes in preferences (for example, changing public awareness and increases in perceived value of forested environments). The utility curves are not parallel because marginal non-timber values may increase relatively faster than timber values as income increases. Intuitively, the land value for timber production is not subject to the owner's income but rather is largely due to management skill and scale. But amenity values are dependent on the owner's income since significant costs prevent the emergence of the market. When income increases, the marginal value curve of timberland shall shift upward. Therefore, the shape of the weighted total marginal value curve varies. The concave shape of the curve is consistent with the findings of other research (e.g. Butler and Leatherberry 2004).

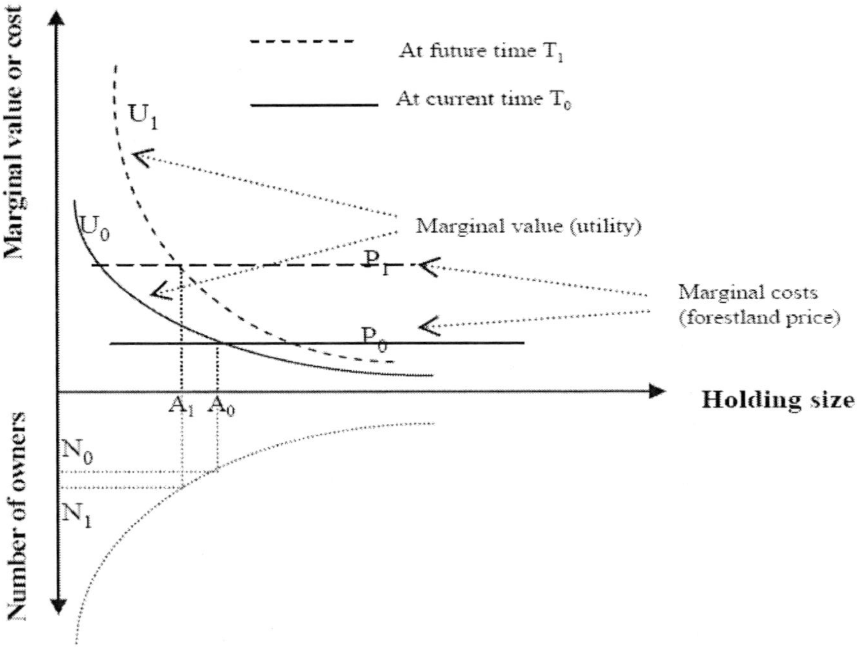

Figure 1. Shrinking holding size and increasing number of small family owners.

The optimal landholding size for a family forest owner is the holding size for which marginal value is equal to marginal timberland price. An increase in the number of people with small holdings occurs when land becomes more costly, or in other words, when marginal timberland cost rises more sharply and holding larger tracts of land becomes less affordable. Under these circumstances (increasing number of wealthy people creating a demand for the non-wood value of forests and increasing costs), as figure 1 indicates, the landholding size will become smaller. Although larger numbers of people with more wealth are owning forestland, a substantial rise in forestland values in the past decades means that they can only afford smaller tracts of land. As a consequence, the total number of small family forest owners increase from N_0 to N_1.

When the primary ownership objective is timber production, the optimal holding size is likely based on the efficiency of timber production (figure 2). Holding sizes tend to be larger because of technological advances that favor larger scale of operations for forest management. Of course, the ultimate choice of size of forest holdings will be mitigated by capital and land availability.

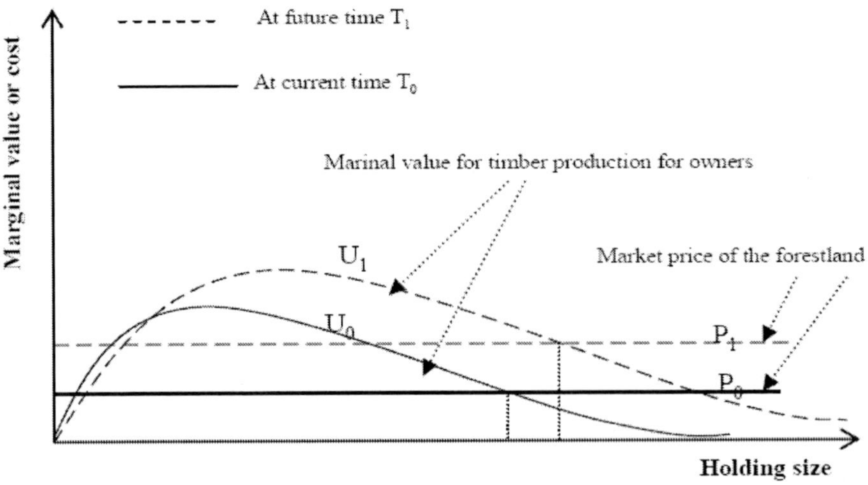

Figure 2. Holding size versus cost of timberland for timber production.

3. Econometric Models

Socio-economic and biophysical conditions determine possible management objectives as well as timber and non-timber values. Therefore, there should be observable relationships between the distribution of forest land holding sizes and the biophysical characteristics of the forest land and the socio-economic characteristics of the forest landowners. *Following Mehmood and Zhang (2001) and Pan (2006), an empirical analysis has been conducted of these factors. For this analysis, it is hypothesized that holding size is a function of a group of variables representing the factors that ultimately determine the holding size.*

It is hypothesized that, as population increases, more forest and agricultural land will be converted to residential and commercial uses (Nagubadi and Zhang 2005). Therefore population density influences demand and price for forest land. Changes in the spatial distribution of the population, such as urbanization, could also be expected to affect the structure of forest holding sizes. People living in cities, suburbs or urban-rural interfaces have different goals for their land than rural people. They are more likely to own land as part of their house site or for recreation purposes, and less likely to have timber production as a major objective.

Demographics and economic factors may also be associated with changes in holding size (Gobster and Rickenbach 2003). For example, an aging population might result in more land transfers, subsequent parcelization, and hence an increase in the number of smaller holdings. Income levels and number of retirees may affect the utility function and owners' demands for various mixes of products and services from their forestland, and thereby influence the distribution of forest landholding size. As income increases, living in or around the forest has become a growing lifestyle choice (DeCoster 1998). Distribution of income may also be an important determinant of land property size. Several researchers (e.g. Sisock 1998; Pan 2006) have found the Gini index, which quantifies income disparity, to be a significant predictor of parcelization and consolidation. Private forestland availability may also affect average holding size and holding size structure.

To put all these variables together, a model has been hypothesized of the form:

$$Y_m = \alpha + \beta_1 POPD + \beta_2 OLDP + \beta_3 URBAN + \beta_4 INCP + \beta_5 GINI + \beta_6 PFP + \varepsilon \quad (1)$$

$$Y_i = \alpha + \beta_{i1} POPD + \beta_{i2} OLDP + \beta_{i3} URBAN + \beta_{i4} INCP + \beta_{i5} GINI + \beta_{i6} PFP + \varepsilon_i \quad (2)$$

where Y_m is the mean holding size in a state (ha), and i represents the category of holding size (small, medium and large). Equation (2) has three sub-equations: Y_i is the percentage of family forestland in holdings between 1 and 19 ha (i=small), the percent of forestland in holdings between 20 and 199 ha (i=medium), and the percentage of forestland in holdings 200 ha or larger (i=large); POPD is population density (number of persons per square mile); OLDP is the percentage of the population who are at least 65 years of age; URBAN is the percent of the population living in urban areas; INCP is per capita income ($US 1000); GINI is the Gini index of income disparity; and PFP is the average holding size (ha/owner). The data sources for these variables are listed in table 1.

Table 1. Description of variables and data sources

Variable	Description	Data sources
MEAN	Average holding size	Birch (1996), Butler et al. (2005)
SMALL	Percent of forestland in holding size less than 20 ha (%)	Birch (1996), Butler et al. (2005)
MEDIUM	Percent of forestland in holding size between 20 and 200 ha (%)	Birch (1996), Butler et al. (2005)
LARGE	Percent of forestland in holding size larger than 200 ha (%)	Birch (1996), Butler et al. (2005)
POPD	Persons per square mile	U.S. Census Bureau 2000 and 1990
OLDP	Persons 65 years old and over	U.S. Census Bureau 2000 and 1990
URBAN	Percent of urban population	U.S. Census Bureau 2000 and 1990
INCP	Per capita income ($1,000)	U.S. Census Bureau 2000 and 1990
GINI	Gini index of family income	Volscho (2004)
PFP	Per capita private forestland (ha)	Smith et al. (2004)

Ordinary least square (OLS) has been used to conduct the regression for Equation 1, which has average holding size as the dependent variable for the panel data (two periods for all states). Seemingly unrelated regression (SUR) has been used to jointly estimate the equations represented by Equation 2. SUR is used because the explanatory variables that affect the share of forestland are the same for the three models (small, medium, and large). This method allows individual shares of forestland and residual variances to differ across the models. An advantage is that heteroscedasticity across equations is explicitly incorporated, thus strengthening the testing of joint hypotheses (Binder 1985). Varying levels and log transformations are explored for these equations.

Ordinary least square (OLS) has been used to conduct the regression for Equation 1, which has average holding size as the dependent variable for the panel data (two periods for all states). Seemingly unrelated regression (SUR) has been used to jointly estimate the equations represented by Equation 2. SUR is used because the explanatory variables that affect the share of forestland are the same for the three models (small, medium, and large). This method allows individual shares of forestland and residual variances to differ across the models. An advantage is that heteroscedasticity across equations is explicitly incorporated, thus strengthening the testing of joint hypotheses (Binder 1985). Varying levels and log transformations are explored for these equations.

4. Data

To conduct the regressions of Equations 1 and 2, the data were collected in 1993 (for more information, see Birch 1996) and 2002-04 by the USDA Forest Service (for more information, see Butler et al. 2005). The data from 2002-04 are simply interpreted as year 2003. Data from the 1978 national study of private forest owners as described in Birch et al. (1982) could not be used because of figures were not reported for size of holding for family forest owners.

The USDA Forest Service Forest Inventory and Analysis program conducts the National Woodland Owner Survey (NWOS), which is the nation's survey of forest owners. On a recurring basis (previously conducted every five years, but now annually), the survey collects data from a random sample of forest owners across the United States to obtain information about their forestland, ownership objectives, forest use, forest management practices, sources of information, concerns and issues, and demographics. The NWOS uses remotely sensed imagery to determine whether each of a random set of sample points across the United States is forested. For the forested points, ownership information is collected from property tax offices and other public sources. These forest owners are sent a postal questionnaire as the primary data collection instrument, and telephone interviews are used to increase response rates. More detailed information on the data collection and processing procedures are described in Birch (1996) and Butler et al. (2005).

The data set used in this study has 92 observations for 46 states and two points in time (1993 and 2003). Four states – Hawaii, Nevada, Alaska, and Idaho – are excluded from the analysis due to either missing data or too few observations. Table 1 describes the dependent and independent variables and data sources. For each state, landowners are placed into six size classes: less than 4, 4-19, 20-39, 40-199, 200-399, and more than 400 ha. Means have been calculated as the weighted average size by states.

Forest holdings are further collapsed into three groups based on previous research, data availability, and the objective of an approximately uniform distribution of the sample among the groups. Accordingly, the percent of forestland in holdings less than 20 ha was used to indicate owners with smaller holdings and, presumably, non-timber ownership objectives. An increase in the percentage of forestland in this group is considered an indicator of forest parcelization. The group of owners with holdings between 20 and 199 ha was labeled medium, and the group with owners of holdings of 200 ha or more was labeled large. The specific break points are arbitrary, but are judged to be appropriate for measuring the structure of forestland holdings.

Socio-economic variables including POPD, OLDP, URBAN, and INCP were obtained from the US Census Bureau for 1990 and 2000 (US Census Bureau 1990, 2000). Data for the Gini index on income disparity was drawn from Volscho (2004). Per capita private forestland area is used as a proxy for private forestland availability and was obtained from Smith et al. (2004).

5. RESULTS

As indicated in Table 2, the mean holding size is largest in the South (21 ha), followed by the West (14 ha) and the North (11 ha). About 90% of the family forest owners in the United States own less than 20 ha , for a total amount of one third of family forestland.

From 1993 to 2003, the average size of family forest holdings decreased marginally (by 1%), but the share of small and large holdings increased by 4% and 13% respectively and the share of medium sized holdings decreased by 8%. This is consistent with the findings of Ripatti (1996) and the argument made by Zhang et al. (2005) about average holding size change. It highlights the importance of examining the structure of forest land ownership, because this change would not have been evident through an analysis of average holding sizes only.

Table 2. Mean holding size and structure of family forest holdings by state, 2003

Region, State	Mean (ha)	Small (%)	Medium (%)	Large (%)
North				
Connecticut	5.72	55	43	2
Delaware	4.48	80	10	10
Illinois	10.39	46	47	6
Indiana	8.34	51	45	4
Iowa	9.02	46	51	3
Maine	15.12	29	53	18
Maryland	5.12	51	40	9
Massachusetts	3.83	56	40	4
Michigan	11.25	48	45	7
Minnesota	18.53	33	62	5
Missouri	19.48	29	59	13
New Hampshire	11.41	36	49	15
New Jersey	5.34	77	19	4
New York	11.61	43	52	5
Ohio	8.83	60	35	5
Pennsylvania	11.92	45	46	9
Rhode Island	4.59	65	35	0
Vermont	30.40	23	66	10
West Virginia	17.64	27	62	11
Wisconsin	16.13	31	63	6
North Average	11.46	47	46	7
South				
Arkansas	14.51	23	58	19
Alabama	19.64	22	40	38
Florida	7.74	38	41	21
Georgia	15.20	24	44	32
Kentucky	13.33	32	57	10
Louisiana	32.05	23	41	36
Mississippi	40.23	14	48	38
North Carolina	14.70	38	46	15
Oklahoma	49.53	14	63	24
South Carolina	16.58	25	51	24

Table 2. (Continued).

Region, State	Mean (ha)	Small (%)	Medium (%)	Large (%)
Tennessee	13.47	35	46	18
Texas	18.83	30	49	21
Virginia	16.07	33	51	16
South Average	20.91	27	49	24
Rocky Mountain				
Arizona	12.92	33	33	33
Colorado	12.58	26	33	41
Kansas	10.00	45	50	5
Montana	109.96	6	44	50
Nebraska	11.02	34	36	30
New Mexico	29.77	11	11	79
North Dakota	10.84	39	46	14
South Dakota	14.36	35	42	23
Utah	17.36	15	33	51
Wyoming	58.03	6	36	58
Rocky Mountain Average	28.68	25	37	38
Pacific Coast				
California	20.74	21	37	42
Oregon	14.17	38	34	28
Washington	8.24	51	33	16
Pacific Coast Average	14.38	36	35	29
National Average	18.06	36	44	20

Data Source: USDA Forest Service (Birch 1996, Butler et al. 2005).

Table 3 presents the estimated OLS (for the mean) and SUR (for structure) models. The impacts of individual variables can be compared directly because the coefficients have an interpretation as the elasticity of a log-form model. Overall, the independent variables explain the four dependent variables measuring forestland holding structure quite well, with R^2 values between 0.40 and 0.61. All of the independent variables have showed the expected sign of relationship with the dependent variables.

As expected, the significance of the coefficient for the population density variable is consistent with the literature in that, as it increases, the mean size and share of large holdings decreases and the share of small holdings increases. It is likely that population density is a determinant of the holding of small parcels. As population density increases and area of forestland remains fixed, the demand for forestland increases and drives the conversion from larger land owners who are more interested in timber production to small owners who are more interested in non-timber activities, such as aesthetics and privacy (often associated with a home site) and recreational opportunities like hunting.

Our findings show that the percentage of the population older than 65 years has a positive effect on the share of medium and small holdings, but a negative effect on the share of large holdings. It is likely that advancing age portends an increase in the transfer of larger holdings through inheritance, and parcelization is associated land going to multiple heirs or being subdivided for development. For obvious reasons, the percentage of population older than 65 years has a strong correlation with increased death rates and previous studies have shown that death rate is correlated with smaller holdings (DeCoster 1998; Mehmood and Zhang 2001).

Table 3. Regression results on factors influencing forest landholding size and distribution

Variable	Mean Coeff. (S.E)	Small Coeff. (S.E)	Medium Coeff. (S.E)	Large Coeff. (S.E)
Constant	6.752*** (2.277)	–4.943*** (2.224)	–5.716*** (1.642)	13.510*** (3.555)
POPD	–0.271*** (0.040)	0.193*** (0.039)	0.032 (0.028)	–0.398*** (0.062)
OLDP	0.045 (0.059)	0.100* (0.057)	0.177*** (0.042)	–0.229** (0.094)
URBAN	0.678** (0.301)	-0.800*** (0.295)	–0.796*** (0.219)	1.903*** (0.463)
INCP	–0.196 (0.162)	0.030 (0.159)	0.066 (0.117)	–0.467* (0.250)
GINI	0.425 (0.858)	–1.220 (0.839)	–1.677*** (0.621)	6.223*** (1.326)
PFP	0.306*** (0.059)	–0.279*** (0.058)	0.061 (0.043)	0.352*** (0.090)
R^2	0.61	0.53	0.40	0.60
N	92	92	92	90

Note: *, ** and *** denote significance at the 10%, 5%, and 1% levels.

Urban population versus rural population refers to the share of urban and rural residents. The percentage of a state's population that lives in an urban area, an appropriate identifier of urban versus rural states, has a significantly positive relationship with the mean size and share of large holdings, but is negatively related to the shares of medium and small holdings. This finding indicates that, other things being equal, population concentration in cities is positively correlated with large forest land holdings. This contradicts findings from pervious studies (Befort et al 1998; Mehmood and Zhang 2001). The results indicate that, for states with higher percentages of urban population, overall mean holding size is larger.

Income per capita has a significant negative relationship with the share of larger holdings, which is in agreement with the literature (e.g. Goldschmidt 1978; Mehmood and Zhang 2001). The estimate for the income variable indicates that, as income increases, people devote more money to non-timber activities. However, this result should be considered preliminary; the shares of medium-sized and small holdings show the expected signs, but neither of the coefficients is statistically significant.

It is interesting to note that income inequality, which is measured by the Gini Index, has a significantly positive impact on the share of large holdings, but a significantly negative impact on the share of medium-sized holdings. This implies that income inequality is a driving force for large landholdings. A plausible interpretation is that states where income inequality is greatest have smaller shares of small and medium holdings and a greater proportion of larger holdings, a reflection of rurality. These results should be considered with caution because the negative impact of small holding sizes is not significant.

Not surprisingly, per capita private forestland area had significant positive impact on mean family forestland holding size and share of large holdings, but a negative impact on the share of small holdings; it also has a positive, but statistically non-significant, impact on the share of medium-sized holdings. This finding suggests that per capita forestland is a driving force behind large landholdings, which are more likely when forestland is abundant and in less populated regions. A possible explanation is that, with increasing availability of private forestland, the possibility of owning larger parcels increases.

6. CONCLUSION

The results of the statistical analysis indicate that population density, percentage of older people, percentage of urban population, per capita income, income variation, and per capita private forestland correlate with structural differences in the sizes of family forest holdings. These results strongly support the hypothesized link between the variables and the structure of forestland, and help explain why some states have relatively higher numbers of smaller family forests. For example, higher population density, aging population and per capita income are associated with more small family forest owners (i.e., more owners with smaller parcels), whereas percentage of urban population, income variation and per capita private forestland are positively associated with larger holdings.

To understand the impact of percentage of urban population on overall mean holding size, suppose two states have the same population density, one being highly urbanized and another highly rural. The average holding size in an urbanized state is likely larger than in a rural state where the population is more evenly distributed across the state. It is likely that more people own land in the rural state while in the urbanized state a smaller number of people living in rural areas own larger parcels. This is counter to conventional thinking – most people think that urbanization will increase parcelization. But in fact, urbanization is not necessary population growth but more about spatial changes in population.

Due to data limitations, it has not been possible to address directly the process of parcelization. But the statistical results reveal some general trends associated with family forestland holdings. Family forestland under current circumstances is tending to shift from medium-sized holdings to smaller parcels that are primarily used for non-timber purposes. The results suggest that growth in population causes parcelization when measured either by the average holding size or as share of small holdings, but urbanization measured by the share of population which lives in cities does not. These results indicate that states with a higher share of urban population, all else equal, are more likely to have larger average holding sizes and larger proportions of forestland in large holdings.

Parcelization, or an increasing number of small owners with smaller parcels, is often presumed to be related to forest fragmentation, but this may not always be the case. Examining the structure of forest holding size instead of the mean has proved a useful step towards understanding the drivers of parcelization. A further research step would be to develop transition matrices that include transitions between sizes of family forest holdings along with gains and losses to and from other types of land uses.

The USA is not unique in being subject to forestland parcelization. As countries in Asia become wealthier, similar changes can be expected. Harrison et al. (2002) observed, 'throughout the world, there appears a trend to move from industrial forestry towards landholder-based forest management and community forestry and small-scale (often referred to as 'smallholder') forestry is of growing importance.' There are important policy implications of this change; hence it is an important research topic.

Acknowledgments

This research was financially supported by the McIntire-Stennis program of the USDA (Project number ALAZ00032), the USDA Forest Service Southern Research Station, and Auburn University's Center for Forest Sustainability. We appreciate comments and suggestions from Dr. Indrajit Majumdar and Dr. Steve Harrison.

References

Allen, D.W, &Lueck. D. (1998). The nature of the farm. *Journal of Law and Economics.* 41 (2): 343-386

Befort, W.A., Luloff, A.E., & Morrone, M. (1988). Rural land use and demographic change in a rapidly urbanizing environment. *Landscape and Urban Planning.* 16 (4):345-56

Binder, J.J. (1985*). Measuring the effects of regulation with stock price data.* Rand Journal of Economics. 16(2): 167-183.

Birch, T.W. (1996). Private forestland owners of the United States, 1994. *USDA For. Serv. Northeast For. Exp. Stat. Resour. Bull.* NE-138, Delaware, OH.

Birch, T.W., Lewis, D.G., Kaiser, H.F. (1982). The private forest-land owners of the united states. *U.S. Department of Agriculture, Forest Service. Resour. Bull. WO-1.* Washington DC.

Bliss, J.C., Sisock, M.L., & Birch, T.W. (1998). Ownership matters: forestland concentration in rural Alabama. *Society and Natural Resources:* 11 (4): 401-410.

Butler, B.J., Leatherberry, E.C. (2004). America's family forest owners. *Journal of Forestry.* 102 (7): 4-14.

Butler, B.J, Leatherberry, E.C., Williams, M.S. (2005). Design, implementation, and analysis methods for the National Woodland Owner Survey. U.S. Department of Agriculture, Forest Service, Northeastern Research Station, Newtown Square, PA. *Gen. Tech. Rep.* NE-GTR-336. P. 43.

Decoster, L.A. (1998). The boom in forest owners—A bust for forestry? *Journal of Forestry.* 96(5):25-28.

Gilles, J.L., Michael, D. (1988). Rural Well-Being and Agricultural Change in Two Farming Regions. *Rural Sociology.* 53 (1):40-55.

Gobster, P.H., Rickenbach, M.G. (2003). Private forestland parcelization and development in Wisconsin's Northwoods: perceptions of resource-oriented stakeholders. *Landscape and Urban Planning.* 69 (2/3): 165-182.

Goldschmidt, W. (1978). As you sow: three studies in the social consequences of Agribusiness. Montclair, NJ: Allenhead, Osmun.

Harrison, S., Herbohn, J., Niskanen, A. (2002). Non-industrial, smallholders, small-scale and family forestry: what's in a name? *Small-scale Forest Economics, Management and Policy.* 1 (1): 1-11.

Hartsell, A.J., Brown, M.J. (2002). Forest Statistics for Alabama, 2000. Resource Bulletin SRS-67. USDA Southern Research Station. http://vm.uconn.edu/~twv00001. Viewed 15 July 2006.

Joshi, M. L., Bliss, J.C., Bailey, C., Teeter, L.J., Ward, K.J. (2000). Investing in industry, underinvesting in human capital: Forest-based rural development in Alabama. *Society and Natural Resources*. 13(4): 291-319.

Fortmann, L., J. Kusel, Y. Kawamura, M. Olson and C. Danks 1991. The effects of forest control and use on county well-being. Pp. 49-67 in Jonathan Kusel and Louise Fortmann, *Well-Being in Forest-Dependent Communities*, Vol. 1. September 1991. Berkeley, California: Department of Forestry and Resource Management, University of California, Berkeley.

Labao, LM. and Michael DS (1991). Farming patterns, rural restructuring, and poverty: A comparative regional analysis. *Rural sociology.* 56 (4): 565-602.

Larson, K. (2004). Family forests—the bigger picture. *Journal of Forestry.* (Oct./Nov. 2004):13-14.

Mather, A.S. (2001). Forests of consumption: postproductivism, postmaterialism, and the postindustrial forest. *Environment and Planning C: Government and Policy.* 19(2): 249 – 268.

MacCleert, D.W. (1990). Brief overview of timber conditions and trends of U.S. forests. USDA, Washington, DC.

Mehmood, S.R., Zhang, D. (2001). Forest Parcelization in the United States. Journal of Forestry 99(4): 30-34.

Nagubadi, R.V., Zhang, D. (2005). Determinants of timberland use by ownership and forest type in Alabama and Georgia. *Journal of Agricultural and Applied Economics.* 37(1):173-186

Pan, Y. (2006). Analysis of Holding Size Distribution of private Forestland Ownership. *Master Thesis,* Auburn University. Auburn.

Ripatti, P. (1996). Factors Affecting Partitioning of Private Forest Holdings in Finland: A Logit Analysis. *Acta Forestalia Fennica.* 252 (monograph).

Rosen, J.F., Doolittle, L. (1987). Profiles of Midsouth Non-industrial private forests and owners. *Resource Bulletin.* SO-125. United States Department of Agricultural, Forest Service, Southern Forest Experiment Station. New Orleans, LA:

Siry, K. P. (2002). Chapter 14: Intensive Timber Management Practice. In Wear D. and J. G. Greis (eds): Southern Forest Resource Assessment. Department of Agriculture, Forest Service, Southern Research Station, Asheville, NC. *General Tech. Rep.* SRS-53.

Sisock, M.L. (1998). Unequal shares: forest land concentration and well-being in the rural Alabama. *Master Thesis,* Auburn University. Auburn.

Smith, W.B., Miles, P.D., Vissage, J.S. et al (2004). Forest resources of the United States, 2002. US Department of Agriculture, Forest Service, North Central Research Station, St. Paul, MN. *Gen. Tech. Rep.* NC-241.

Swanson, L.E. (1988). Agriculture and community change in the U.S. Boulder. CO: Westview Press.

US Census Bureau (1990). Census 1990 Data for the United States http://factfinder.census.gov/servlet. Viewed 1 Aug 2006

US Census Bureau (2000). Census 2000 Data for the United States http://factfinder.census.gov/servlet. Viewed 1 Aug 2006

Volscho, T.W. (2004). Gini Index of Family Income by U.S. County, 2000. University of Connecticut, Dept of Sociology: http://vm.uconn.edu/~twv00001/counties.htm. Viewed 15 July 2006.

Wang, S., van Kooten, G.C. (2000). Forestry and New Institutional Economics: *An Application of Contract Theory to Forest Silvicultural Investment.* Ashgate, Aldershort.

Zhang, Y. (2001) Economics of transaction costs saving forestry. *Ecological Economics.* 36 (2): 197–204.

Zhang, Y., Zhang, D., Schelhas, J. (2005). Small-scale non-industrial private forest ownership in the United States: Rational and Implications for forest management. *Silva Fennica.* 39 (3):443-454.

In: Global Change and Forestry
Editors: J. Gan, S. Grado and I. A. Munn

ISBN: 978-1-60876-262-0

Chapter 16

PREDICTING MANAGEMENT BEHAVIORS ON FAMILY FORESTS USING FIA AND CENSUS DATA

Indrajit Majumdar*[*1], Larry Teeter[2] and Brett J. Butler[3]

[1] Ontario Ministry of Natural Resources, Sault Ste. Marie, Ontario, Canada
[2] Forest Policy Center, Auburn University, AL, USA
[3] USDA Forest Service, Northern Research Station, PA, USA

ABSTRACT

Understanding family forest owners has been an important focus of study for researchers in the field of forestry. In this chapter we investigated the classification of *multiple-objective*, *timber* and *non-timber* motivated family forest landowner groups in the three southeastern states of Alabama, Georgia, and South Carolina. Using non-parametric discriminatory analysis procedures we found that the bio-physical, socio-economic, and demographic attributes best differentiate the three landowner groups. Results further indicated that 84% of landowners across the three landowner groups were correctly classified based on their bio-physical, socio-economic and demographic attributes. With all variables used to develop the classification scheme known, *a-priori*, before the landowners on a Forest Inventory and Analysis (FIA) plot location were contacted for the National Woodland Owner Survey (NWOS), we suggested the possibility of predicting what attitudinal type of landowner was likely to own forest land at a particular location with known woodlot (FIA) and demographic (U.S. Census Bureau) attributes. Results suggested that forest lands which are closer to population centers with high population densities and counties that have a higher median household income were likely to be more appealing to *non-timber* motivated owners relative to the *timber* or *multiple-objective* motivated owners. Pine stands which have higher commercial value than oak-pine or hardwoods and better land quality parcels were likely to be owned by *timber* owners relative to *non-timber* owners or *multiple-objective* owners.

* Email: Indrajit.majumin@ontario.ca, Ontario Ministry of Natural Resources, Sault Ste. Marie, Canada.

Keywords: Discriminant analysis; Family forest; FIA; Landowner motivation.

1. INTRODUCTION

Family forest owners[1], a term almost used synonymously with the traditional nonindustrial private forest (NIPF) landowner nomenclature, is the dominant ownership group in the United States holding every four out of 10 forest land acres (Butler and Leatherberry 2004). The motivations, intentions, and management behavior of these owners have always been a subject of interest given their diversity. Timber supply has been a major emphasis for studying family forest owners in the past as is evident from the fact that the two most researched areas documented within the NIPF literature are timber harvest (Binkley 1981, Boyd 1984, Dennis 1990, Hyberg and Holthausen 1989, Pattanayak et al. 2003) and reforestation behavior (Alig 1986, Newman and Wear 1993, Kline et al. 2002).[2] Some recent studies have focused on ecosystem and other non-timber amenity values of family forests (Belin et al. 2005, Jacobson 2002, Kline et al. 2000, Rickenbach et al. 1998). It is, however, widely acknowledged that these owners, with similar backgrounds and when facing similar choices often have different objectives and as such make different management decisions (Wear and Greis 2002). One of the emerging fields of interest among family forest researchers is the development of ownership typologies to aid in understanding this diverse group of owners, implement effective policies (Emtage et al. 2007) and extend educational needs to these landowners according to their motivations and goals for their forestland (Kittredge 2004, Majumdar et al. 2008). A number of recently published studies are examples of the development and implications of forest landowner typologies (Kline et al. 2000, Kluender and Walkingstick 2000, Kendra and Hull 2005, Finley et al. 2006; Butler et al. 2007, Majumdar et al. 2008). While the above studies provided valuable insights to the application of multivariate statistical techniques such as *cluster analysis* for grouping landowners, they all required information/data on the survey questions used in the particular typological study. Survey data, while providing valuable detailed information about respondents that are otherwise not available from secondary data sources, are expensive, time consuming, and generally non-replicable. Moreover, survey methodology relies on the subjective evaluation of the researcher to develop survey questions which are general enough to be minimally appropriate for all respondents. The objective of our research was to predict the likelihood of forest land to be owned by a particular motivational group by using demographic (U.S. Census Bureau) and forest characteristics [i.e., Forest Inventory and Analysis (FIA)] data on non-surveyed forest landholders. The use of extant data for developing information on management behavior of family forest owners is not only less expensive and less time consuming but also can provide a more generalized account of the likely behaviors of owners at larger scales, a tenet important for meaningful policy analysis (Wear and Bolstad 1998).

---

[1] 'Family forests include lands that are at least one acre in size, 10% stocked, and owned by individuals, married couples, family estates and trusts, or other groups of individuals who are not incorporated or otherwise associated as a legal entity' (Butler and Leatherberry 2004).

[2] For a review of literature of the studies on NIPF management behavior refer to Beach et al. 2005.

In an earlier work the family forest owners from three southeastern states of Alabama, Georgia, and South Carolina were grouped according to their stated reasons for owning their forest land (Majumdar et al. 2008). Three attitudinal groups were identified and named as *multiple-objective*, *timber,* and *non-timber* oriented owners. The objective of this chapter is three-fold; first to identify characteristics that discriminate/differentiate best the three above mentioned owner types using discriminant analysis procedures; second to develop a classification scheme that will help in predicting the motivational type of landowner likely to own forest land for a given vector of socio-demographic and resource characteristics; and third to use logistic regressions to explore the relationship between the explanatory variables (discriminatory variables) to the objectives of the landowner (dependent variable). Results from this chapter will be helpful in making a connection between policy makers and family forest owners and for developing effective policy prescriptions and educational programs targeting the stewardship goals of the forest owner. Our work also has important implications for estimating the availability of family forest land for timber production. This information will be beneficial to policy makers interested in sustaining family forests, so that they continue to provide the stream of timber and non-timber goods in the future.

2. Data

The diverse nature of family owned forests in terms of their economic and social surroundings indicated that socio-demographic factors, forest characteristics, and spatial location will be likely correlates of landowner motivations. As such, information on the socio-demographic characteristics, economic surroundings, and bio-physical characteristics of the forests of each landowner belonging to one of the three family forest ownership types (determined previously by Majumdar et al. 2008) were used to differentiate them. The bio-physical data came from the FIA program of the United States Department of Agriculture Forest Service (USDA FS). FIA forest resources inventory collects forest resources data annually from a sample of standard plots each representing roughly 6,000 acres in the eastern U.S. The social counterpart of the FIA forest resource inventory is the National Woodland Owner Survey (NWOS) which is conducted on a sub-sample of NIPF landowners of FIA plots already inventoried. The socio-demographic and economic data were incorporated by linking U.S. Census Bureau data with the FIA plot location.

Biological and physical characteristics of forest properties delineate to some extent what the forest plot's efficient use would be and choice of a particular plot with its associated land characteristics may reflect upon the preferences and motives of the landowner. For example, high soil quality and flatter terrain forest plots are likely to be owned by timber production oriented owners due to their high productivity and relatively (to a steep terrain plot) lower cost of performing harvest operations. As such, these plots are also less likely to be owned by those who value forest-based amenities more than the timber value of the forest land. Social and demographic factors are also likely to be important indicators of landowner preferences. For example, timber oriented owners are likely to own forest land in rural areas when compared to owners in cities or near urban centers who primarily are interested in non-timber amenity values. The opportunity cost for forest landowners to own and manage forest land for timber production in urban and exurban settings, where there is a greater demand for aesthetic and forest-based amenities, would be high. Also, timber harvest operations would be less popular in these settings (Xu et al. 2003). Empirical studies on timber harvest behavior by NIPF landowners have shown that timber harvest activity and increasing urbanization were negatively correlated (Barlow et al. 1998; Munn et al. 2002).

Bio-physical variables used to characterize NIPF landowner motivations included percent slope of the forest plot (SLOPE), dummy variables for forest type as pine (PINE), site quality (SITE) and average stand age (AGE). Socio-demographic and economic variables incorporated through linkages between the FIA forest plot location and the U.S. Census Bureau data included a population gravity index (PGI)[3] and median household county income (INC). Sources and descriptions of all variables used were presented in table 1.

Table 1. Mean and standard deviations of the explanatory variables along with their description and source

Variable	Description	Source	Mean	Std. dev
PGI	Number of persons/km2 around each FIA plot within a 100km radius	FIA plot and U.S. Census Bureau	522.34	1432.23
INC	Median household income by county in 2005 dollars	Economic Research Service (ERS) unit of USDA	32,852.98	7164.54
SLOPE	Angle of slope in percent	FIA Cond	6.69	9.26
PINE	Forest type dummy with value of 1 for pine and 0 for all other species	FIA Cond	2.01	0.92
AGE	Average stand age	FIA Cond	31.87	23.86
SITE	Site productivity class code taking values from 1 to 6 with 6 representing the best site	FIA Cond	4.41	0.90

Note: FIA Cond represents the multiple conditions and is defined by heterogeneity in reserved status, owner group, forest type, stand-size class, regeneration status and stand density within FIA plots (for details on FIA data description and collection methods see Alerich et al. 2004).

3. METHODS

Discriminant Analysis (DA) is a statistical technique that allows researchers to study differences between two or more groups of objects with respect to several variables simultaneously (Klecka 1988, Johnson and Wichern 2002). DA has been recently used in a number of studies related to NIPF landowner behavior. For example, in their study of 829 Mississippi forest landowners Arano et al. (2004) used canonical discriminant function analysis to discriminate landowner groups based on ownership characteristics, attitudes

[3] PGI, a measure of the influence of population on landowner preferences was calculated as:

$$PGI_i = \sum_k \frac{P_k}{D_{ki}^2} \forall k : D_{ki} \leq 100 \text{ miles},$$

where PGI_i is the population gravity index for forest parcel *i*, P_k is the population of the Census populated place *k* within 100 miles from each forest plot and D_{ki} is the distance between the U.S Census Bureau populated place *k* and forest plot *i*. U.S Census Bureau populated place data was obtained from ESRI Data and Maps (ESRI 2005).

towards timber investment, awareness of assistance programs, and attendance in educational programs. Finley et al. (2006) used multiple discriminant analysis to associate predefined groups of Massachusetts NIPF landowners, namely, general cooperators, conservation cooperators, neutralists and non-cooperators, with forest land ownership reasons, attitudes and actions towards forest management and perceived barriers to cooperation. Greene and Blatner (1986) successfully classified nearly 80% of respondents in a study of Arkansas private forest landowner groups, timber managers, and nonmanagers, based on a set of 47 variables related to reasons for owning forest land, owner characteristics, forest characteristics, and their attitudes towards forest management. Though the studies mentioned above provide insights about factors or variables that best differentiate predefined groups of NIPF landowners, they do not report on the accuracy of classification results.

Our research suggested that DA can be used to predict group membership of owners with a high degree of accuracy. However, since the entire set of variables in the classification models comes from survey instruments administered to respondents, it was not possible to predict group memberships for landowners (and their forests) who have not been administered the survey. This revealed the importance of exogenous discrimination of landowner groups (i.e., where all discrimination variables were exogenous and not dependent on survey responses), an issue that we explored in this chapter. In summary, our focus was to be able to use easily available secondary information on landowner decision-making environments from sources other than the primary NWOS[4] to predict their membership in one of the predefined landowner motivational groups established using NWOS previously. Majumdar et al. (2008) found that NWOS data can be used to classify landowners as timber, multiple-objective, and non-timber forest managers. Our aim was to investigate ways to classify landowners into those groups using secondary data sources exogenous to the NWOS survey.

Two assumptions for conducting parametric discriminant analysis were multivariate normality and homogeneity of variances. Surprisingly, researchers frequently have ignored the statistical limitations and assumptions required to be fulfilled to perform parametric techniques. Under such circumstances it was difficult to assess whether the results were accurate or not and statements made about the findings were thus unreliable. For example, employing parametric discriminant analysis when the above mentioned assumptions were not satisfied could lead to an overestimation of degree of association between predetermined classes (Hosmer et al. 1983, Hosmer and Lameshow 1989). Our test results (table 2) from analysis of the NWOS data indicated a rejection of both assumptions. Alternatives to parametric discriminant analysis are non-parametric discriminant analysis and logistic regression analysis. Since one of the primary goals of this paper was prediction of landowner membership into one of the predetermined groups based on a vector of predictor variables, and the relevant assumptions for linear discriminant analysis could not be met, we adopted the non-parametric analytical technique.

[4] For details on the NWOS survey design, implementation, and analysis methods see Butler et al. (2005).

Table 2. Multivariate normality and homogeneity of variance-covariance test results for the data

Test	Chi-square statistic	p-value
Mardia's Skewness[a]	197.6	<0.0001
Mardia's Kurtosis[a]	328.5	<0.0001
Levene Homogeneity[b]	2338.8	<0.0001

[a] H_0: Multivariate normality.
[b] H_0: Homoscedasticity.

K-nearest neighbor classification (KNN), also known as nearest neighbor discriminant analysis, introduced by Fix and Hodges (1951), was used for our analysis. This method is based on distances from 'immediate neighbors' eliminating the need for a probability density estimation based on a distribution assumption. It was used to predict the response of an observation using a non-parametric estimate of the response distribution of its 'K' nearest (i.e., in predictor space) neighbors. Consequently, this technique was relatively flexible and, unlike traditional classifiers such as discriminant analysis and generalized logit models, does not require an assumption of multivariate normality or a strong assumption implicit in specifying a link function (e.g., the logit link which assumes the distribution of the dependent variable to be within the exponential family of distributions, such as normal, poisson, binomial, gamma). KNN was based on the simple logic that members that share characteristics were likely to be representatives of a common group. This method assigns a new observation to a group to which the majority of the k-nearest neighbors of known predetermined groups belong. For illustration purposes, consider a family forest owner with known vector of attributes $\mathbf{x}$ but unknown group membership. Let $\mathbf{x}_1$, $\mathbf{x}_2$, $\mathbf{x}_3$..... $\mathbf{x}_n$ be the attribute vectors of 'n' family forest owners with predetermined membership to a particular group '*i*' (where '*i*' is the number of predetermined groups, in this case 3 for the *multiple-objective*, *timber,* and *non-timber* owners). Given that the variance-covariance matrix for the vector of attributes is $\mathbf{v}_i^{-1}$ the generalized (Mahalanobis) squared distance ($d_i^{\,2}$) between two observations $\mathbf{x}$ and $\mathbf{x}_n$ is $d_i^2(\mathbf{x},\mathbf{x}_n) = (\mathbf{x}-\mathbf{x}_n)'\mathbf{v}_i^{-1}(\mathbf{x}-\mathbf{x}_n)$

Based on the squared distance defined above and specified parameter '*k*', a positive integer that denoted the number of nearest neighbors to be considered, '*k*' observations that were closest to $\mathbf{x}$ (least squared distance) were identified and $\mathbf{x}$ was assigned to the group that the majority of the '*k*' nearest neighbors belonged to. KNN classification performance was evaluated using the apparent error rate. Percentage of correct classification within each group (cluster) and, for the whole population, based on predictions of KNN classification and is reported in table 3. Classification accuracy peaked within each ownership group and for all NIPF landowners taken together when two nearest neighbors (K = 2) were considered.

Table 3. Classification results for the apparent-error-rate of the K-Nearest Neighbor method

Percentage Correct				
k	Multiple-objective	Non-timber	Timber	Total
2	87.4	77.5	82.9	84.1
3	70.5	58.6	59.8	65.0
4	65.8	43.7	52.5	57.5
5	76.1	43.1	58.1	64.2
6	75.3	39.1	50.9	60.9
7	74.9	42.1	53.1	62.0
8	71.4	36.1	49.5	58.0
9	73.2	34.4	48.6	58.3

Results

This chapter examined whether it was possible to predict group membership of family NIPF owners in the Southeast U.S. using variables which were not collected during a survey of landowners. In other words, was it possible to assign landowners to the *multiple-objective*, *non-timber,* or *timber* oriented group *ex-ante* (i.e. without acquiring primary data from the landowners). Table 1 provides summary statistics for the variables selected as discriminators of the three landowner groups using step-wise procedures. The socio-demographic (i.e., PGI, INC) and bio-physical characteristics (i.e., SLOPE, SITE, AGE, PINE) describing each group of landowners was used to classify a previously unclassified landowner (unknown owners with known FIA plot information) into one of the three groups.

Three binary logistic models were used to investigate the relationship of the discriminatory variables and landowner motivations. Likelihood ratio (LR) tests confirmed that all explanatory variables in the three models were jointly significant in explaining the heterogeneity of landowner motivations as indicated by their membership in the *non-timber*, *multiple-objective,* or *timber* motivated owner group. The LR test statistic for the logistic models explained the difference in landowner motivations: Timber vs. Non-timber (table 4, LR = 116.4), Non-timber vs. Multiple-objective (table 5, LR = 106.9) and Timber vs. Multiple-objective (table 6, LR = 23.9) indicated that *timber* motivated owners were most separated from *non-timber* motivated owners while the *timber* and *multiple-objective* motivated owners were the most overlapping.

Table 4. Logistic regression analysis results for *timber* relative to *non-timber* motivated owners

	Estimate	Standard error	Odds ratio
Intercept	0.0735	0.5573	
PGI	-0.0011	0.0002*	0.998
INC	-0.00001	0.00001	1.000
SLOPE	-0.0282	0.0077*	0.972
SITE	0.2822	0.0878*	1.326
AGE	0.0023	0.0036	1.002

Table 4. (Continued).

	Estimate	Standard error	Odds ratio
PINE	0.7271	0.1788*	2.069
Log-likelihood function	909.988		
Restricted log-likelihood function	1026.377		
Chi-square	116.389		
Number of observations	765.0		
Percent correct	66.3		

* Significant at greater than 5% accuracy level.

Table 5. Logistic regression analysis results for *non-timber* relative to *multiple-objective* motivated owners

	Estimate	Standard error	Odds ratio
Intercept	-0.9214	0.4468	
PGI	0.00019	0.00006*	1.000
INC	0.00002	0.0000095*	1.000
SLOPE	0.0467	0.00794*	1.048
SITE	-0.1729	0.0706*	0.841
AGE	-0.00826	0.00329*	0.992
PINE	-0.8268	0.1630*	0.437
Log-likelihood function	1175.052		
Restricted log-likelihood function	1281.988		
Chi-square	106.936		
Number of observations	1083.0		
Percent correct	74.1		

* Significant at greater than 5% accuracy level.

Table 6. Logistic regression analysis results for *timber* relative to *multiple-objective* motivated owners

	Estimate	Standard error	Odds ratio
Intercept	-0.1306	0.4255	
PGI	-0.00028	0.00014*	1.000
INC	-0.00000008	0.000009	1.000
SLOPE	0.0152	0.00751*	1.015
SITE	0.1875	0.0659*	1.206
AGE	-0.00621	0.00282*	0.994
PINE	-0.0352	0.1283	0.965
Log-likelihood function	1618.474		
Restricted log-likelihood function	1642.351		
Chi-square	23.8775		
Number of observations	1244.0		
Percent correct	63.3		

* Significant at greater than 5% accuracy level.

Timber Relative to Non-Timber Owners

Bio-physical characteristics such as SLOPE, SITE, and PINE and socio-demographic PGI variables were significant in differentiating *timber* from *non-timber* oriented owners (table 4). Consistent with our *a-priori* expectation, the negative coefficient on PGI indicates that forest plots close to population centers were less likely to be owned by *timber* oriented landowners relative to *non-timber* motivated owners. Results suggested that forest lands which were closer to population centers, with high population densities and counties that have higher median household income were likely to be more appealing to the *non-timber* motivated owners relative to the *timber* or *multiple-objective* objective owners. Further, we found that pine stands which have higher commercial values than oak-pine or hardwoods and better land quality parcels were likely to be owned by the *timber* relative to *non-timber* motivated owners. Consistent with our expectation, steep terrain forest plots with high slope were less likely to be owned by *timber* motivated owners relative to *non-timber* owners as indicated by negative coefficient on the SLOPE variable.

Non-Timber Relative to Multiple-Objective Owners

All model variables were statistically significant in terms of distinguishing the *non-timber* oriented owners from the *multiple-objective* owners (table 5). The positive coefficient on PGI and INC indicated that forest plots with a high population influence (i.e., close to cities and urban centers) and richer counties with high median household income were likely to belong to *non-timber* objective owners relative to *multiple-objective* owners. The positive SLOPE coefficient, as expected, indicated that steep terrain forest plots were more likely to be owned by *non-timber* relative to *multiple-objective* owners, a result which agreed with our prior expectation. The negative coefficients on AGE and PINE suggested that mature stands and pine stands with high commercial values were less attractive to the *non-timber* relative to the *multiple-objective* motivated owners.

Timber Relative to Multiple-Objective Owners

In table 6 we presented the estimated parameters, standard errors, and the odds ratios for the model comparing landowners that were *timber* oriented relative to *multiple-objective* owners given the socio-demographic and bio-physical characteristics of the forest plot. The negative coefficient on PGI confirmed earlier comparisons of landowner preferences along the timber to non-timber continuum. NIPF owners motivated by profit such as *timber* relative to *multiple-objective* (owners who were only partially motivated by timber production) were less likely to own forest land with an increase in PGI. The positive coefficient on SITE indicated that better quality lands were more likely to belong to *timber* motivated owners relative to *multiple-objective* owners. The negative coefficient on AGE variable indicated that mature stands were less likely to be managed by *timber* relative to *multiple-objective* owners. This could be due to *multiple-objective* owners who prefer a mix of both mature stands (which have high aesthetic value) and young stands (managed for timber production) relative to *timber* only motivated owners who could be managing pine plantations with lower average stand age. The positive coefficient on SLOPE indicates that steeper plots were more likely to be owned by a *timber* relative to *multiple-objective* NIPF owner, a result that was contrary to our prior expectation.

CONCLUSION

Given their increasing numbers and forest land acres they managed, family forest owners have become increasingly important stewards of productive timberlands in the Southern U.S. Thus, it was important to critically examine their motivations for managing their land either for timber and/or non-timber forest products. In this chapter, we have shown that landowners can be accurately classified into heterogeneous attitudinal groups (*multiple-objective*, *non-timber,* and *timber*) using predetermined demographic data from the U.S. Census Bureau and woodlot data from FIA variables using the KNN technique. Overall accuracy rate of classification of landowner groups was 84.1% (table 3). We found that bio-physical (i.e., SLOPE, SITE, PINE, AGE) and socio-demographic (i.e., PGI, INC) variables had a strong association with landowner group profiles. Given the increasing number of family forest owners and increasing proportion of timberland they own and manage as an ownership class, results from this research can effectively help in estimating the different adjustment factors for diversely motivated landowner groups to more accurately project future timber supply.

Family forest owners who owned forest land in close proximity to population centers were likely to be motivated by the non-timber amenity values, a result that was consistent with several past studies (Tremblay and Dunlap 1978, Brunson et al. 1997, Ribe and Matteson 2002, Xu et al. 2003). Rural areas with less population pressure on the other hand were more appealing to family forest owners who were motivated mainly by production of timber for profit. This chapter served as an example of the vast body of recent research focusing on understanding family forest owners, their future intentions, and management behavior and motivations and emphasizes the use of existing data in predicting goals and management preferences of family forest owners given their demographic and bio-physical characteristics.

REFERENCES

Alerich, C. L., Klevgard, L., Liff, C., & Miles, P. D. (2004). The Forest Inventory and Analysis Database: Database Description and Users Guide Version 1.7.

Alig, R. (1986). Econometric analysis of the factors influencing forest acreage trends in the Southeast. *Forest Science, 32(1)*, 119-134.

Arano, K. G., Munn, I. A., Gunter, J. E., Bullard, S. H., & Doolittle, M. L. (2004). Comparison between regenerators and non-regenerators in Mississippi: A discriminant analysis. *Southern Journal of Applied Forestry, 28(4)*, 189-195.

Barlow, S. A., Munn, I. A., Cleaves, D. A., & Evans, D. L. (1998). The effect of urban sprawl on timber harvesting: A look at two southern states. *Journal of Forestry, 96(12),* 10-14.

Beach, R. H., Pattanayak, S. K., Yang, J. C., Murray, B. C., & Abt, R. C. (2005). Econometric studies of non-industrial private forest management: A review and synthesis. *Forest Policy and Economics, 7*, 261-281.

Belin, D. L., Kittredge, D. B., Stevens, T. H., Dennis, D. C., Schweik, C. M., & Morzuch, B. J. (2005). Assessing private forest owner attitudes toward ecosystem-based management. *Journal of Forestry, 103(1)*, 28-35.

Binkley, C. S. (1981). Timber supply from non-industrial forests: A microeconomic analysis of landowner behavior. Yale University Press, New Haven, CT.

Boyd, R. (1984). Government support of non-industrial production: The case of private forests. *Southern Economic Journal*, *51(1)*, 89-107.

Brunson, M. W., Shindler, B., & Steel, B. S. (1997). Consensus and dissention among rural and urban publics concerning forest management in the Pacific Northwest. In Public lands management in the West: Citizens, interest groups, and values, ed. B. Steel, 83-94. Westport, CT: Praeger.

Butler, B. J., & Leatherberry, E. C. (2004). America's family forest owners. *Journal of Forestry. 102(7)*, 4-9.

Butler, B. J., Leatherberry, E. C., & Williams, M. S. (2005). *Design, implementation, and analysis methods for the national woodland owner survey*. USDA Forest Service General Technical Report NE-336, Newtown Square, PA. 43pp.

Butler, B. J., Tyrrell, M., Fienberg, G., VanManen, S., Wiseman, L., & Wallinger, S. (2007). Understanding and reaching family forest owners: lessons from social marketing research. *Journal of Forestry, 105(7)*, 348–357.

Dennis, D. (1990). A probit analysis of the harvest decision using pooled time-series and cross-sectional data. *Journal of Environmental Economics and Management, 18*, 176-187.

Emtage, N., Herbohn, J., & Harrison, J. (2007). Landholder profiling and typologies for natural resource–management policy and program support: potential and constraints. *Environmental Management, 40(3)*, 481-492.

ESRI Data & Maps [CD-ROM]. (2005). Redlands, CA: Environmental Systems Research Institute.

Finley, A. O., Kittredge, D. B., Stevens, T. H., Schweik, C. M., & Dennis, D. C. (2006). Interest in cross-boundary cooperation: Identification of distinct types of private forest owners. *Forest Science, 52(1)*, 10-22.

Fix, E., & Hodges, J. (1951). Discriminatory analysis, nonparametric discrimination: consistency properties. Technical Report, Randolph Field, Texas: USAF School of Aviation Medicine.

Greene, J. L., & Blatner, K. A. (1986). Identifying woodland owner characteristics associated with timber management. *Forest Science, 32(1)*, 135-146.

Hosmer, D. W., & Lameshow, S. (1989). *Applied logistic regression*. Wiley-Interscience, New York. 307pp.

Hosmer, T., Hosmer, D. W., & Fisher L. L. (1983). A comparison of maximum likelihood and discriminant function estimators of the coefficients of logistic regression model for mixed continuous and discrete variables. *Communications in Statistics, B12,* 577-593.

Hyberg, B., & Holthausen, D. (1989). The behavior of non-industrial private forest owners. *Canadian Journal of Forest Research, 19*, 1014-1023.

Jacobson, M. G. (2002). Factors affecting private forest landowner interest in ecosystem management: linking spatial and survey data. *Environmental Management, 30(4)*, 577-583.

Johnson, R. A., & Wichern, D. W. (2002). *Applied multivariate statistical data analysis*. Prentice Hall, 5th edition, Upper Saddle River, NJ. 767pp.

Kendra, A., & Hull, R. B. (2005). Motivations and behaviors of new forest owners in Virginia. *Forest Science, 51(2)*, 142-154.

Kittredge, D. B. (2004). Extension/outreach implications for America's family forest owners. *Journal of Forestry, 102(7)*, 15-18.

Klecka, W. R. (1988). *Discriminant Analysis*. Sage University Paper series on Quantitative Applications in the Social Sciences, series no. 07-019. Beverly Hills and London: Sage Publications.

Kline, J. D., Alig, R. J., & Johnson, R. L. (2000). Fostering the production of nontimber services amongst forest owners with heterogeneous objectives. *Forest Science, 46(2)*, 302-311.

Kline, J. D., Butler, B. J., & Alig, R. J. (2002). Tree planting in the south: what does the future hold? *Southern Journal of Applied Forestry, 26(2)*, 99-107.

Kluender, R. A., & Walkingstick, T. L. (2000). Rethinking how nonindustrial landowners view their lands. *Southern Journal of Applied Forestry, 24(3)*, 150-158.

Majumdar, I., Teeter, L. D., & Butler, B. J. (2008). Characterizing Family Forest Owners: A Cluster Analysis Approach. *Forest Science, 54(2)*, 176-184.

Munn, I. A., Barlow, S. A., Evans, D. L., & Cleaves, D. (2002). Urbanization's impact on timber harvesting in the south central United States. *Journal of Environmental Management. 64(1)*, 65-76.

Newman, D., & Wear, D. (1993). Production economics of private forestry: a comparison of industrial and non-industrial forest owners. *American Journal of Agricultural Economics, 75*, 674-684.

Pattanayak, S., Abt, K., & Holmes, T. (2003). Timber and amenities on non-industrial private lands. In: Sills, E. and K. Abt, (eds.), Forests in a market economy. Forestry sciences series. Kluwer academic publishers, Dordrecht.

Ribe, R., & Matteson, M. (2002). Views of old forestry and new among reference groups in the Pacific Northwest. *Western Journal of Applied Forestry, 17(4)*, 173-182.

Rickenbach, M. G., Kittredge, D. B., Dennis, D., & Stevens, T. (1998). Ecosystem management: capturing the concept for woodland owners. *Journal of Forestry, 96(4)*, 18-24.

Tremblay, K. R., & Dunlap, R. (1978). Rural–urban residence and concern with environmental quality. *Rural Sociology, 43*, 474–491.

Wear, D. N., & Bolstad, P. (1998). Land-use changes in southern Appalachian landscapes: spatial analysis and forecast evaluation. *Ecosystems, 1*, 575-594.

Wear, D. N., & Greis, J. G. (2002). *Southern Forest Resource Assessment*: Technical Report. GTR-SRS-053. USDA Forest Service Southern Research Station. 635pp.

Xu, W., Lippke, B. R., & Perez-Garcia, J. (2003). Valuing biodiversity, aesthetics, and job losses associated with ecosystem management using stated preferences. *Forest Science. 49(2)*, 247-257.

In: Global Change and Forestry
Editors: J. Gan, S. Grado and I. A. Munn

ISBN: 978-1-60876-262-0

Chapter 17

KNOWLEDGE OF THREE REGENERATION PROGRAMS AND APPLICATION BEHAVIOR AMONG MISSISSIPPI NONINDUSTRIAL PRIVATE FOREST LANDOWNERS: A TWO-STEP SAMPLE SELECTION APPROACH[1]

Xing Sun[1], Changyou Sun[*2], Ian A. Munn[2] and Anwar Hussain[2]

[1] School of Forestry and Wildlife Sciences,
Auburn University, Auburn, AL 36849, U.S.A.
[2] Forest and Wildlife Research Center, Department of Forestry,
Mississippi State University, Mississippi State University,
363 Thompson Hall, Mississippi, MS 39762, U.S.A.

ABSTRACT

Various public financial assistance programs are available to nonindustrial private forest (NIPF) landowners in the United States. Many landowners have limited knowledge of these programs and have not utilized them. This chapter employed a two-step sample selection model to examine the determinants of NIPF landowner knowledge of three regeneration programs, and conditional on their knowledge, the determinants of the application frequency to these programs since 1996. Data were collected using a phone survey of NIPF landowners in Mississippi. The survey revealed that among these qualified landowners who had clearcut in the past 10 years, about 50% knew of one or more of the programs. Their application frequencies to individual programs ranged from zero to five. Landowner knowledge of the programs was positively related to land size, regeneration experience, gender, and membership in forestry associations. For landowners who knew of these programs, application frequencies were higher for those

---

[1] This chapter is reprinted with permission from the article originally published in the Journal of Forest Economics (in press).

[*] Corresponding author. Tel.: +1 662 325 7271. E-mail address: csun@cfr.msstate.edu (C. Sun).

that had less acreage, had lower education or income, were fully employed, were female, or had no membership in forestry organizations.

1. INTRODUCTION

Nonindustrial private forest (NIPF) landowners have been a major participant in forestry in the United States. In 2002, NIPF landowners owned 58% of total timberlands, forest industry 13%, and the public (e.g., U.S. Forest Service) 29%; they accounted for 63%, 29%, and 8% of total timber harvested, respectively (Smith, Miles, Vissage & Pugh 2004). Furthermore, NIPF landowners not only produce timber as raw material for the forest products industry, but also provide numerous environmental amenities, including soil conservation, carbon storage, wildlife habitats, maintenance of air and water quality, and recreational opportunities (Alig 2003; Wear & Greis 2002). Therefore, for a long time, a variety of public assistance programs have been justified and created for NIPF landowners in the United States. These programs have been designed with the intent of helping landowners to achieve their management goals and also meet societal needs.

Forestland management can be capital intensive and a long period of growth is usually needed before income can be generated. Consequently, when forest landowners harvest their land, they face a critical decision of whether to reforest the land, to let the land regenerate naturally, or to use the land for agriculture, commercial development, or other alternative purposes (Gunter, Bullard, Doolittle & Arano 2001a). Public assistance programs can influence the management of NIPF lands, compensate landowners for high costs of tree planting, and encourage better forest stewardship (Wear & Greis 2002). Many public regeneration programs have been specifically designed to reduce the financial burden from regeneration on NIPF landowners and encourage them to replant their lands after harvest. For example, in 1973, the Forestry Incentives Program (FIP) was authorized by the U.S. Congress to share the cost of tree planting and timber stand improvement with private landowners. The share by FIP has ranged up to 65% of these costs for regeneration or improvement (Gaddis,, New, Cubbage, Abt & Moulton 1995).

Many studies have been conducted to analyze the silvicultural decisions by NIPF landowners and their participation in public assistance programs (Amacher, Conway & Sullivan 2003). These studies also examined the economic efficiency of these programs, practice retention rate, capital substitution, equity, and subsidiary effects (Gaddis et al. 1995). Previous studies generally agreed that these programs have successfully influenced the management of NIPF lands and stimulated more planting activities (Boyd 1984; Mehmood & Zhang 2001; Nagubadi, McNamara, Hoover & Mills 1996; Sun 2007). However, in spite of the benefits, these studies also revealed that NIPF landowners have not always utilized these programs. For example, Gunter, Bullard, Doolittle and Arano (2001b) found that among the 829 landowners that responded to a survey in Mississippi, more than 60% of them did not know of these public financial assistance programs covered. The majority (54.3%) of the landowners who regenerated their timber stands after harvesting from 1994 to 1998 did not receive any financial assistance from these programs (Gunter et al., 2001). At present, many questions related to public assistance programs still have not been fully addressed in the literature. For example, what determines landowner knowledge of a public assistance program, and furthermore, what is the impact of availability of program information on the application behavior of landowners?

The objective of this chapter was to examine NIPF landowner knowledge of three regeneration programs in Mississippi and their application behavior after harvesting their lands from 1996 to 2006. Mississippi is a typical southern state in the United States where timber and the related forest products industry have been important. In 2002, NIPF landowners owned 72% of forest lands in Mississippi and produced 67% of the state's timber outputs (Smith et al. 2004). Three programs were included in this chapter: Mississippi Forest Resource Development Program (FRDP), Forestry Incentives Program (FIP), and Mississippi Reforestation Tax Credit (RTC). They have been the major programs that provide public financial assistance to NIPF landowners in Mississippi for regeneration activities over the past several decades. An innovative two-step sample selection model was developed to examine what factors were associated with landowner knowledge of these programs, and conditional on landowner knowledge, what factors affected their application frequency. The results have important policy implications for designing and promoting public financial assistance programs for NIPF landowners.

2. Major Regeneration Programs and the Response of Landowners

2.1. Major Public Financial Programs for Regeneration

A number of public programs have provided financial assistance to NIPF landowners to share reforestation costs (Mehmood & Zhang 2001; Wear & Greis 2002). Of these programs, FRDP, FIP, and RTC are the major ones available to NIPF landowners in Mississippi for reforestation activities. FRDP is a Mississippi cost-share program, FIP is a federal cost-share program, and RTC is a Mississippi tax incentive program. Their history, eligibility, application and approval criteria, and financial assistance arrangements are briefly described here.

FRDP was established in Mississippi as a state cost-share program for reforestation and timber stand improvement in 1974 (Wear & Greis 2002). FRDP provides financial assistance to eligible landowners for establishing and improving forest lands. This program offsets landowner expenses by sharing the cost of implementing specific forest practices to produce timber and enhance wildlife development. FRDP requires that applicants submit a management prescription for the desired treatment area, comply with Mississippi Forestry Commission standards during operations, and maintain practices for 10 years. Cost-share payments of FRDP cover 50-75% of the total cost of implementing forest practices, with a maximum annual assistance of $7000. The funding for this program has been generated through a timber severance tax in Mississippi. Approximately, 2-4 million dollars each year have been distributed to forest landowners in the past 10 years. From 1974 to 2005, FRDP had a total expenditure of $72 million (Mississippi Forestry Commission 2007).

FIP is a major federal program related to regeneration. Authorized in 1973, the main purpose of FIP has been to increase timber production and encourage good forest management on NIPF lands by sharing the cost of tree planting, timber stand improvement, and site preparation (Wear & Greis 2002). To participate in FIP, eligible lands owned by NIPF landowners can range from 10 to 1000 acres, and with special authorization up to 5000 acres. Assistance provided to NIPF landowners can be up to 65% of actual costs, with a maximum annual cost-share payment of $10,000. In the 2002 Farm Bill, the FIP program was replaced by the Forest Land Enhancement Program; however, some FIP contracts were funded through 2004. From 1974 to 1994, FIP provided more than $200 million to fund 3.32 million acres of tree planting, 1.45 million acres of timber stand improvement, and 0.27 million acres of site preparation for natural regeneration in the United States (Gaddis et al. 1995). The southern states accounted for 90% of the program's allocation. Annual appropriations for FIP have ranged from $10 to $15 million in the last decade (Sun 2007).

RTC was initiated in 1999 to promote reforestation on nonindustrial private lands in Mississippi. The credit is applied to Mississippi state income taxes. RTC is favorable to NIPF landowners because cost-share payments for reforestation and some other practices can be excluded from their taxable income (Wear & Greis 2002). RTC has provided Mississippi NIPF landowners tax credit up to 50% of the cost of qualified reforestation practices. Before 2007, both the annual and lifetime credit limits were $10,000 and landowners were allowed to carry forward any unused credit. In 2007, the Mississippi state legislature made the program even more attractive by raising the lifetime credit limit to $75,000 (Mississippi Forestry Commission 2007). Compared with cost-share programs like FRDP and FIP, RTC as a tax incentive program has no budget constraint and does not need approval by administrative agencies.

2.2. Response of NIPF Landowners to Public Assistance Programs

As reviewed by Amacher et al. (2003), numerous studies have examined the participation behavior of NIPF landowners in public assistance programs in the United States. Typically, these studies have relied on binary choice models (e.g., Bell, Roberts, English & Parks 1994; Nagubadi et al. 1996). The dependent variable was a binary dummy indicating participation behavior. Independent variables usually included landowner characteristics (e.g., income, education) and land features (e.g., acreage). Landowner participation in public assistance programs has been found to be positively associated with total acres owned, membership in forestry organizations, interest in timber production, income, and residence on the forest lands (Konyar & Osborn 1990; Nagubadi et al. 1996; Straka, Wisdom & Moak 1984).

Unfortunately, this type of binary model may be inadequate in analyzing landowner participation in public programs. As revealed in studies like Gunter et al. (2001b), many NIPF landowners have been unaware of these existing public programs. A binary choice model is derived from an individual's utility maximization by comparing two choices: participation or no participation. If an individual is not aware of a public assistance program and has not made the assessment, the dependent variable is actually a missing value, instead of zero. In other words, zero values for the dependent variable in previous studies might come from two sources: individuals who knew of the program and decided not to participate, and individuals who were not aware of the program and did not consider the participation choice at all.

The problem with previous studies has originated from the oversimplified assumption in the binary choice model with regard to the behavior of NIPF landowners. A more suitable approach would be employing a two-step decision model to examine the response of NIPF landowners to public assistance programs. As demonstrated below, the innovation of the two-step decision process is to recognize the reality in forestry that many NIPF landowners are not aware of these public programs. Similar sample selection models (Greene 2003) have been widely applied in the literature to other comparable issues (e.g., Katchova & Miranda 2004; Lee, Lee & Eastwood 2003; Starbuck, Alexander, Berrens & Bohara 2004).

3. METHODOLOGY

Two-step sample selection models have been widely used in the literature to analyze cost-share programs, hunting lease markets, and other forestry-related issues (Ovaskainen, Hannine, Mikkola & Lehtonen 2006; Starbuck et al. 2004; Zhang, Hussain,& Armstrong 2006). In this chapter, a two-step sample selection model is employed to examine the determinants of landowner knowledge of individual public programs and their application frequencies. It is assumed that their applications to a public program are contingent upon whether these landowners know of the program. In the selection stage, landowner knowledge of a specific program is modeled as a function of variables, wi, that are composed of land features, management experiences, and landowner characteristics. In the outcome stage, landowner application frequency in the program over the study period is specified as a function of similar explanatory variables, x_i. Conceptually, the model is expressed as follows:

Selection equation: $z_i = g(w_i)$ (1a)

Outcome equation: $y_i = f(x_i)$ (1b)

where z_i is a binary variable that indicates whether landowner i knows of an individual program (i.e., FRDP, FIP, RTC); z_i equals to one if the landowner is aware of the program, and zero otherwise. y_i is the frequency that the landowner applies to the program during the study period (i.e., 1996-2006). Some landowners may have harvested timber multiple times over the study period and they may have applied to the program several times. y_i is observed only when $z_i = 1$. The variables of knowledge (z_i) and application (y_i) are related but may be influenced by different explanatory variables, or by a same set of factors to a different degree. Therefore, w_i may be different from x_i.

The nature of the dependent variables, z_i and y_i, requires a combination of binary probit model and count data model. The logic of sample selection model is that an outcome variable is only observed if some criterion, defined with respect to a selection variable, is met (Greene 2003). In estimating the model, a correction factor for sample selection is computed from the binary probit model. It is then used in the count data regression to analyze application frequency. Mathematically, the two-step sample selection model can be expressed as follows:

$$z_i^* = w_i\gamma + e_i \quad (2a)$$

$z_i = 1$ if $z_i^* > 0$; 0 otherwise

$$\Pr(z_i = 1) = \Phi(w_i\gamma)$$

$$\Pr(z_i = 0) = 1 - \Phi(w_i\gamma)$$

$$y_i = x_i\beta + \varepsilon_i \qquad (2b)$$

$$\Pr(y_i \mid x) = \frac{e^{-\lambda_i}\lambda_i^{y_i}}{y_i!}$$

$$\lambda_i = e^{x\beta}$$

where γ and β are parameters to be estimated, Φ is the normal cumulative distribution function, e_i and ε_i are error terms, and λ_i is the Poisson distribution parameter. In the selection equation, z_i is a realization of an unobserved continuous variable (z_i^*) with a normally distributed, independent error, e_i. In the outcome equation, y_i is observed only when $z_i = 1$. The two error terms are correlated such that (e_i, ε_i) ~ NID (0, 0, σ_e^2, σ_ε^2, ρ).

The binary probit model, used for the selection equation, is a standard regression method (Greene 2003). For the outcome equation, modeled in the second stage, the choice among various count data models is worthy of some elaboration here. The Poisson regression model is usually central to the development of various models for count data analysis. As a unique feature of the Poisson distribution, its conditional mean is equal to its conditional variance (also known as equidispersion), i.e., $E(y_i \mid x) = Var(y_i \mid x) = e^{x\beta} = \lambda_i$.

In practice, the Poisson regression model is inappropriate if the assumption of equidispersion is not met. The conditional variance is often greater than the conditional mean so overdispersion is quite common. If that is the case, the estimates from the Poisson regression model are consistent but inefficient. The negative binomial regression (NBR) model can be constructed to remedy the limitation. The NBR model allows the variance of the dependent variable to differ from its mean (Long 1997, Greene 2003) by replacing λ_i with a random variable $\tilde{\lambda}_i$:

$$\tilde{\lambda}_i = \exp(x\beta + u_i) = \lambda_i\delta_i \qquad (3)$$

where $\delta_i = \exp(u_i)$. To compute $\Pr(y_i \mid x)$ without conditioning on δ_i, the form of the probability density function of δ_i needs to be specified and used. The most common assumption is that δ_i has a gamma distribution $\Gamma(v_i)$ with parameter v. By utilizing the Poisson distribution and the gamma distribution together, the NBR model can generate the probability of the dependent variable y_i conditional on covariates x as follows:

$$\Pr(y_i \mid x) = \frac{\Gamma(y_i + v_i)}{y_i!\Gamma(v_i)} \left(\frac{v_i}{v_i + \mu_i}\right)^{v_i} \left(\frac{\mu_i}{v_i + \mu_i}\right)^{y_i} \tag{4}$$

The conditional mean and variance are now different from each other:

$$E(y_i \mid x) = \exp(x\beta) = \lambda_i \tag{5a}$$

$$Var(y_i \mid x) = \lambda_i(1 + \frac{\lambda_i}{v_i}) = \lambda_i + \alpha\lambda_i^2 \tag{5b}$$

where $\alpha = 1/v_i$ is known as the dispersion parameter. When $\alpha = 0$, the conditional variance equals to the conditional mean so the NBR model reduces to the Poisson regression model. In practice, if $H_0 : \alpha = 0$ is rejected, the NBR model should be employed.

There are several ways to estimate the selection and outcome equations jointly as a system (Greene 2003). A nonzero correlation between the error terms is intrinsic to the model. If simply regressing y_i on x_i using those observations for which $z_i = 1$, the estimates of β will be both biased and inconsistent. In dealing with the issue, commonly used approaches include full information maximum likelihood (FIML) and two-step method (Murphy & Topel 1985; Starbuck et al. 2004). FIML approach may be inappropriate when defining the joint distribution is problematic. FIML also frequently has convergence problems, which occurred during the preliminary data analyses in this chapter. In contrast, two-step method can be used under more general conditions because there is no need to determine a joint-density function for the errors, and as long as each separate function is estimable, it does not have convergence problems (Greene 2003). Given these considerations, the two-step approach was finally employed in this chapter.

The two-step sample selection model consists of two marginal distributions: $g(z \mid w, \gamma)$ and $f(y \mid w, x, \gamma, \beta)$. First, estimate the binary probit model through maximum likelihood and denote the estimated parameter as $\hat{\gamma}$. Then, estimate the count data model with $\hat{\gamma}$ inserted in place of γ as if it were known. The predicted value is commonly known as the inverse Mills ratio (IMR) and is specified for the correction of selectivity in the count data model:

$$IMR = \frac{\phi(z_i^*)}{1 - \Phi(z_i^*)} \tag{6}$$

where $\phi(.)$ and $\Phi(.)$ are the density and distribution function for the selection equation, respectively. When the coefficient of estimated IMR is significant, it implies the parameter estimators for the second stage of outcome would be biased if the two-step estimation procedure was not used. This entails modeling y_i as dependent upon variables x but considering the fact that y_i is only observed when $z_i = 1$.

The key part of the two-step approach is to correct the estimated asymptotic covariance matrix for the estimator in the outcome equation for the randomness of the estimator carried forward from the selection equation (Greene 2002). Let V_1 be the estimator of the asymptotic covariance matrix for the parameter estimates from the selection equation. Let V_2 be the uncorrected covariance matrix computed from the outcome equation, using the parameter estimates obtained in the selection equation as if they were known. Both of these estimators are based on the log likelihood functions. In addition, define:

$$C = \sum_{i=1}^{n} \left(\frac{\partial \log f(x_i)}{\partial \beta} \right) \left(\frac{\partial \log f(x_i)}{\partial \gamma'} \right) \tag{7a}$$

$$R = \sum_{i=1}^{n} \left(\frac{\partial \log f(x_i)}{\partial \beta} \right) \left(\frac{\partial \log g(w_i)}{\partial \gamma'} \right) \tag{7b}$$

where n is the number of observations. The corrected covariance matrix for the estimator of the outcome equation, V_2^*, can be computed as

$$V_2^* = V_2 + V_2 \left[CV_1C' - RV_1C' - CV_1R' \right] V_2 . \tag{8}$$

Finally, the two sets of explanatory variables, *w* and *x*, can be the same or different. If *w* is equal to *x*, or *w* is a subset of *x*, then it may still be possible to identify the parameters of the outcome equation because of the nonlinearity of the model (Breen 1996). Given the nonzero correlation between the error terms, the model would not be identified if both equations are linear. In practice, reliance on the nonlinearity of the probit model can result in unstable parameter estimates. As a general rule, it is not recommended to rely on the model nonlinearity for identification; instead, it is much better to place restrictions on coefficients (Breen 1996). This will ensure model identification, although which restrictions are appropriate will depend upon the conceptual model that underlines the analysis. In this chapter, with limited guidance from economic theory for variable selection, several insignificant variables were excluded from the second stage to ensure model identification.

4. Data Collection

A survey was designed and conducted in Mississippi to examine NIPF landowner knowledge of three regeneration programs and their application behavior. FRDP, FIP, and RTC were selected because of their focus on forest regeneration and their availability in recent years to NIPF landowners in Mississippi. Information was collected for each of them. Application to these programs was differentiated from participation. For FRDP and FIP, landowners may apply but not be able to participate in the programs for various reasons (e.g., unavailability of funds). In contrast, RTC is a tax incentive code for state income tax, and for qualified landowners application is synonymous with participation. This chapter focused on the response of NIPF landowners to public assistance programs so their application behavior was examined. Finally, considering that clearcut occurs infrequently for many NIPF landowners with small acreages, the survey covered ten years from 1996 to 2006 to increase the likelihood that respondents had harvested during the study period. For specific programs, the time span varied slightly because not all of them were available over the whole study period. The survey covered 10 years for FRDP (1996-2006), 8 years for FIP (1996-2004), and 7 years for RTC (1999- 2006).

The Social Science Research Center at Mississippi State University conducted the telephone survey in August 2006 and collected the data used in this chapter. The population of NIPF landowners in Mississippi is similar to other southern states. In 2006, there were 245,000 NIPF landowners in Mississippi, and they owned 13.5 million acres of forest lands in total, with an average of 55 acres per landowner (Butler 2007). Among them, 32,000 landowners (13% of all the landowners) owned at least 100 acres, and altogether, they owned 9.8 million acres (73% of the total acreage). In this chapter, the survey sample was drawn from a database of landowner records in Mississippi with several filters. All counties except Hinds in Mississippi were covered (there was a lack of records for Hinds). NIPF landowners were the chapter focus so companies and partnerships were excluded. Only NIPF owners with at least 100 acres of land were selected to eliminate small landowners with infrequent forest management activities. Then, names and addresses of these landowners were used to find their phone numbers through a commercial service agency. After this filtering, a random sample of 9925 landowners was selected and used in the telephone survey. Furthermore, during the survey, several questions were asked to select relevant landowners. The chapter objective focused on the application behavior of landowners in public programs. A landowner was supposed to have timber harvesting during the study period so a question was used to exclude those landowners without harvests. Another question was also asked about the types of harvesting activities (e.g., clearcut, thinning, and selective cut). A landowner was included only when the landowner conducted a clearcut because other harvesting types usually do not need regeneration.

Of the 9925 landowners contacted by phone, 2126 owned less than 100 acres (the database was inaccurate about acreage), and 2132 did not harvest timber in the past 10 years, so these landowners were excluded from the survey at the beginning. There were also 1110 incorrect telephone numbers. Other reasons for unsuccessful calls included communication problems, refusal to participate, and deceased landowners. As a result, 2229 landowners completed the survey for a completion rate of 49.8%, i.e., 2229/(9925 - 2126 - 2132 - 1110). Furthermore, among those with complete records, most of them only reported thinning or selective harvesting activities during the study period so they were excluded from the statistical analysis. At the end, there were 934 observations for FRDP, 833 for FIP, and 725 for RTC.

The survey questionnaire was designed to collect information on variables needed for the empirical analysis (table 1). There were two sets of dependent variables, i.e., z_i and y_i. One set measured landowner knowledge of the individual programs. The other set recorded their application frequencies in each program during the study period. Independent variables contained in w_i and x_i were divided into three groups: land features, management experiences, and landowner characteristics. First, three variables were used to represent land features: *acreage*, *land type*, and *forest type*. Second, three variables were constructed to represent management experiences of the landowner: *year*, *timber*, and *regeneration*. Finally, eight variables were used to represent the demographic characteristics of individual landowner: *age*, *education*, *income*, *employment*, *race*, *gender*, *membership*, and *residence*. *Membership* equaled to one if a landowner was a member of any major forestry organization available to Mississippian landowners, and zero if not. In Mississippi, major organizations available to NIPF landowners are the Mississippi Forestry Association, the Mississippi County Forestry Association, and the Society of American Foresters.

Table 1. Definitions and descriptive statistics for the variables used in the two-step sample selection regressions by program

Variables	Definitions	Mean (Std. dev.)		
		FRDP (n = 934)	FIP (n = 833)	RTC (n = 725)
Z_i	Dummy = 1 if the landowner knew of the program; 0 otherwise	0.457 –	0.496 –	0.502 –
Y_i	Application frequencies for the program over the study period	0.143 (0.450)	0.155 (0.472)	0.263 (0.564)
acreage	Total acreage owned by the landowner	491.410 (889.970)	487.526 (917.826)	493.247 (855.369)
land type	Dummy = 1 if forest lands; 0 otherwise	0.777 –	0.776 –	0.763 –
forest type	Dummy = 1 if pine forests; 0 otherwise	0.550 –	0.547 –	0.537 –
year	Years of land ownership till 2006	33.604 (19.098)	33.473 (18.876)	33.091 (19.216)
timber	Dummy = 1 if the landowner was interested in timber production; 0 otherwise	0.903 –	0.900 –	0.894 –
regeneration	Regeneration frequencies over the study period	0.744 (0.680)	0.712 (0.713)	0.644 (0.715)
age	Landowner age in 2006	65.430 (11.262)	65.456 (11.401)	65.294 (11.070)
education	Dummy = 1 if the landowner only had high school education or lower	0.337 –	0.343 –	0.353 –
income	Household income before taxes in 2005 ($1,000)	64.507 (27.755)	63.974 (27.577)	64.234 (28.218)
employment	Dummy = 1 if the landowner was fully employed; 0 otherwise	0.392 –	0.382 –	0.403 –
race	Dummy = 1 if Caucasian; 0 otherwise	0.961 –	0.957 –	0.954 –
gender	Dummy = 1 if male; 0 otherwise	0.749 –	0.753 –	0.756 –
membership	Dummy = 1 if the landowner was a member of a forestry association; 0 otherwise	0.269 –	0.262 –	0.257 –
residence	Dummy = 1 if the landowner resided on the land; 0 otherwise	0.455 –	0.455 –	0.457 –

5. RESULTS

5.1. Descriptive Statistics and Model Selection

The descriptive statistics for the variables used in the analysis are presented in table 1. Landowner knowledge of individual programs was measured by a dummy variable so its mean also revealed the percentage-45.7% of the 934 qualified landowners knew of FRDP, 49.6% of the 833 qualified landowners knew of FIP, and 50.2% of the 725 qualified landowners knew of RTC. Overall, about 50% of these qualified landowners were not aware of these public programs. This was consistent with findings from a previous survey in Mississippi (Gunter et al. 2001b). Furthermore, the application frequencies by qualified landowners varied slightly among programs. Among the 934 qualified landowners for FRDP, for instance, one applied five times during the study period, two applied four times, two applied three times, nine applied twice, 97 applied once, and 823 did not apply. The maximum application frequencies were five for FIP and four for RTC, and the distribution patterns were similar to that for FRDP. The variable mean was 0.143 for FRDP, 0.155 for FIP, and 0.263 for RTC so the application frequency was the highest for RTC.

The descriptive statistics for the independent variables were similar across the three programs. Taking FRDP as an example, the average acreage owned for the 934 qualified landowners was 491.4 acres. Forest lands were the predominant land use for 77.1% qualified landowners. Pine was the predominant forest type for 55.0% landowners while the remainder had either hardwood or mixed forest types. Average length of ownership was 33.6 years and 90.3% of them were interested in timber production. The mean of regeneration frequency was 0.7 times over the study period. For demographic characteristics, the qualified landowners who applied for FRDP were 65.4 years old on average; 33.7% had a high school diploma or lower education; and their household income in 2005 was $64,507. In addition, about 39.2% of the qualified landowners who applied for FRDP were fully employed, 96.1% were Caucasian, 74.9% were male, 26.9% were members of a forestry association, and 45.5% resided on their lands.

Regression results from the two-step sample selection model are reported in table 2 for FRDP, table 3 for FIP, and table 4 for RTC. The coefficients on the IMR were significant for all of the three programs and they indicated that the use of the two-step sample selection model was appropriate. Two variables (i.e., regeneration and race) were not significant in the second stage, and with the confirmation of a Wald test, they were excluded from the second stage to ensure model identification. In addition, for the count data model, the dispersion parameter (Alpha) was significant for FRDP at the 10% level and for FIP at the 5% level. Consequently, the NBR model was employed for FRDP and FIP at the second stage. However, the dispersion parameter was not significant for RTC so the results from a Poisson regression model were reported. In the subsequent presentation, to address the chapter objectives directly, the determinants of landowner knowledge were first examined for the three public programs together. They are followed by the determinants of landowner applications in these programs.

Table 2. Results of two-step sample selection estimation of NIPF landowner knowledge of Mississippi Forest Resource Development Program (FRDP) and their application behavior from 1996 to 2006

Variables	First step: binary probit model		Second step: negative binomial model	
	Coefficient (*t*-ratio)	Marginal effect	Coefficient (*t*-ratio)	Marginal effect
constant	-0.593 (-1.348)	-0.235	-10.074*** (-8.685)	-0.961
acreage	2.309E-4*** (2.798)	9.161E-5	-0.001*** (-5.563)	-8.246E-5
land type	-0.123 (-1.162)	-0.049	0.909*** (3.448)	0.087
forest type	-0.211** (-2.414)	-0.084	1.398*** (5.289)	0.133
year	0.001 (0.533)	0.001	-0.006 (-1.046)	-0.001
timber	0.212 (1.409)	0.083	-0.834 (-1.437)	-0.079
regeneration	0.116* (1.719)	0.046	–	–
age	-0.001 (-0.275)	-0.001	0.020* (1.750)	0.002
education	-0.071 (-0.733)	-0.028	0.512** (2.055)	0.049
income	0.001 (0.739)	0.001	-0.010** (-2.433)	-0.001
employment	-0.169 (-1.498)	-0.067	1.389*** (5.120)	0.132
race	0.025 (0.115)	0.010	–	–
gender	0.274*** (2.704)	0.108	-1.645*** (-4.983)	-0.157
membership	0.435*** (4.382)	0.172	-2.439*** (-5.549)	-0.233
residence	0.029 (0.329)	0.012	-0.421** (-2.013)	-0.040
Inverse Mills ratio	–	–	18.240*** (7.709)	–
Dispersion alpha	–	–	0.589* (1.826)	–
Log Likelihood	-607.734		-349.485	
χ^2	72.470		5.667	

***Significance at 1% level.
**Significance at 5% level.
*Significance at 10% level.

Table 3. Results of two-step sample selection estimation of NIPF landowner knowledge of Forestry Incentive Program (FIP) and their application behavior from 1996 to 2004

Variables	First step: binary probit model		Second step: negative binomial model	
	Coefficient (*t*-ratio)	Marginal effect	Coefficient (*t*-ratio)	Marginal effect
constant	-0.906* (-1.923)	-0.361	-9.539*** (-7.378)	-0.888
acreage	1.832E-4** (2.435)	7.307E-5	-0.001*** (-5.435)	-7.156E-5
land type	-0.057 (-0.513)	-0.023	0.360 (1.366)	0.034
forest type	-0.061 (-0.654)	-0.024	0.631*** (2.886)	0.059
year	-0.002 (-0.834)	-0.001	0.017*** (2.808)	0.002
timber	0.486*** (3.006)	0.188	-2.983*** (-3.555)	-0.278
regeneration	0.115* (1.671)	0.046	–	–
age	-0.001 (-0.159)	-3.363E-4	0.014 (1.110)	0.001
education	-0.111 (-1.084)	-0.044	0.592** (1.951)	0.055
income	0.001 (0.550)	4.071E-4	-0.012*** (-2.700)	-0.001
employment	-0.105 (-0.856)	-0.042	1.145*** (4.026)	0.107
race	0.234 (1.036)	0.092	–	–
gender	0.254** (2.369)	0.101	-1.885*** (-4.971)	-0.175
membership	0.428*** (4.012)	0.169	-2.956*** (-5.932)	-0.275
residence	-0.039 (-0.416)	-0.016	0.269 (1.210)	0.025
Inverse Mills ratio	–	–	20.866*** (7.244)	–
Dispersion alpha	–	–	0.889** (2.428)	–
Log Likelihood	-541.146		-323.204	
χ^2	72.433		11.344	

***Significance at 1% level.
**Significance at 5% level.
*Significance at 10% level.

Table 4. Results of two-step sample selection estimation of NIPF landowner knowledge of Reforestation Tax Credit (RTC) and their application behavior from 1999 to 2006

Variables	First step: binary probit model		Second step: poisson model	
	Coefficient (*t*-ratio)	Marginal effect	Coefficient (*t*-ratio)	Marginal effect
constant	-1.064** (-2.125)	-0.424	-5.508*** (-8.127)	-0.862
acreage	2.394E-4** (2.269)	9.548E-5	-0.001*** (-4.807)	-1.106E-4
land type	0.241** (2.006)	0.096	-0.884*** (-4.538)	-0.139
forest type	0.056 (0.551)	0.022	-0.262** (-1.917)	-0.041
year	0.001 (0.366)	4.212E-4	0.003 (0.899)	0.001
timber	0.275 (1.609)	0.109	-0.632 (-1.555)	-0.099
regeneration	0.282*** (3.663)	0.112	–	–
age	-0.004 (-0.635)	-0.001	0.018** (2.128)	0.003
education	-0.107 (-0.954)	-0.043	0.315* (1.761)	0.049
income	0.002 (0.994)	0.001	-0.014*** (-4.746)	-0.002
employment	-0.157 (-1.207)	-0.063	0.895*** (5.295)	0.141
race	0.135 (0.567)	0.054	–	–
gender	0.357*** (3.015)	0.142	-1.008*** (-4.451)	-0.158
membership	0.663*** (5.466)	0.255	-2.064*** (-9.685)	-0.324
residence	-0.124 (-1.187)	-0.050	0.491*** (3.455)	-0.078
Inverse Mills ratio	–	–	11.157*** (14.647)	–
Dispersion alpha	–	–	0.001 (0.182)	–
Log Likelihood	-443.232		-365.574	
χ^2	118.588		216.794	

***Significance at 1% level.
**Significance at 5% level.
*Significance at 10% level.

5.2. Determinants of Landowner Knowledge of Public Programs

The first stage of binary probit model generated similar results across the three programs. Among land features variables, the coefficients for acreage were positive and significant for all the three programs. Thus, qualified landowners with more land acreage were more likely to know of these programs. Land type was positive and only significant for RTC, implying that qualified landowners with a majority of their land in forestry were more likely to know of the tax incentive program. Forest type was negative and only significant for FRDP so these qualified landowners with predominantly pine forest types were less likely aware of this program.

Among the three land management variables, the coefficients for regeneration were positive and significant for FRDP and FIP at the 10% level, and for RTC at the 1% level. This suggested that the regeneration experience usually helped qualified landowners learn about these programs. Timber was positive for FIP only, so qualified landowners were more likely to know of FIP if they were interested in timber production. The length of ownership as measured by year was not significant for any of them. Finally, among the eight demographic variables, two had consistent impacts across all the three programs. Both gender and membership were positive and significant at the 5% level or better. Thus, male landowners or those with membership in forestry associations had higher probability of knowing of these programs. Other demographic variables did not show any significant impacts on the awareness of the three programs.

Overall, landowner knowledge was positively influenced by land size, regeneration experience, gender, and membership in forestry associations, and the impact was consistent across all the three programs. Land types, forest types, and interest in timber production demonstrated different degrees of influence on landowner knowledge of some programs. Among these significant variables, membership had the largest marginal effect at 0.172 for FRDP, 0.169 for FIP, and 0.255 to RTC.

5.3. Determinants of Landowner Applications to Public Programs

The second stage count data model examined the determinants of landowner applications to individual programs. The empirical results were comparable across the three programs. Among land features, acreage showed negative and significant signs for all the three programs at the 1% level. This revealed that when qualified landowners knew of these programs, their application frequency was higher for those with less land. Results for land type were mixed with a positive impact on FRDP and negative on RTC. Forest type had a positive impact on the application frequency for FRDP and FIP at the 1% level, but negative for RTC at the 5% level. Furthermore, among the two variables representing management experience, year was positive for FIP and timber was negative for FIP. Thus, when qualified landowners knew of FIP, longer ownership of the land and weaker interest in timber production were associated with a higher application frequency in FIP. However, both year and timber were not significant for FRDP and RTC.

Most of the demographic variables had significant impacts and the results were similar across programs. Five variables had the same signs for all three programs: positive for education and employment, and negative for income, gender, and membership. This suggested that when qualified landowners knew of these programs, their application frequency was higher if they were fully employed, female or members of forestry organizations, and was lower as education and income increased. In addition, age had positive impact on the application behavior for FRDP and RTC but was not significant for FIP. Residence also had mixed results with a negative impact on FRDP, a positive impact on RTC, but no significant impact on FIP.

Overall, the results from the second stage revealed that application frequency was higher for qualified landowners that need more public assistance. Among these landowners with knowledge of the programs, application frequency was higher for landowners that in general had less acreage, lower education and income, or were not members in forestry associations. The marginal effect for membership had the largest impact on application frequency, with values of -0.233 for FRDP, -0.275 for FIP, and -0.324 for RTC.

CONCLUSION

This chapter examined how land features, management experiences, and landowner characteristics have impacted NIPF landowner knowledge of public financial assistance programs and their application behavior. The three programs considered in this chapter (i.e., FRDP, FIP, and RTC) have provided financial assistance to forest landowners in Mississippi for a long time. The analysis was conducted on the landowners that clearcut some or all of their forest lands from 1996 to 2006. A two-step sample selection model was employed to analyze their application behavior conditional on their knowledge of these public programs. Modeling the application frequency conditional on landowner knowledge yields more inspiring results than simple binary regressions typically employed in the literature.

The survey revealed that among those NIPF landowners who conducted clearcuts in the past 10 years, about 50% knew of FRDP, FIP, and RTC. Their application frequencies to individual programs ranged from 0 to 5 times. The two-step sample selection model generated several clear results. From the first stage of selection and binary probit model, landowner knowledge of all three programs was positively related to land size, regeneration experience, gender, and membership in forestry associations. In other words, qualified NIPF landowners that had larger ownerships and more regeneration experience, were male, or were members of forestry associations were more likely to know of these assistance programs. Land type, forest type, and interest in timber production demonstrated different degrees of impact on their knowledge of individual programs. In the outcome stage, the application behavior of NIPF landowners in these programs was modeled conditional on their knowledge of the programs. When landowners knew of these programs, their application frequencies were higher for those with smaller land area, lower education or income, and full employment, that were female, or without membership in forestry associations. Therefore, the application frequency for these public programs was higher for landowners that need more public assistance.

These results have several policy implications for designing, promoting, and implementing public assistance programs. First, information asymmetry for small NIPF landowners is severe and it is clearly a vital issue that merits more attention and analysis. Half of the surveyed Mississippi NIPF landowners in this chapter had no knowledge of major public assistance programs for regeneration. Considering all NIPF landowners, including those that did not qualify for this survey such as landowners who owned less than 100 acres or had not clearcut any of their forest land during the study period, the proportion of landowners that have no knowledge of these programs is likely much higher. Thus, considering the similarity of nonindustrial private forestland ownership in the southern states, the assumption of perfect information for landowners as assumed in many previous studies is invalid. This information barrier also has been a concern for forest landowners in timber markets (Amacher et al. 2003; Munn 1998). To improve the efficiency of public assistance programs, more effort should be made to disseminate information among the forest landowner community. Extension outreach can be more effective through forestry organizations. Forestry organizations typically provide information and technical guidance and thus affect the involvement of NIPF landowners in public financial assistance programs. Therefore, a useful long-term strategy is to encourage NIPF landowners to join forestry organizations, and furthermore, attend workshops and meetings on a regular basis. Although membership costs for forestry associations in Mississippi are generally quite low, where these costs are burdensome for landowners, appropriate measures to reduce them could be introduced.

Second, the chapter results also address some equity concerns related to public assistance programs. Many public assistance programs, including FIP, are not designed as an income redistribution or regional development program (Gaddis et al. 1995). Nonetheless, these programs are often involved in debates related to equity. Equity consideration for public assistance programs has several aspects: distribution of funds among states or regions, income and wealth of program participants, secondary program benefits, and transitional equity. For example, a question that has often been asked is whether these programs are subsidizing wealthy landowners who could well afford to invest in reforestation without these public assistance programs. This chapter revealed that when disadvantaged landowners (i.e., with less acreage, less education, or lower income) were aware of these public assistance programs, they were more likely to apply to them. This is clearly different from the simplified conclusion in the literature that the richer a landowner, the higher the participation probability. Our results are probably indicative of the difference in perceptions of marginal costs and marginal benefits between advantaged and disadvantaged landowners. This also suggests that in designing public financial programs, the needs of small landowners should be emphasized through administrative procedures and regulations.

Finally, the difference between cost-share and tax incentive programs needs more examination. Compared with tax incentive programs, cost-share programs usually are more challenging and expensive for landowners to apply and for pubic agencies to manage. For example, about 10% of FIP appropriations in the past have been spent as administration costs for providing technical assistance, writing plans, and monitoring practices (Gaddis et al. 1995). Thus, other things being equal or similar, tax incentive programs should be more appealing to NIPF landowners than cost-share programs. However, cost-share programs may be better than tax incentive programs in program implementation and evaluation. In this chapter, NIPF landowners did demonstrate a higher application frequency for the state tax incentive program (i.e., RTC) than for cost-share programs (i.e., FRDP and FIP). The determinants of landowner knowledge of these programs and determinants of their application behavior were nevertheless similar across the three programs. Future research will need to further examine the different impacts of cost-share and tax incentive programs on both governments and forest landowners in achieving their objectives.

In summary, public financial assistance will continue to be a vital concern to governments and NIPF in the United States. These assistance programs have faced many challenges over time. These challenges include how to handle the fluctuation of budgets and how to allocate funds effectively to achieve the program goals. Future research needs to examine how to improve program design and implementation to improve program efficiency.

ACKNOWLEDGMENTS

The authors wish to thank an associate editor and two anonymous reviewers of Journal of Forest Economics and Stephen C. Grado for their valuable comments, and William H. Greene for his great help in model selection and data analysis.

REFERENCES

Alig, R. J. (2003). U.S. landowner behavior, land use and land cover changes, and climate change mitigation. *Silva Fennica. 37*(4), 511-527.

Amacher, G. S., Conway, M. C., & Sullivan, J. (2003). Econometric analyses of nonindustrial forest landowners: is there anything left to study? *Journal of Forest Economics. 9*(2), 137-164.

Bell, C. D., Roberts, R. K., English, B. C., & Park, W. M. (1994). A logit analysis of participation in Tennessee's Forest Stewardship Program. *Journal of Agricultural and Applied Economics. 26*(2), 463-472.

Boyd, R. G. (1984). Government support of non-industrial production: The case of private forests. *Southern Economic Journal. 51*(1), 89-107.

Breen, R. (1996). *Regression models: Censored, sample-selected, or truncated data*. Sage University Paper series on Quantitative Applications in the Social Science, series no. 07-111. Thousand Oaks, CA: Sage:

Butler, B. J. (2007). National Woodland Owner Survey 2006. Forest Inventory and Analysis, USDA Forest Service, 20.

Gaddis, D. A., New, B. D., Cubbage, F. W., Abt, R. C., & Moulton, R. J. (1995). Accomplishments and economic evaluations of the Forestry Incentives Program: a review. Southeastern Center for Forest Economics Research Working Paper No. 78. Research Triangle Park, NC, 83.

Greene, W. H. (2002). *LIMDEP Version 8.0 Econometric Modeling Guide*. Econometric Software, Inc. 15 Gloria Place, Plainview, New York, United States:

Greene, W. H. (2003). *Econometric Analysis*. Pearson Education, Inc., Delhi, India.

Gunter, J. E., Bullard, S. H., Doolittle, M. L., & Arano, K. G. (2001a). Reforestation of harvested timberlands in Mississippi: behavior and attitudes of non-industrial private forest landowners. FWRC Research Bulletin #FO 172, Forest and Wildlife Research Center, Mississippi State University, 25.

Gunter, J. E., Bullard, S. H., Doolittle, M. L., & Arano, K. G. (2001b). Reforestation of harvested timberlands in Mississippi: behavior and attitudes of non-industrial private forest landowners. *Forest and Wildlife Research Center, Bulletin #FO172, Mississippi State University.*, 25pp.

Katchova, A. L., & Miranda, M. J. (2004). Two-step econometric estimation of farm characteristics affecting marketing contract decisions. *American Journal of Agricultural Economics. 86*(1), 88-102.

Konyar, K., & Osborn, C. T. (1990). A national-level economic analysis of conservation reserve program participation: A discrete choice approach. *The Journal of Agricultural Economics Research. 42*(2), 5-12.

Lee, E., Lee, J., & Eastwood, D. (2003). A two-step estimation of consumer adoption of technology-based service innovations. *Journal of Consumer Affairs. 37*(2), 256-282.

Mehmood, S. R., & Zhang, D. W. (2001). Causes for continuation of state cost-share programs for nonindustrial private forest landowners. *Forest Science. 48*(3), 471-478.

Mississippi Forestry Commission (2007). Forest Management. http://www.mfc.state.ms.us (Accessed in September 2007),

Munn, I. A. (1998). The impact of the forest products industry on the Mississippi economy: an input-output analysis. FWRC Research Bulletin #FO 087. Forest and Wildlife Research Center, Mississippi State University, 12.

Murphy, K. M., & Topel, R. H. (1985). Estimation and Inference in Two-Step Econometric Models. *Journal of Business & Economic Statistics. 3*(4), 88-97.

Nagubadi, V., McNamara, K. T., Hoover, W. L., & Mills, W. L. (1996). Program participation behavior of nonindustrial forest landowners: a probit analysis. *Journal of Agricultural and Applied Economics. 28*(2), 323-336.

Ovaskainen, V., Hannine, H., Mikkola, J., & Lehtonen, E. (2006). Cost-sharing and private timber stand improvements: A two-step estimation approach. *Forest Science. 52*(1), 44-54.

Smith, W. B., Miles, P. D., Vissage, J. S., & Pugh, S. A. (2004). Forest resources of the United States 2002. *USDA Forest Service Gen. Tech. Rep. NC-241*,

Starbuck, C. M., Alexander, S. J., Berrens, R. P., & Bohara, A. K. (2004). Valuing special forest products harvesting: a two-step travel cost recreation demand analysis. *Journal of Forest Economics. 10*(1), 37-53.

Straka, T. J., Wisdom, H. W., & Moak, J. E. (1984). Size of forest holding and investment behavior of nonindustrial private owners. *Journal of Forestry. 82*, 495-496.

Sun, C. (2007). Variation of federal cost-share programs in the United States and the inducement effects on tree planting. *Journal of Forest Economics. 12*(4), 279-296.

Wear, D. H., & Greis, J. G. (2002). Southern forest resource assessment. *Southern Research Station. Gen. Tech. Rep. SRS. 53*, 175-223.

Zhang, D., Hussain, A., & Armstrong, J. B. (2006). Supply of hunting leases from non-industrial private forest lands in Alabama. *Human Dimensions of Wildlife. 11*, 1-14.

In: Global Change and Forestry
Editors: J. Gan, S. Grado and I. A. Munn

ISBN: 978-1-60876-262-0

Chapter 18

How Promptly Nonindustrial Private Forest Landowners Regenerate Their Lands after Harvest: A Duration Analysis[1]

***Xing Sun*[1], *Ian A. Munn*[*2], *Changyou Sun*[2] *and Anwar Hussain*[2]**

[1] School of Forestry and Wildlife Sciences,
Auburn University, Auburn, AL 36849, U.S.A.
[2] Forest and Wildlife Research Center, Department of Forestry,
Mississippi State University, Mississippi State University,
363 Thompson Hall, Mississippi, MS 39762, U.S.A.

Abstract

Understanding factors that influence how promptly landowners regenerate their timberlands after harvest, if at all, is critical to developing policies to improve forest productivity. Mississippi forest landowners with over 100 acres of forestland were surveyed in 2006 to collect harvest and regeneration data from 1996 to 2006. This chapter investigated the length of the time interval between harvest and reforestation. Non-parametric duration analysis was used to examine how long nonindustrial private forest (NIPF) landowners waited to reforest after harvesting. Parametric duration analysis was used to examine factors that influenced the length of this period. The average time elapsed from harvest to regeneration was 11 months for landowners that regenerated their lands. The instantaneous probability of regeneration reached its highest value in the 16th month after harvest and thereafter decreased steadily until the 28th month, after which the probability of regeneration was essentially nil. Interest in timber production, employing a

---

[1] This chapter is reprinted with permission from the article originally published in the *Canadian Journal of Forest Research* 38(8):2109-2117.

[*] Corresponding author. Tel.: +1 662 325 4546. E-mail address: imunn@cfr.msstate.edu (I.A. Munn). The authors wish to thank the associate editor and two anonymous reviewers of Canadian Journal of Forest Research for their valuable comments.

consultant, and ownerships that were predominantly pine forest types were factors associated with substantially shorter reforestation times. Lower stumpage prices and higher reforestation costs were associated with substantially longer reforestation times.

1. INTRODUCTION

Regenerating forestland after final harvest is fundamental to sound forest management. Providing for adequate regeneration, either through artificial or natural methods, is essential for maintaining productive timberlands. Landowners benefit monetarily from increased timber production, and at the same time, society benefits from restored forest-based environmental services such as providing aesthetic landscapes, clean water, enhanced wildlife habitat, and recreational opportunities. Given that regeneration is fundamental to sound forest management, it follows that the timing of regeneration is also critical. Site preparation and planting costs may increase with time if these activities are delayed (Sessions, Bettinger, Buckman, Newton & Hamann 2004). Furthermore, if harvested sites are not regenerated promptly, water, soil, and amenity values may deteriorate and wildlife habitats may degrade (Wear & Greis 2002). Therefore, the time elapsed before reforestation is a critical indicator of good forest management.

Many studies have investigated landowner reforestation behavior (Amacher, Conway & Sullivan 2003; Beach, Pattanayak, Yang, & Muray 2005). In general, these studies modeled landowner reforestation decisions as a binary choice: regenerate or not regenerate. Explanatory variables usually included land characteristics (e.g., ownership size, land type), owner demographics (e.g., household income, education, residence), and market factors (e.g., sawtimber price, pulpwood price, reforestation costs). Royer (1987), for example, employed logistic regression to model reforestation probabilities. Income, reforestation costs, government cost-sharing, technical assistance, and pulpwood price were highly significant determinants of reforestation (Hyberg & Holthausen 1989). They also used logistic regression to investigate harvest timing and reforestation investment decisions of private landowners and obtained similar results. Other modeling approaches included Zhang and Flick (2001), who estimated a two-step selectivity model, and found income and government financial assistance programs positively related with increased reforestation probabilities. Gunter, Bullard, Doolittle and Arano (2001), using non-parametric techniques, found landowners who regenerated were more likely to have larger ownerships, higher income levels, and more education. Also they were more likely to reside in urban areas, be white males, and be employed in professional or business occupations.

In aggregate, the literature on NIPF regeneration suggests that the likelihood of replanting decreases as the time since harvest completion increases. Site preparation and replanting costs increase with time (Sessions et al. 2004) and higher reforestation costs result in lower reforestation probabilities (Hyberg & Holthausen 1989; Royer 1985); thus, it is reasonable to infer that landowners are less likely to replant as the time since harvesting increases. Previous studies, however, have not investigated this time dimension as a factor influencing landowner replanting decisions. Given that landowners do not always replant promptly (Amacher et al. 2003), important unanswered questions include: how long do NIPF landowners wait to replant after harvesting; what factors affect the time interval between harvest and replanting; and does the probability of replanting change over time? Answers to these questions would be extremely useful in fine tuning policies intended to improve reforestation rates among NIPF landowners. Currently, federal incentive programs such as the Forestland Enhancement Program (http://www.fs.fed.us/spf/coop/programs/loa/flep.shtml accessed March 18, 2008) and state incentive programs such as Mississippi's Forest Resources Development do not specifically address or consider time delays in the allocation of benefits (Wear & Greis 2002). Therefore, the objective of this chapter is to determine how long NIPF landowners wait to reforest after harvesting (if at all), what factors affect this interval, how much each factor contributes to the probability of reforestation, and to what extent each factor affects the interval length.

2. METHODS

We assume that NIPF landowners are utility maximizers and we model their reforestation decisions accordingly. This approach has been commonly adopted in the forestry literature (Amacher et al. 2003). (See, for example, Hyberg & Holthausen 1989; Nagubadi, McNamara, Hoover & Mills 1996) Specifically, landowners maximize their expected utility by choosing to replant their harvested forestland at time T, where T can take on values from 1 to ∞; with ∞ implying that the landowner chooses never to replant. The value of T that maximizes landowner utility is a function of a vector of variables, x_i, that is composed of land and ownership characteristics, landowner demographics, harvest characteristics, and market factors. Conceptually, the model is expressed as:

$$T_i = f(x_i) + \varepsilon_i \tag{1}$$

where i denotes the ith individual and ε is the associated error term.

Unlike most typical utility models, T is not a binary variable and thus the model cannot be estimated using standard logit or probit regression techniques. Instead, we employ duration analysis, which is a class of statistical methods that investigates survival times, i.e., the occurrence and timing of events (Allison 1995; Greene 2003). The phrase "survival time" derives from early applications in medical research when the interval of interest was typically that between a medical treatment and the patient's demise. Duration analysis techniques can accommodate situations where the interval may not be complete at the end of the study period. In the case of medical research, the patient may still be living at the end of the study. When this occurs, the observation is censored in the sense that the survival time is at least the observed interval within the study period. To accommodate such situations, an additional variable, δ, is defined. δ is an indicator variable equal to 1 if the interval ends within the study period; 0 otherwise. Estimation needs to account for the censored nature of the data.

There are four equivalent ways to describe the continuous probability distribution for T. Two of these, the probability density function (p.d.f.) denoted as $f(t)$ and the cumulative distribution function (c.d.f.) denoted as $F(t)$, are used to estimate model parameters. Mathematically, they can be expressed as:

$$f(t) = \frac{dF(t)}{dt} = \lim_{\Delta t \to 0} \frac{\Pr(t \le T < t + \Delta t)}{\Delta t}, \tag{2}$$

$$F(t) = \Pr(T \le t) = \int_0^t f(x)dx \tag{3}$$

In addition to these two functions, the survivor function $S(t)$ and hazard function $h(t)$ are used in duration analysis. $S(t)$ is an unconditional probability distribution defined as the probability that the interval of interest will be greater than *t*. It is expressed mathematically as follows:

$$S(t) = \Pr(T > t) = 1 - F(t) = \int_t^\infty f(x)dx \tag{4}$$

In this chapter, the survivor function provides the probability that a harvested site will not be regenerated by time *t*. $S(t)$ is at a maximum when t equals 0.

The hazard function, $h(t)$ is a conditional density distribution and in this chapter represents the instantaneous rate of regeneration at time *t*, given that regeneration has not occurred prior to time *t*. This function is a popular and useful way of describing *T*'s distribution in duration analysis (Allison 1995). It is defined as follows:

$$h(t) = \lim \frac{\Pr(t \le T < t + \Delta t \mid T \ge t)}{\Delta t} = \frac{f(t)}{S(t)}. \tag{5}$$

In this chapter, the hazard function provides the probability that regeneration occurs in the small interval between t and $t + \Delta t$ conditional on $T \geq t$. Overall, these four functions provide the theoretical basis for empirical duration analysis. Both non-parametric and parametric analytical techniques were employed in this chapter. Non-parametric analysis was used to examine the probability of regeneration occurring over time after the harvest has been completed. Parametric analysis was employed to examine the effect of explanatory variables on timely reforestation behavior.

2.1. Non-Parametric Duration Analysis

Non-parametric techniques can be used to compute survival time and plot the survival probability. In this chapter, survival time is the time elapsed between completion of harvest and beginning of regeneration. $S(t)$ is the probability that reforestation has not occurred by time *t*, where *t* can be any nonnegative number. Kaplan-Meier estimation (KM) is the most widely used method for estimating survival functions to obtain exact survival proportions and survival times and has a solid theoretical justification (Allison 1995). If all NIPF landowners regenerated their harvested lands within the study period, the KM estimator $\hat{S}(t)$ is just the sample proportion of observations that have not been regenerated by t. The KM estimator is then defined as:

$$\hat{S}(t) = \prod_{j:t_j \le t} \left[1 - \frac{d_j}{n_j}\right], \tag{6}$$

where at each time t_j, there are n_j observations. Of these, d_j have been regenerated (Allison 1995).

$$\hat{\mu} = \sum_{j=1}^{k} \hat{S}(t_{j-1})(t_j - t_{j-1}), \tag{7}$$

where t_0 is defined to be zero and k represents distinct event times. When the largest observed time is censored, this sum underestimates the mean. The standard error of $\hat{\mu}$ is estimated as

$$\hat{\sigma}(\hat{\mu}) = \sqrt{\frac{m}{m-1} \sum_{j=1}^{k-1} \frac{A_j^2}{n_j s_j}} \tag{8}$$

where prior to t_j, s_j have been not regenerated.

$$A_j = \sum_{i=j}^{k-1} \hat{S}(t_i)(t_{i+1} - t_i) \tag{9}$$

$$m = \sum_{j=1}^{k} d_j \tag{10}$$

The KM estimator provides a sound basis for analysis and does not depend on the grouping of the data into a certain number of intervals. The disadvantage of the KM method is that to accurately interpret the results, the study period must be sufficiently long as to include all the interval times. If the study period is not long enough to include all interval times, some observations will be censored, i.e., the interval of interest was not completed before the end of the study period and cannot be assigned a value.

The Life Table (LT) method, used to estimate the survivor and hazard functions in this chapter, avoids this problem, and accommodates censored observations. In the LT method, the study period can be as long or short as needed. Consider intervals $[t_j, t_{j+1}]$, as for KM estimator, except all intervals to be relatively long. Let c_j be the number of landowners who do not regenerate during the interval. Since intervals are long, replace n_j by $n_j - \frac{c_j}{2}$ when estimating the survival function. The LT estimator is defined as:

$$\hat{S}(t_i) = \prod_{j=1}^{i-1} \left(1 - \frac{d_j}{n_j - \frac{c_j}{2}} \right), \quad (11)$$

where for interval i, t_i is the start time.

Non-parametric analysis, as the name suggests, drops the formal modeling framework (Greene 2003). Furthermore, it does not consider the impact of other variables on the dependent variable. Parametric analysis can provide a more complete characterization of the relationship between T and various variables influencing the regeneration interval.

2.2. Parametric Duration Analysis

Parametric duration analysis produces estimates of parametric regression models based on the accelerated failure time (AFT) model (Allison 1995). The AFT model describes a relationship between survival functions of any two individuals. If $S_i(t)$ is the survival function for individual i, $S_j(t)$ for another individual j, the AFT model is expressed as:

$$S_i(t) = S_j(\phi_{ij} t) \text{ for all } t \quad (12)$$

where ϕ_{ij} is a constant that is specific to the pair (i, j).

Parametric duration analysis estimates a model similar to an ordinary linear regression model (Allison 1995). Let T_i be a random variable denoting the time interval, and $x_{i1}, \ldots, x_{ij}$ be j explanatory variables for the ith individual. Then, the model is:

$$\log T_i = \beta_0 + \beta_1 \chi_{i1} + \ldots + \beta_k \chi_{ik} + \sigma\varepsilon_i \quad (13)$$

where $\beta_0, \ldots, \beta_k$, and σ are parameters to be estimated; ε_i is a random disturbance term with variance σ. The log transformation of T is to ensure that predicted values of T are positive, regardless of the values of x's and β's. Thus, exponentiating both sides of the above equation gives an alternative way of expressing the model:

$$T_i = \exp\{\beta_0 + \beta_1 \chi_{i1} + \ldots + \beta_k \chi_{ik} + \sigma\varepsilon_i\} \quad (14)$$

Depending on the data, ε can take on a number of different distributions. Regardless of the distribution selected, all model forms are estimated by the method of maximum likelihood (Allison 1995). Because this chapter included censored observations, the likelihood function is expressed as follows:

$$L = \prod_{i=1}^{n} \left[f_i(t_i)\right]^{\delta_i} \left[S_i(t_i)\right]^{1-\delta_i} \tag{15}$$

where n is the number of observations; t_i is the time interval of interest or the time of censoring; and δ_i is the indicator variable referenced previously with a value of 1 if t_i is uncensored or 0 if censored. The functions $f(t_i)$ and $S(t_i)$ represent the probability density function and survival function, respectively.

It is generally easier to work with the natural logarithm of the likelihood function to maximize Eq. [15] because the logarithm is an increasing function, so whatever maximizes the logarithm also maximizes the original function (Greene 2003). Thus, taking the logarithm of both sides of Eq. [15], the likelihood function is expressed as follows:

$$\log L = \sum_{i-1}^{n} \delta_i \log\left[f_i(t_i)\right] + \sum_{i-1}^{n} (1-\delta_i) \log\left[S_i(t_i)\right]. \tag{16}$$

Once a particular model is chosen, appropriate expressions related to β can be substituted for $f(t_i)$ and $S(t_i)$.

In parametric survival analysis, coefficient signs reveal the direction of the relationship (Allison 1995). For example, for binary variables, positive coefficients indicate that those with a value of 1 take longer to regenerate than those with a value of 0, whereas negative coefficients indicate the opposite. In contrast, the numerical values of the estimated coefficients are not directly informative. Simple transformations of the coefficients provide the odds ratios which do have interpretive values. For dummy variables, the odds ratio, e^{β}, provides the estimated ratio of the expected (mean) survival times for the two groups. For quantitative variables, the odds ratio, 100(e^{β}-1), gives the percent increase in expected survival time for each one-unit increase in the variable (Allison 1995).

2.3. Data

Mississippi NIPF landowners were surveyed in 2006 to collect harvest and regeneration data from 1996 to 2006. Mississippi is a typical southern state where timber plays an important role in the state economy and most timberland is owned by NIPF landowners. The Social Science Research Center at Mississippi State University conducted the telephone survey during July and August of 2006. The survey sample was drawn from a database of landowner records in Mississippi. The database covered 81 of the 82 counties in Mississippi. NIPF landowners were the chapter focus so companies and partnerships were excluded. In addition, only NIPF landowners with at least 100 acres of land were selected to eliminate small landowners with infrequent forest management activities. That yielded a list of about 20,000 landowners. Telephone numbers were provided by a commercial service agency. Finally, among landowners with telephone numbers, a random sample of 9,925 landowners was selected and used in the survey.

At the beginning of the survey, two questions were asked to eliminate landowners not relevant for this chapter. As land records are sometimes incorrect, landowners were asked if they owned one hundred or more acres in Mississippi. Landowners were also asked if they had conducted a timber harvest in the period 1996 to 2006. Respondents who replied no to either of these questions were thanked and the survey was terminated.

QUESTIONNAIRE AND VARIABLES

The survey questionnaire was designed to collect information about the harvesting and regeneration behavior of Mississippi landowners during the period 1996-2006. Questions about each harvesting and regeneration event, land and ownership characteristics, and landowner demographics supplied much of the information needed for the empirical analysis (table 1). The two dependent variables, T and δ, were constructed from the reported dates of harvest and regeneration. T was the interval length in months between the completion of harvest and the beginning of regeneration. If the landowner provided only the season and not an exact month for either of these events, the mid-point of the season was used (i.e., March for Spring, June for Summer, September for Fall, and December for Winter). If the harvest had not been regenerated during the study period, δ was set equal to 0; otherwise, δ was set equal to 1.

Table 1. Summary statistics of variables used in the duration analysis of nonindustrial private forest landowner regeneration behavior in Mississippi from 1996 to 2006 (n = 655)

Variable	Definitions	Mean	Std. dev.
Dependent variables			
T	Time from finishing harvest to beginning regeneration (month)	28.246	35.928
δ	Dummy = 1 if the landowner replanted; 0 otherwise	0.603	--
Land & ownership characteristics			
Acreage	Total acreage owned by the landowner	559.669	938.201
Forest Type	Dummy = 1 if pine is the dominant forest type; 0 otherwise	0.331	--
Ownership Length	Years of land ownership	32.266	18.624
Timber	Dummy = 1 if the landowner is strongly interested in timber production; 0 otherwise	0.795	--
Landowner demographics			
Age	Landowner age	64.570	12.008
Education	Dummy = 1 if the landowner has a bachelor degree or better; 0 otherwise	0.505	--
Income	Household income before taxes in 2005 ($1,000)	66.382	28.908
Race	Dummy = 1 if Caucasian; 0 otherwise	0.960	--
Gender	Dummy = 1 if male; 0 otherwise	0.795	--

Table 1. (Continued).

Variable	Definitions	Mean	Std. dev.
Membership	Dummy = 1 if the landowner is a member of the Mississippi Forestry Association or any of its associated County Forestry Associations; 0 otherwise	0.295	--
Residence	Dummy = 1 if the landowner resides on forestland; 0 otherwise	0.501	--
Harvest characteristics			
Harvest Acreage	Harvested acreage for each harvest activity	98.208	123.768
Harvest Date	Time (month) from beginning harvest to end of the study period	65.413	40.901
Consult	Dummy = 1 if a consultant handled the harvest; 0 otherwise	0.545	--
Market factors			
Sawtimber Price	Sawtimber real price (base = 1996)	40.769	4.671
Pulpwood Price	Pulpwood real price (base = 1996)	8.866	3.289
Reforestation Cost	Reforestation real cost (base = 1996)	104.229	9.439

The independent variables were selected to mirror factors identified as significant determinants of replanting behavior in previous research, as discussed in the Introduction. These variables consisted of four groups: land and ownership characteristics, landowner demographics, harvest characteristics, and market factors. Variable definitions are presented in table 1. Four variables were used to represent land and ownership characteristics: *Acreage* (acres owned), *Forest Type* (pine or other), *Ownership Length* (years), and *Timber* (a measure of interest in timber production). Seven variables represented landowner demographics: *Age* (years), *Education* (Bachelors degree or not), *Income* (dollars), *Race* (white or other), *Gender* (male or female), *Membership* (membership in forestry organizations), and *Residence* (residing on/off the forested property). Three variables represented characteristics of the harvest: *Harvest Acreage* (acres), *Harvest Date*, and *Consult* (participation of a consulting forester). Finally, three variables were constructed from non-survey data to represent market factors: *Sawtimber Price*, *Pulpwood Price*, and *Reforestation Cost*. Nominal prices for sawtimber and pulpwood at the time of harvest were obtained from Timber-Mart South from 1996 to 2006. Nominal costs for forestry practices in the South at the time of harvest were obtained from the Cost and Cost Trends series produced on two-year intervals (Dubois, McNabb, Straka & Watson 1995; Dubois, McNabb & Straka 1997; Dubois, McNabb & Straka 1999; Dubois, McNabb & Straka 2001; Dubois, Straka, Crim & Robinson 2003; Smidt, Dubois & Folegatti 2005). For the unreported years, cost was calculated by averaging the costs over adjacent years. Reforestation costs included chemical site preparation and hand planting. Real prices and costs (adjusted for inflation, expressed in 1996 dollars) were calculated by dividing their nominal values by the Producer Price Index. Thus, *Sawtimber Price*, *Pulpwood Price*, and *Reforestation Cost* were expressed in real terms.

3. Results

Of the 9,925 landowners contacted by phone, 2,126 owned less than 100 acres, another 2,132 did not harvest timber in the past 10 years. These landowners were excluded from the survey. There were also 1,110 wrong telephone numbers. Other reasons for unsuccessful calls included communication problems, refusal to participate, and deceased owners. In total, 2,229 landowners completed the survey for a response rate of 48.9%, i.e., 2,229/ (9,925-2,126-2,132-1,110).

There were 1,081 final harvests conducted by these 2,229 landowners. Of these harvests, 695 had been replanted by the end of the study period and 386 had not. For 264 of the harvests that had been replanted, the landowner did not recall either the harvest date or regeneration date and another 36 reported the harvest date after the regeneration date, so these observations were excluded from the analysis. For 121 of the harvests that had not been replanted, the landowner did not recall the harvest date and another 5 reported a harvest date outside the survey period. Hence, these observations were also excluded from the analysis. At the end, 655 harvests were available for statistical analysis. Of these, 395 had been regenerated within the study period and 260 harvests had not been regenerated.

The average time elapsed before regeneration (T) was 11 months for uncensored observations, i.e., harvests that were regenerated within the survey period (n = 395). Including censored observations, i.e., harvests that were not regenerated before the end of the study period, the average interval length was 28 months for all harvests (n = 655). As described above, 60.3% of the 655 harvests were replanted during the study period. Conversely, about 39.7% were not replanted. The average acreage owned was 560 acres. For 33.1 % of respondents, pine was the predominant forest type. The average length of ownership was 32 years. Most landowners (79.5%) were interested in timber production.

On average, respondents were 65 years old with household incomes of $66,382. About half of respondents (50.5%) had a bachelor's degree or better. In addition, 96% were Caucasian and 79.5% were male. About 29.5% belonged to a forestry organization and 50.1% resided on their forestland. The average harvest size was 98 acres. A total of 357 landowners hired consultants to handle the harvest. Finally, over the study period, sawtimber averaged $40.77/ton; pulpwood averaged $8.87/ton; and reforestation costs averaged $104.23/acre in real terms.

3.1. Results from Non-Parametric Duration Analysis

The graph of the survival function $S(t)$ depicts the probability that a harvested site had not been regenerated by time t (figure 1). This probability decreased rapidly initially, then flattened out at approximately 28 months. The graph of the hazard function $h(t)$ depicts the probability a harvest site that has not been regenerated will be regenerated at time t (figure 2). This probability increased until the 16th month and thereafter decreased rapidly until the 28th month and remained less than 1% from then on. Along this prediction track, the probability of regeneration approached zero as the time since harvest increased.

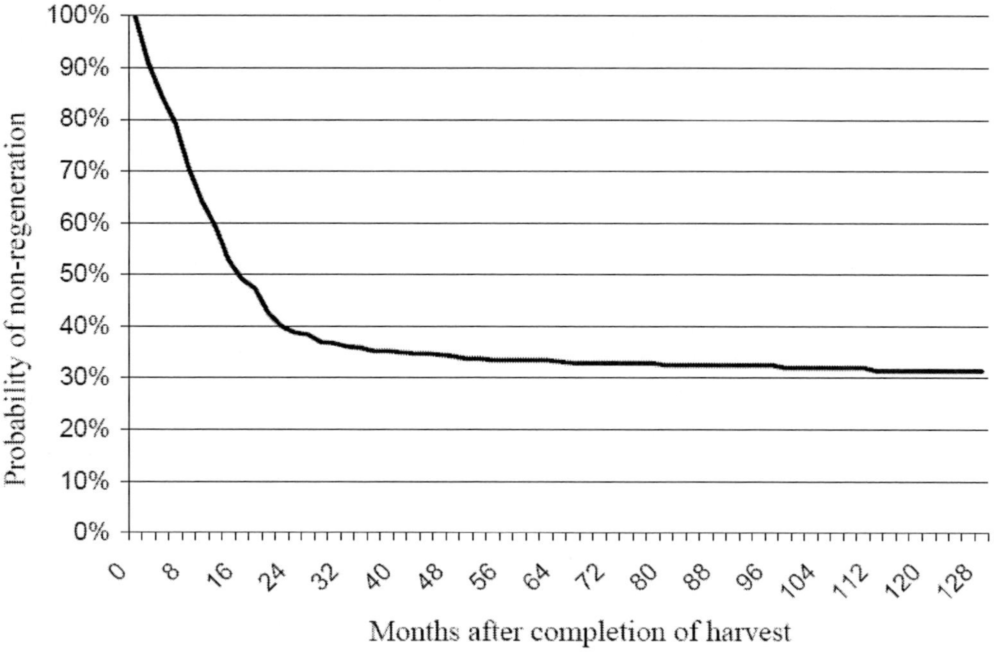

Figure 1. The probability over time that nonindustrial private forest landowners have not regenerated after a timber harvest, based on Mississippi NIPF timber harvests from 1996 to 2006.

3.2. Results from Parametric Duration Analysis

The log-normal distribution was used for estimation as it best fits the distribution for T. Through preliminary analysis, the shape parameter of the generalized gamma model was very close to 0 (i.e., 0.072), which indicated that the log-normal model should be employed (Allison 1995). The results of parametric duration analysis are reported in table 2.

Among land and ownership characteristics, only *Forest Type* and *Timber* were significant and both were negative. If pine was their predominant forest type, landowners took less time to regenerate than others. Landowners interested in timber production regenerated more rapidly than non-interested landowners. The odds ratios for these two variables were 0.566 and 0.185, respectively. Therefore, the predicted regeneration period for landowners whose predominant forest type was pine was 43.4% shorter than landowners with other forest types predominant. The predicted regeneration period for landowners interested in timber production was 81.5% shorter than other landowners.

Among landowner demographic variables, only *Race* and *Residence* were significant and both had negative coefficients, with odds ratios of 0.314 and 0.605, respectively. Caucasian landowners waited 68.6% less time to regenerate than other landowners. The predicted regeneration period for those living on their forestland was 39.5% shorter than for those who did not. Among the harvest characteristics, only *Consult* was significant. The odds ratio was 0.438; thus, for landowners who used a consultant, the regeneration period was 56.2% shorter than those who did not.

Finally, all market factors had significant impacts on the regeneration period. *Sawtimber Price* and *Pulpwood Price* negatively influenced the regeneration period, and *Reforestation Cost* had a positive effect, with odds ratios equal to 0.962, 0.926, and 1.046, respectively. Therefore, each additional dollar increase in *Sawtimber Price* was associated with a 3.8% decrease in predicted regeneration period. Each additional dollar increase in *Pulpwood Price* was associated with a 7.4% decrease in the predicted regeneration period. Each additional dollar increase in R*eforestation Cost* was associated with a 4.6% increase in the predicted regeneration period.

Table 2. A parametric survival model of factors that influence how long Mississippi NIPF landowners take to regenerate their forestland after harvesting during the period 1996 to 2006

Variable	Coefficient	*t*-ratio	Odds Ratio
Constant	4.247***	8.230	69.895
Land & ownership characteristics			
Acreage	-1.000E-4	1.620	1.000
Forest Type	-0.570***	12.340	0.566
Ownership Length	-0.007	2.540	0.993
Timber	-1.689***	58.660	0.185
Landowner demographics			
Age	0.003	0.140	1.003
Education	-0.001	0.000	0.999
Income	-7.000E-4	0.050	0.999
Race	-1.157**	6.500	0.314
Gender	-0.018	0.010	0.982
Membership	-0.086	0.250	0.918
Residence	-0.503***	9.750	0.605
Harvest characteristics			
Harvest Acreage	-2.000E-4	0.090	1.000
Harvest Date	-0.001	0.020	0.999
Consult	-0.825***	25.660	0.438
Market factor			
Sawtimber Price	-0.039**	4.650	0.962
Pulpwood Price	-0.077**	5.130	0.926
Reforestation Cost	0.045***	8.880	1.046
LOG Likelihood	-954.942		

***, **, and * indicate significance at the 1%, 5%, and 10% level, respectively.

4. Conclusion

This chapter employed non-parametric and parametric duration analysis to investigate how quickly landowners regenerated their timberlands after harvest and what factors influenced the length of this interval. This chapter is the first to use duration analysis to examine the timing of silvicultural operations. In doing so, it reveals interesting aspects of reforestation behavior with important policy implications that were previously unknown.

The non-parametric results paint an interesting picture of reforestation behavior. Consistent with previous research (Gunter et al. 2001), about forty percent of the surveyed NIPF landowners did not replant after harvesting. Unlike previous research, however, this chapter sheds light on the timing of reforestation activities. For those landowners who replanted, most did so within a year and virtually all replanted within two. Given that planting is largely confined to the winter months, landowners really have one window of opportunity a year. There are many unavoidable reasons that landowners may miss this window during the first year. For example, if harvesting is completed late in the year, site preparation may not be complete before winter arrives. Likewise, where pales weevils, *Hylobius pales* (Herbst), are a problem, delaying replanting until the second winter after harvest is a commonly accepted control practice. Even with the best of intentions, a sizeable portion of landowners would still be replanting in the second winter after harvesting. These results indicate that, in general, landowners who intend to replant, do so promptly. The implication for policymakers is that programs intended to promote replanting should be streamlined to avoid bureaucratic delays. Landowners waiting on assistance typically delay reforestation by a year or more (Nodine 1992). An unintended consequence of assistance programs is that, because of the long delays, the probability of replanting by landowners who ultimately do not receive assistance decreases substantially.

The parametric results are largely consistent with previous studies in that factors associated with shorter intervals between harvesting and replanting are those previously found associated with higher replanting probabilities, e.g., interest in timber production, employing a consulting forester, having predominantly pines, higher stumpage prices, residing on forestlands, and race (Hyberg & Holthausen 1989; Royer 1985). Similarly, reforestation costs lengthened the replanting interval, consistent with lower replanting probabilities found in previous studies (Brooks 1985; Hyberg & Holthausen 1989; Royer 1985). The estimated coefficients and associated odds ratios for some of the significant variables have important policy implications for decreasing planting delays and improving overall reforestation rates.

Stumpage prices and reforestation costs potentially have the greatest impact on timely reforestation. While the marginal impacts of a one unit change in these variables are relatively small compared to the other significant variables, stumpage prices and reforestation costs are relatively volatile[2] compared to the other explanatory variables, thus large changes in prices and costs over time are more likely than for the other explanatory variables. A four dollar increase in the per ton price of sawtimber, a 10% shift from the mean value over the study period, translates to a four month increase in the mean planting interval of 28 months. An additional four month delay for tracts not planted by 20 months essentially decreases the probability that these tracts will be replanted to virtually zero (See figure 2). Coupled with the non-parametric results, that tracts not replanted within two years are very unlikely to be replanted at all, the implication is that if prices remain low over a period of more than two years, tracts not planted because of adverse prices are unlikely to be replanted when prices improve. From a policy perspective, promoting landowner assistance programs during periods of low stumpage prices is critical.

---

[2] For an illustration of this volatility, see the Market News Quarterly report in any issue of Timber Mart-South for quarterly changes in stumpage prices. Sawtimber prices ranged from $51.37/ton to $31.30/ton over the study period

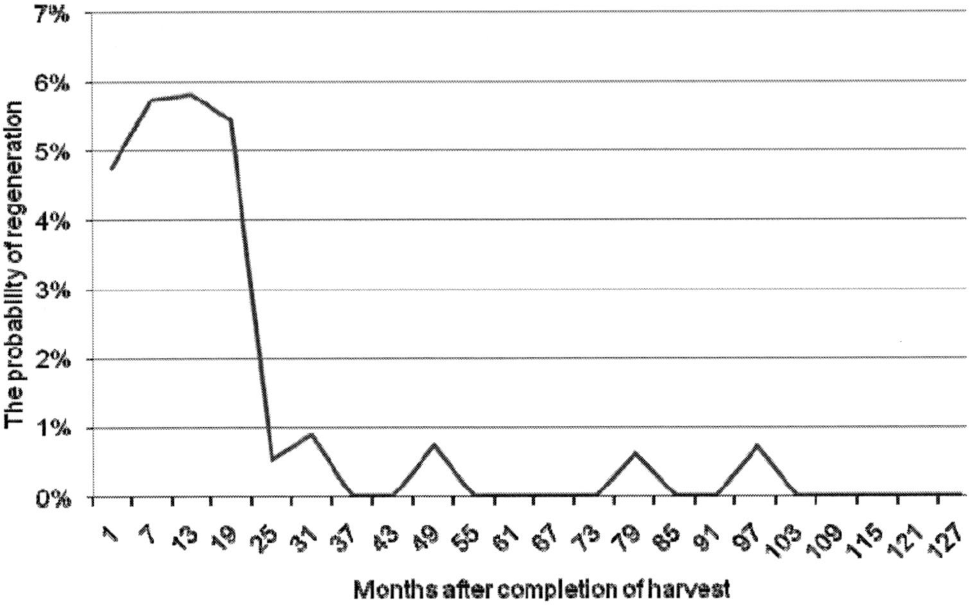

Figure 2. The probability that nonindustrial private forest landowners will regenerate a harvested site at a specific time, given the site has not yet been regenerated, based on Mississippi NIPF timber harvests from 1996 to 2006.

Consult is another key variable. Landowners employing consultants are more likely to replant and do so substantially faster than landowners who do not employ consultants. Given that 45.5% of landowners did not employ consultants to handle their timber sales, promoting consultant involvement to landowner groups would be an effective policy tool to encourage prompt reforestation among NIPF landowners.

The relatively large odds ratio for *Timber* suggests that generating interest in timber production among landowners would be a fruitful policy approach to enhance reforestation. However, the percentage of landowners currently not interested in timber production is very small (20.5%). Identifying those landowners would be difficult. In addition, many landowners may be unlikely to change in this regard because of preferences for recreation, aesthetics, or other non-market values.

Similarly, *Forest Type,* despite its relatively large odds ratio, has limited policy value. To decrease planting delays by changing the predominant forest type to pine would require converting mixed pine hardwoods and upland hardwood stands to pines. Landowners interested in timber production have probably already done so and the remaining landowners are unlikely to adopt conversion practices for reasons mentioned previously.

Race and *Residence* also have large marginal impacts on reforestation times but do not suggest viable policy venues. Non-Caucasians represent a very small portion of the timberland-owning population and encouraging reforestation within this segment of the population may be justifiable on equity grounds, however, the impact on timely reforestation at the landscape level would be minor. Landowners that do not reside on their forestland represent a substantially larger segment of the population; however, the reasons that resident landowners reforest more quickly than non-residents is very likely linked to issues associated with their continuous presence on their lands, e.g., aesthetics. Short of encouraging landowners to reside on their lands, designing policies to improve replanting rates by addressing differences between residents and non-residents is not likely to be a productive venue for improving reforestation efforts.

In conclusion, duration analysis has provided some key insights into the reforestation period and factors that impact its length. Timing is a critical component of reforestation behavior that has been ignored in previous research. To be effective, policies must emphasize prompt reforestation and avoid imposing delays on potential participants. Periods of low stumpage prices, in particular, warrant particular attention. Involvement of consultants, or other professional foresters, appears to be a highly effective venue to decrease planting delays. Future research should examine regional and/or species differences, other potentially influential factors such as government support programs, and the effect of key dates such as when harvests are completed. Duration analysis is a very effective tool for analyzing forestry issues where time is a concern. Reforestation delays are an obvious application. Silvicultural treatments where the duration of the response varies, such as fertilization and herbicide applications, are a potentially fruitful area. There are a number of other socio-economic issues where the technique would also be relevant. Rotation lengths, ownership tenure, hunting lease turn-over are some examples.

References

Allison, P. D. (1995). *Survival Analysis Using SAS: A practical guide*. Cary, North Carolina. SAS Publishing.

Amacher, G. S., Conway, M. C., & Sullivan, J. (2003). Econometric analyses of nonindustrial forest landowners: is there anything left to study? *Journal of Forest Economics*, *9*(2), 137-164.

Beach, R. H., Pattanayak, S. K., Yang, J. C., & Muray, B. C. (2005). Econometric studies of non-industrial private forest management: a review and synthesis. *Forest Policy and Economics*, *7*(3), 261-281.

Brooks, D. J. (1985). Public policy and long-term timber supply in the South. *Forest Science*, *31*(2), 342-357.

Dubois, M. R., McNabb, K., Straka, T. J., & Watson, W. F. (1995). Costs and cost trends for forestry practices in the South. *Forest Farmer 30th Manual Edition, March/April*, 10-17.

Dubois, M. R., McNabb, K., & Straka, T. J. (1997). Costs and cost trends for forestry practices in the South. *Forest landowner 31st Manual Edition, March/April*, 7-13.

Dubois, M. R., McNabb, K., & Straka, T. J. (1999). Costs and cost trends for forestry practices in the South. *Forest Landowner 32nd Manual Edition, March/April*, 3-8.

Dubois, M. R., McNabb, K., & Straka, T. J. (2001). Costs and cost trends for forestry practices in the South. *Forest Landowner 33rd Manual Edition, March/April*, 3-9.

Dubois, M. R., Straka, T. J., Crim, S. D., & Robinson, L. J. (2003). Costs and cost trends for forestry practices in the South. *Forest Landowner 34th Manual Edition, March/April*, 3-10.

Greene, W. H. (2003). *Econometric Analysis*. Pearson Education, Inc., Delhi, India.

Gunter, J. E., Bullard, S. H., Doolittle, M. L., & Arano, K. G. (2001). Reforestation of harvested timberlands in Mississippi: behavior and attitudes of non-industrial private forest landowners. FWRC Research Bulletin #FO 172, Forest and Wildlife Research Center, Mississippi State University, 25.

Hyberg, B. T., & Holthausen, D. M. (1989). The behavior of non-industrial private forest landowners. *Canadian Journal of Forest Research, 19*, 1014–1023.

Nagubadi, V., McNamara, K. T., Hoover, W. L., & Mills, W. L. (1996). Program participation behavior of nonindustrial forest landowners: a probit analysis. *Journal of Agricultural and Applied Economics, 28*(2), 323-336.

Nodine, S. K. (1992). Reforestation delay tolerance in the Forestry Incentives Program. *Canadian Journal of Forest Research, 23*, 414-426.

Royer, J. P. (1985). The effects of markets and public policies on the reforestation behavior of southern landowners. Southeastern Center for Forest Economics Research WP 12, Research Triangle Park, NC,

Royer, J. P. (1987). Determinants of reforestation behavior among southern landowners. *Forest Science, 33*(3), 654-667.

Sessions, J., Bettinger, P., Buckman, R., Newton, M., & Hamann (2004). Hastening the return of complex forests following fire: The consequences of delay. *Journal of Forestry, 102*(3), 38-45.

Smidt, M., Dubois, M. R., & Folegatti, B. S. (2005). Costs and cost trend for forestry practices in the south. *Forest Landowner 35th Manual Edition, March/April*, 25-31.

Wear, D. H., & Greis, J. G. (2002). Southern forest resource assessment. *Southern Research Station. Gen. Tech. Rep. SRS, 53*, 175-223.

Zhang, D., & Flick, W. A. (2001). Sticks, carrots, and reforestation investment. *Land Economics, 77*(3), 443-456.

In: Global Change and Forestry
Editors: J. Gan, S. Grado and I. A. Munn
ISBN: 978-1-60876-262-0

Chapter 19

Reforestation Tax Incentives under the American Jobs Creation Act of 2004[1]

Thomas J. Straka*[*1] and John L. Greene[2]
[1] Department of Forestry and Natural Resources, Clemson University, SC, USA
[2] USDA Forest Service, Southern Research Station, NC, USA

Abstract

The American Jobs Creation Act of 2004 made significant changes in the reforestation tax incentives available to private forest owners. Owners can now deduct outright reforestation costs up to $10,000 per year for each qualifying timber property and amortize any additional amount over 8 tax years. To assess the financial benefit the new incentives provide to forest owners, the authors developed spreadsheets that calculate after-tax Bare Land Value (BLV) for a representative southern pine management plan under three tax situations: no reforestation incentives, the incentives under previous law, and the incentives under the current law. They found that compared to no tax incentive, the current law chiefly benefits owners with high non-timber income, increasing BLV by an amount equivalent to a reforestation cost share of roughly 25–30% as opposed to 5–15% for owners with low or median income. Compared to previous law, the current law chiefly benefits owners of large forest holdings, increasing BLV by an amount equivalent to a reforestation cost share of roughly 10–20%. For owners of small forest holdings, however, BLV decreased by an amount equivalent to a 5–10% increase in reforestation costs. These findings are significant as Congress likely intended that the new incentives continue to benefit primarily "small woodland owners" with modest incomes and forest holdings.

Keywords: Reforestation; Income tax; Incentive; Cost share; Financial analysis.

---

[1] This chapter is reprinted with permission from the article originally published in the *Southern Journal of Applied Forestry* 31(1): 23–27. This is Technical Contribution Number 5195 of the Clemson University Experiment Station.

[*] E-mail: tstraka@clemson.edu, Department of Forestry and Natural Resources, Clemson University, Clemson, SC 29634-0317.

1. Introduction

The American Jobs Creation Act of 2004 (PL 108-357) made significant changes in the reforestation tax incentives available to private forest owners. Under the previous law (PL 96-451) owners could take a 10% tax credit on and amortize (write off) reforestation costs up to $10,000/year over 8 tax years.[2] Beginning on Oct. 23, 2004, the day after President Bush signed the Act into law, owners could deduct outright reforestation costs up to $10,000/year for each qualified timber property and amortize any additional amount, again, over 8 tax years. The reforestation tax credit is eliminated (RIA 2004).

With its $10,000 cap on both the tax credit and the amortization provisions, the previous law was intended to benefit primarily "small woodland owners." In contrast, the current law benefits owners of forest holdings of all sizes, large and small. With its large, up-front deduction, the current law is comparable with a reforestation cost share. The size of the cost share it is equivalent to, however, varies with the size of the forest holding and amount of the owner's nontimber income.

2. Study Approach

The approach presented in this study was used to develop a measure of the net financial benefit that both the previous and the current reforestation tax incentives provide to private forest owners and to show the differences between them. To accomplish this, we developed spreadsheets that modeled a representative southern pine management plan under three tax situations. The first was the no incentive situation, with reforestation costs established as "basis" until timber is harvested; the second was the previous law, with its limited reforestation tax credit and amortization provisions; the third was the current law, with its reforestation deduction and unlimited amortization provisions.

Each spreadsheet calculates on a year-by-year basis the net financial effect of owning and managing a forest holding under the associated tax situation, including the costs of site preparation and planting; property tax; the effect on federal and state income taxes of deducting forest management expenses and using any reforestation incentives; and the returns, costs, harvest taxes, and federal and state capital gain taxes resulting from timber harvests. The annual net cost and return figures were discounted to the beginning of the rotation using the owners' personal discount rate (see in the following paragraphs) and summed to calculate after-tax net present value (NPV) per acre and bare land value (BLV) per acre. The underlying equations used in the spreadsheets to calculate NPV and BLV are shown in table 1. The spreadsheets also were capable of modeling real annual increases in reforestation costs, forest product prices, and property tax, but that feature was not used for this analysis.

[2] The regulations for the amortization provision required that forest owners reduce the amount amortized by half of any reforestation tax credit taken.

Table 1. Equations used in the study spreadsheets to calculate Net Present Value (NPV) per acre and Bare Land Value (BLV) per acre.

Net Present Value (NPV):

$$NPV = \sum_{n=0}^{r} \frac{-SP + (HR - AC) - PT - HT - (FT_{WO} - FT_{W}) - (ST_{WO} - ST_{W}) - FC - SC}{(1+i)^{r}}$$

Bare Land Value (BLV):

$$BLV = \frac{NPV * (1+i)^{r}}{(1+i)^{r} - 1}$$

where: r = Rotation length in years
SP = Site preparation and planting cost
HR = Harvest return
AC = Harvest administration cost
PT = Property tax
HT = Harvest tax
FT_{WO} = Federal tax on the owners' ordinary, non-timber income, without including the effect of the forest holding
FT_{W} = Federal tax on the owners' ordinary, non-timber income, including the effect of the forest holding
ST_{WO} = State tax on the owner's ordinary, non-timber income, without including the effect of the forest holding
ST_{W} = State tax on the owners' ordinary, non-timber income, including the effect of the forest holding
FC = Federal tax on timber capital gain
SC = State tax on timber capital gain
I = Owners' personal discount rate
A = Number of acres in the forest holding

The spreadsheets were used to calculate after-tax NPV and BLV for ownership scenarios representing five different combinations of forest size and nontimber income:

- Low-income owners with a small forest holding.
- Median-income owners with a small forest holding.
- Median-income owners with a large forest holding.
- High-income owners with a small forest holding.
- High-income owners with a large forest holding.

In each ownership scenario, the forest owners were assumed to be married couples who file jointly. Nontimber income was assumed to be \$60,000/year for the median income scenarios – a figure that closely approximates average 2005 disposable personal income for a two-person household (Council of Economic Advisors 2006) – \$20,000/year for the low-income scenarios and \$180,000/ year for the high-income scenarios. It was further assumed that the owners qualify as material participants in their forest enterprise, have itemized deductions equal to the standard deduction, and a personal discount rate of 4% real (with inflation factored out).

The forest holding was assumed to consist of a single even-age stand constituting one qualified timber property. Stand size was assumed to be 40 ac for the small-holding ownership scenarios and 400 ac for the large-holding scenarios (table 2a). Timber management costs were adapted from the *Forest Landowner 34th Manual Edition* (DuBois et al. 2003). The sawtimber and pulpwood stumpage prices used were 5-year regional averages from the *Timber Mart-South Market Newsletter* (Timber Mart-South 2001–2005). Sale administration costs were assumed to equal 10% of the gross stumpage price (table 2b).

Taxes the forest owners were assumed to be subject to include a property tax of $5/ac per year, a harvest tax equal to 2.5% of the gross stumpage price, federal income and capital gain taxes at 2005 rates, and state income and capital gain taxes at 25% of the federal rates (table 2c). The forest management plan was taken from Busby et al. (1990; table 2d).

This marginal approach enabled us to show the effect of the change in reforestation tax incentives on private forest owners with various sizes of forest holdings and income levels. The remainder of this chapter presents and discusses the study findings.

Table 2. Assumptions about the forest owners and ownership, costs and returns, taxes, and forest management plan used in the analysis.

a. Forest Owners and Ownership Married couple filing jointly, with	$20,000of non-timber income per year, $60,000 of non-timber income per year, or $180,000 of non-timber income per year
Qualify as material participants	
Itemized deductions equal to the standard deduction	
Personal discount rate	4% real
Tract size	40 acres, or 400 acres
b. Timber Costs and Returns	
Site preparation and planting	270.00 per acre, 0% annual increase
Pulpwood stumpage price	18.00 per cord, 0% annual increase
Chip-n-saw stumpage price	62.50 per cord, 0% annual increase
Sawtimber stumpage price	277.00 per MBF, 0% annual increase
Sale administration cost	10% of gross stumpage price
c. Federal, State, and Local Taxes	
Annual property tax	5.00 per acre, 0.00 annual increase
Harvest tax	2.5% of gross stumpage price
Federal income and capital gain tax rates	2005 schedules
State income and capital gain tax rates	25% of federal rates
d. Forest Management Plan	
Year 0: Site preparation and planting	
Year 15: Commercial thinning	3.85 cords per acre pulpwood
	0.75 cords per acre chip-n-saw
	0.00 MBF per acre sawtimber
Year 30: Final harvest	12.21 cords per acre pulpwood
	25.44 cords per acre chip-n-saw
	2.89 MBF per acre sawtimber

3. RESULTS

3.1. No Reforestation Incentives

Even in the absence of reforestation tax incentives after-tax BLV varied with owner income and forest size, from $504.48 in the low-income, small-holding scenario to $438.73/ac in the medianincome, large-holding scenario (table 3, tax situation 1). Two factors contributed to the high BLV in the low-income, small-holding scenario, both related to the owners' low level of nontimber income. First, the deductions for property and harvest taxes removed a much larger fraction of the owners' nontimber income from taxable income than in the other scenarios. Second, because capital gains are allocated between the 5 and 15% tax rates based on total income, most of the owners' timber capital gains were taxed at the lower rate (table 3, tax situation 1).

Table 3. Comparison of the financial effect of reforestation tax incentives on owners under differing assumptions about forest size and non-timber income (all after-tax BLVs are on a per-acre basis)

	Tax Situation					
	1: No Reforestation Incentives		2: Previous Law (P.L. 96-451)		3: Current Law (P.L. 108-357)	
		Increase Over		Increase Over		Increase Over
Ownership Scenario	After-Tax BLV	Tax Situation 1	After-Tax BLV	Tax Situation 1	After-Tax BLV	Tax Situation 1
Low income, small holding	504.48	--	566.81	12.4%	526.54	4.4%
Median income, small holding	458.54	--	538.07	17.3%	507.74	10.7%
Median income, large holding	438.73	--	446.55	1.8%	474.84	8.2%
High income, small holding	453.69	--	574.73	26.7%	557.17	22.8%
High income, large holding	453.69	--	465.79	2.7%	543.44	19.8%

Comparing the two median-income scenarios – median income, small holding and median income, large holding – the higher forest acreage under management in the large-holding scenario resulted in higher timber capital gains in years with a thinning or final harvest. But, again, because capital gains are allocated between the 5 and 15% tax rates based on total income, a substantially higher proportion of the large-holding owners' timber capital gains were taxed at the higher rate, reducing BLV (table 3, tax situation 1). Had there been only one capital gain tax rate, BLV would have been the same for both scenarios.

Comparing the two-high income scenarios – high income, small holding and high income, large holding – in both cases, the owners' nontimber income was above the ceiling for the 5% tax rate for capital gains. All their timber capital gains were taxed at the 15% rate with the result that BLV was the same for both scenarios (table 3, tax situation 1).

Comparing the two large-holding scenarios – median income, large holding and high income, large holding – property and harvest taxes were deducted against nontimber income in the 28% federal tax bracket in the high-income scenario, as opposed to income in the 15%

bracket in the median-income scenario. This resulted in a greater reduction in federal and state income taxes in the high-income scenario and a higher BLV (table 3, tax situation 1).

BLV was nearly the same in the two remaining small-holding scenarios – median income, small holding and high income, small holding – but for different reasons. The advantage to the median income owners of having a proportion of their timber capital gains taxed at 5% rather than 15% slightly exceeded the advantage to the high-income owners of having their property and harvest taxes deducted against nontimber income in the 28% federal tax bracket rather than the 15% bracket (table 3, tax situation 1).

3.2. Previous Law

The previous law (PL 96-451) increased after-tax BLVs over the no-incentive tax situation in all five ownership scenarios (table 3, tax situation 2). The greatest increases occurred in the scenarios characterized by small forest holdings. In these scenarios, the reforestation tax credit provided a dollar-for-dollar reduction in the amount of income tax due. In two scenarios – median income, small holding and high income, small holding – the owners were able to take the full amount of the credit in the first year of a rotation. In the other scenario – low income, small holding – the owners had to spread the credit over 6 tax years because it exceeded their income tax due; thus, the economic value of the credit was reduced for the lowest-income owners (table 3, tax situation 2).

More important from an economic standpoint, the law's amortization provision enabled all three sets of owners to recover nearly all of their reforestation costs during the first 8 years of a rotation. For the low-income, small-holding owners the amortization deduction was taken against nontimber income in the 10% federal tax bracket, providing the smallest increase in BLV of the three. For the median-income, small-holding owners it was taken against nontimber income in the 15% bracket, providing somewhat larger increase in BLV. For the high-income, small-holding owners it was taken against nontimber income in the 28% bracket, providing the largest increase in BLV (table 3, tax situation 2).

In the ownership scenarios characterized by large forest holdings – median income, large holding and high income, large holding – the increases in BLV were substantially lower (table 2, tax situation 2). This is because the $10,000 cap on both the reforestation tax credit and the amortization provisions allowed the owners to recover only a fraction—less than one-tenth— of their reforestation costs in the early years of a rotation. The large majority of the costs had to be carried as "basis" until timber was harvested. Again, the high-income owners took the amortization deduction against nontimber income in the 28% federal tax bracket as opposed to the 15% bracket for the median-income owners, resulting in a larger increase in BLV (table 3, tax situation 2).

3.3. Current Law

The current law (PL 108-357) also increased after-tax BLVs over the no-incentive tax situation in all five ownership scenarios (table 3, tax situation 3). The pattern, however, was quite different from under the previous law, with the greatest increases occurring in the ownership scenarios characterized by high nontimber income. In the high-income, small-holding scenario the owners benefited most from the law's reforestation deduction provision, which enabled them to recover nearly all of their reforestation costs in the year they occurred. Little was left to amortize. In the high-income, large-holding scenario the owners benefited most from the law's unlimited amortization provision, which allowed them to recover reforestation costs above the $10,000 deduction amount during the first 8 years of a rotation. In addition, in these scenarios, both the reforestation and the amortization deductions were taken against nontimber income in the 28% federal tax bracket (table 3, tax situation 3).

In the scenarios characterized by median income, BLV increased by roughly one-half as much as in the high-income scenarios (table 3, tax situation 3). The median-income, small-holding scenario mirrored the high-income, small-holding scenario, mentioned previously, with the owners benefiting most from the law's reforestation deduction provision. The median-income, large-holding scenario mirrored the high-income, large-holding scenario, mentioned previously, with the owners benefiting most from the law's unlimited amortization provision. The increases in BLV were lower than for the high-income owners because in these scenarios the reforestation and amortization deductions were taken against nontimber income in the 15% federal tax bracket (table 3, tax situation 3).

In the low-income, small-holding scenario the owners were not able to make full use of the $10,000 reforestation deduction, because it exceeded their taxable income by a sizeable amount. As a result, they benefited most from amortization of their reforestation expenses. The deductions were taken against nontimber income in the 10% federal tax bracket, resulting in the lowest increase in BLV (table 3, tax situation 3).

It should be noted that after-tax BLVs for the two large-holding scenarios were higher under current law than under the previous law, but those for the three small-holding scenarios were lower (table 3, tax situations 2 and 3). This indicates that regardless of income level, for owners of small forest holdings, the current law's more generous provisions for reforestation and amortization deductions are outweighed by the loss of the previous law's reforestation tax credit.

3.4. Cost Share Equivalents

Two sets of cost share equivalents were calculated; the first was to determine what size reforestation cost share made the no-incentive situation equivalent to the current law and the second was to determine what size reforestation cost share made the previous law equivalent to the current law. The same procedure was used in both cases: reforestation costs were changed incrementally until after-tax BLV equaled that for the current law for each ownership scenario.

The results were again divided by characteristics of the forest ownership. Compared with the no-incentive situation, the current law increased BLV by an amount equivalent to a reforestation cost share of 5–15% in the low- and median-income scenarios and by an amount equivalent to a cost share of roughly 25–30% in the high income scenarios (table 4).

Compared with the previous law, the current law increased BLV by an amount equivalent to a reforestation cost share of roughly 10–20% in the large-holding scenarios. In the small-holding scenarios, however, BLV decreased by an amount equivalent to a 5–10% increase in reforestation costs (table 4). As mentioned previously, this result is chiefly because of the elimination of the previous law's reforestation tax credit.

4. CONCLUSION

The objective of reforestation tax incentives is to encourage reforestation on privately owned forestland. With its limited reforestation tax credit and amortization provisions, the previous law was clearly intended to benefit primarily "small woodland owners." In contrast, the broader reforestation deduction and amortization provisions of the current law benefit owners of both large and small forest holdings.

Even with no reforestation incentives (table 3, tax situation 1), after-tax BLV varies with size of the forest holding and amount of the owner's nonforest income. Allocation of a large part of their timber income to the 5% capital gain tax rate increases BLV for low- and median-income owners. Deduction of property and harvest taxes against income in a high marginal tax bracket increases BLV for high-income owners.

The previous law (PL 96-451; table 3, tax situation 2) increased after-tax BLVs over the no-incentive tax situation in all five ownership scenarios. The largest% increases occurred for the owners of small forest holdings. This is primarily because the $10,000 cap on both the reforestation tax credit and the amortization provisions limits the deductions available to owners of large forest holdings to a fraction of their total reforestation costs.

The current law (PL 108-357; table 3, tax situation 3) also increased after-tax BLVs over the no-incentive tax situation in all five ownership scenarios, but the pattern of benefits is quite different. Compared with the no-tax incentive, the current law primarily benefits owners with high levels of nontimber income, because the tax savings from the $10,000 reforestation deduction and unlimited amortization provisions is greatest for owners in high marginal tax brackets. Compared with the previous law, however, the current law primarily benefits owners of large forest holdings, because removal of the $10,000 cap on benefits enables them to recover all their reforestation costs in the early years of a rotation. For owners of small forest holdings, loss of the previous law's reforestation tax credit outweighs the current law's more generous provisions for deduction and amortization of reforestation costs.

With its large, up-front deduction, the current law is comparable with a reforestation cost share, and a straightforward way to show its effect is to calculate its cost share equivalent. We did this by calculating the% reduction in reforestation costs under the no-incentive situation and under the previous law that made after-tax BLV equal to that under the current law for each ownership scenario. Compared with the no-incentive situation, the current law chiefly benefits owners with high nontimber income, increasing BLV by an amount equivalent to a reforestation cost share of roughly 25–30% as opposed to 5–15% for owners with low or median income. Compared with the previous law, the current law chiefly benefits owners of large forest holdings, increasing BLV by an amount equivalent to a reforestation cost share of roughly 10–20%. For owners of small forest holdings, however, BLV decreased by an amount equivalent to a 5–10% increase in reforestation costs.

The study results indicate that the current law is less favorable than the previous law for owners of small forest holdings. In addition, its provisions are most beneficial to high-income owners who may have less need for a reforestation incentive. These findings are significant as Congress likely intended that the new incentives continue to benefit primarily "small woodland owners" with modest incomes and forest holdings.

REFERENCES

Busby, R.L., Ward, K.B., & Baldwin, Jr., V.C. 1990. COMPUTE_ MERCHLOB: A growth and yield prediction system with a merchandizing optimizer for planted loblolly pine in the West Gulf Region. *Research Paper* SO-255. New Orleans, LA: USDA-FS Southern Forest Experiment Station.

Council of Economic Advisors. 2006. Economic indicators: March 2006 (Prepared for the Joint Economic Committee, 109th Congress, 2nd Session). Washington, D.C.: U.S. Government Printing Office.

DuBois, M.R., Straka, T.J, Crim, S.D., & Robinson, L.J. 2003. Costs and cost trends for forestry practices in the South. *Forest Landowner,* 62 (2), 3–9.

Research Institute of America (RIA). 2004. RIA's complete analysis of the American Jobs Creation Act of 2004: With code sections as amended and committee reports. New York, NY: RIA

Timber Mart-South. 2000–2005. Timber Mart-South Market Newsletter. Athens, GA: The University of Georgia Daniel B. Warnell School of Forest Resources.

In: Global Change and Forestry
Editors: J. Gan, S. Grado and I. A. Munn

ISBN: 978-1-60876-262-0

Chapter 20

Economic Impacts of Two Mississippi Birding Festivals, 2006-2008

Marcus K. Measells [*] ***and Stephen C. Grado***
Forest and Wildlife Research Center, Mississippi State University, MS, USA

Abstract

Birdwatching is a recreational activity that has continually gained in popularity in the United States. The Great River Birding Trail (GRBT) research project conducted at Mississippi State University will measure current and potential social and economic impacts of birdwatching on private and public sites and festivals along the GRBT in Mississippi. This chapter focused on expenditure data and economic impacts from two types of birding festivals. On-site interviews were conducted in 2006 and 2007 at the Mississippi River Nature Festival at TARA Wildlife near Vicksburg, Mississippi and in 2006 and 2008 at the Hummingbird Migration Celebration at Strawberry Plains Audubon Center near Holly Springs, Mississippi. Economic impacts were modeled using the Impact Analysis for Planning (IMPLAN) System software. Attendance at the Mississippi River Nature Festival ranged from 130 to 241 with total economic impacts to the State ranging from $10,347 to $25,612 per year (2008 dollars). The Hummingbird Migration Celebration attendance ranged from 7,970 to 8,500 with total economic impacts from $103,752 to $160,286 per year (2008 dollars). This information will assist natural resource and tourism agencies and non-governmental organizations as they attempt to complete the Trail along the Lower Mississippi River. It will also allow rural land planners and policy makers to estimate benefits accrued from various land management alternatives on areas related to the Trail and will be useful for establishing marketing and policy strategies related to eco-tourism and resource management oriented toward birdwatching.

Keywords: Birdwatching; Economic impacts; Eco-tourism; Expenditures; Festivals: ; Great River Birding Trail; On-site surveys.

---

[*] E-mail: mmeasells@cfr.msstate.edu, Mississippi State University, Forest and Wildlife Research Center, College of Forest Resources, Box 9681, Mississippi State, MS 39762.

1. Introduction

Mississippi is a state endowed with a wealth of natural resources. Among these are its avian resources, which have increasingly been utilized not only for hunting, but also for insect control, ecological diversity, and recreation-related activities such as birdwatching and wildlife photography. An assumption can be made that birdwatching is an important component of eco-tourism. Therefore, proper avian resource management should be of utmost importance to private and public landowners and natural resource and tourism agencies. One component of this overall effort in Mississippi is the Great River Birding Trial (GRBT). Currently, there is a self-guided birding tour on the northern reaches of what is one of America's longest birding trails. The GRBT is named after the federally designated scenic drive called the Great River Road, which runs from Canada to the Gulf of Mexico. In general, the sites along the GRBT include established wildlife refuges, parks, overlooks, and other attractions no more than 25 to 30 miles from the Great River Road. Currently, the GRBT parallels both sides of the upper Mississippi River from the headwaters at Lake Itasca, Minnesota, downstream for 1,366 miles to the confluence with the Ohio River at Cairo, Illinois. It covers both sides of the upper Mississippi River through Minnesota, Wisconsin, Iowa, Illinois, and Missouri.

Unfortunately many of the Mississippi Delta counties along the GRBT are among the poorest economically in the Nation. Many of these counties, and their citizens, are dependent on natural resources for their economic vitality. While some are more heavily dependent on the forest industry others rely primarily upon the agricultural industry. Landowners in these areas are looking for ways to diversify their income while local and state government leaders are trying to do the same for the local economy. As a result, landowners are turning to alternative sources of income such as wildlife recreation including birdwatching while many governmental agencies are focusing on multiple-use management. Birdwatching may provide these landowners, and governmental agencies (e.g., National Wildlife Refuges, Wildlife Management Areas) alternative venues that are compatible with their changing management objectives and policies. Also, with the increasing demand for nature-based tourism and recreation, many local and state governments are now changing policies and are focusing on promoting nature-based tourism on private, state, and federal lands.

The next step in completing the GRBT is to extend it along both sides of the Lower Mississippi River to the Gulf of Mexico to include the states of Mississippi, Louisiana, Arkansas, Tennessee, Missouri, and Kentucky. The driving forces behind this effort are Audubon Mississippi and the U.S. Fish and Wildlife Service. Together, they have identified and delineated birdwatching sites (approximately 300) and pinpointed bird species of interest (T. Pullen, Audubon Mississippi Contract Employee, pers. comm., 2005). Currently, two-thirds of these sites are open to the public with 95% of them publicly owned and only 5% privately owned. Audubon Mississippi would like to increase private landowner participation on this portion of the GRBT (T. Pullen, Audubon Mississippi Contract Employee, pers. comm., 2005). However, to accomplish this and promote the GRBT in general, they need information on potential economic impacts to compliment what they are trying to achieve (M. Lindsey, Executive Director, Audubon Mississippi, pers. comm., 2005). Without this information as a framework, it will be difficult to evaluate the importance of private landowners in providing birdwatching opportunities to the public. This aspect of the project has both significant social and economic implications that could determine the GRBT's success. This information will also provide a monetary measure of return to county economies from the retention, improvement, or expansion of avian habitats. Thus, economic and social assessments of birdwatching activities are extremely important to the promotion and development of the GRBT in the future.

Birding, or birdwatching, is a recreational activity gaining popularity around the United States. Kerlinger (1992) reported that in 1970 only about 4% of Americans were considered birdwatchers; however, by the mid-1980s 25% were considered birders. A study conducted in 2002 indicated the number of birdwatchers nationwide increased by 27% since 1995 and by 225% since 1982 (Scott and Thigpen 2003). Kerlinger (1992) also stated that fewer than 10 site-specific studies of birders had been completed worldwide at that time. To highlight the importance of birdwatching to local economies, he provided the example of Cape May, New Jersey where \$6 to \$10 million was spent on an annual basis. Crandall et al. (1992) reported approximately 38,000 people visited two conservation areas in southeast Arizona between July 1991 and June 1992 and spent \$1.6 million. In a later study, Kerlinger et al. (1997) reported that birders spent from \$100 to \$130 per day while touring for birds. This amount did not include their travel expenses to and from an area. When trip-related expenses and other expenses such as durable goods were considered, the ensuing expenditures, and hence economic impacts, could be sizable. For example, it was determined that an estimated economic impact of \$90 million per year was attributed to birders in the Rio Grande Valley of South Texas (Kerlinger et al. 1997).

Nationwide, 71.1 million U.S. residents participated in wildlife-watching activities (USDI, FWS, and USDC 2007). Of the 23 million participants who observed wildlife away from home; 12.2 million went to public areas only, 2.5 million visited private areas only, and 6.2 million visited both public and private areas (USDI, FWS, and USDC 2007). There was a strong indication that individuals were searching for both public and private locations for wildlife-watching activities. Nearly 47.7 million of these wildlife watchers observed birds around their homes and on trips in 2006 (USDI, FWS, and USDC 2007). Beyond natural settings and habitats, other avian-related activities are also important. For example, Waldrup (1994) reported that 40,000 people participated nationwide in the annual Christmas Bird Count. Nationwide, wildlife watchers, 16 years and older, spent \$45.7 billion which included equipment expenditures of \$23.2 billion, trip-related expenses of \$12.9 billion, and other expenditures of \$9.6 billion (USDI, FWS, and USDC 2007). In 2006, 731,000 people, 16 years and older, participated in wildlife watching in Mississippi (535,000 birdwatchers) and spent \$176 million dollars (USDI, FWS, and USDC 2008).

Expenditures represent dollars spent in an economy of interest; however, economic impacts measure dollars that remain in that economy. Economic impacts, examined through input-output analysis, are especially useful in describing current and potential economic contributions of natural resource-based recreational activities (e.g., birdwatching) to an overall economy (Johnson and Moore 1993, Strauss et al. 1995, Grado et al. 2001). Economic impacts are often modeled using the Impact Analysis for Planning (IMPLAN) System software (Olson and Lindall 2000). These studies provide regions and states with useful information about the social and economic effects of proposed new projects, programs, and current activities (i.e., birdwatching) (Loomis and Walsh 1997). In 1991, five southern states (Alabama, Arkansas, Louisiana, Mississippi, and Tennessee) generated \$270.2 million in retail sales for non-consumptive bird use (Bird Conservation 1997). Mississippi's portion accounted for \$34.9 million; however, this still led to approximately 1,200 full- and part-time jobs in Mississippi supported by non-consumptive bird use (Bird Conservation 1997). This could be due to the lack of birding festivals and established birdwatching locations within Mississippi. One purpose of this overall research project is to increase this number, especially by taking advantage of the large private landownership in the State.

As important as birdwatching expenditures are, it is paramount to understand the residence and demographics of birdwatchers. This is of great value when determining advertising and marketing strategies. Hvenegaard (2002) studied birders in Thailand and determined there were three specialization levels among them that included novice, advanced-active, and advanced-experienced birders. Hvenegaard (2002) also stated that understanding the demographics of birders will allow for improving economic impacts, developing effective communication and educational programs, and implementing effective marketing campaigns. For example, more advanced birders were more interested only in birding activities (Hvenegaard 2002). Most birdwatchers (97%) in South Carolina prefer to enjoy birdwatching within the state (South Carolina Department of Parks, Recreation, and Tourism 2001). The average South Carolinian went birdwatching 63 times per year, ranking second only to walking as the most favored recreational activity (South Carolina Department of Parks, Recreation, and Tourism 2001). In 1992, birders who were readers of the American Birds magazine were almost evenly split between males and females, 70% attended college, and had a median household income 35% greater than the general public (Waldrup 1994). Average birders at wildlife refuges were middle-aged, highly educated, and had higher incomes (Kerlinger et al. 1997). The first objective of this chapter was to determine birdwatching expenditures of participants at two Mississippi birding festivals. The second objective was to quantify the current and potential economic impacts of these birding festivals affiliated with the GRBT on the Mississippi economy.

2. METHODS

Dillman (2000) found face-to-face and telephone surveys more popular than mail surveys because they provided more accurate data. Consequently, on-site, face-to-face interviews were used to gather information from birders at birdwatching festivals to achieve higher response rates and more reliable data. This survey method allowed the interviewer to explain and interpret any questions the interviewee might have about the survey process. These were important considerations when using a detailed on-site survey associated with expenditure data collection.

Two birding festivals were selected for this project. The Mississippi River Nature Festival held annually in August was hosted at TARA Wildlife (a private entity) located 30 minutes northwest of Vicksburg, Mississippi. The Hummingbird Migration Celebration scheduled in September was hosted by Audubon Mississippi at Strawberry Plains Audubon Center (a non-governmental organization) located north of Holly Springs, Mississippi. Face-to-face surveys were conducted in 2006 and 2007 at the Mississippi River Nature Festival and during 2006 and 2008 at the Hummingbird Migration Celebration.

Survey questions pertained to participant birding activities and trip-related expenditures and birding participation during the year. Participants were randomly selected and asked to provide their on-site, trip-related, and equipment expenditures and the purchase location for each item. In-state expenses were cataloged by amount and county of purchase, and out-of-state expenses by amount and state of purchase. Participants were asked to provide on-site and trip-related expenditures for the current 24 hours to minimize recall error. In situations where participants were on day trips, they were asked to estimate their trip expenses for the remainder of the day. Equipment expenditures included durable items related to participation at the site, acquired during the past year, and brought on the trip where the participant was being interviewed. Estimates on annual use in days for durable items for all purposes was were also collected and divided into the expense for that item. All expenses per participant per day were derived for each specific expenditure category and aligned with the corresponding industrial sector in the modeled economy developed using IMPLAN.

Economic impacts of birding-related activities were modeled using IMPLAN. This software package has been used extensively to study economic impacts of activities related to forestry, agriculture, recreation, tourism, commercial development, and the commercial endeavors of specific industries (Olson and Lindall 2000). IMPLAN software uses economic data from an area of interest to construct a model of its economy based on 509 sectors. Associated databases provide information required to construct regional or state IMPLAN models (Olson and Lindall 2000). County and state level models define relationships between industries and account for monetary leakages (i.e., business transactions) outside of an economy of interest. These 2006 data sets were used to analyze the state input-output structure. Expenditures made on behalf of birding-related activities were then organized as final demands on state industries and businesses.

The IMPLAN model was built to identify aggregated (1 Digit SIC) direct and secondary impacts resulting from birdwatcher expenditures (e.g., binocular expenses were entered into sector 409). Direct impacts represented that portion of expenditures retained by an economic entity in the operation of its business such as sales, salaries, wages, and jobs created by initial purchases of participants. Secondary impacts included indirect effects of inter-industry trade within the region and the induced effects of household consumption originating from employment tied to the direct and indirect activities. Economic impacts were measured in terms of aggregated shipment values, value added to the total economy, and employment attributed to direct and secondary activities.

3. Results and Discussion

A total of 152 festival attendees were asked to participate in the survey with 137 surveys completed at the two birding festivals between 2006 and 2008. Overall the response rate reflecting this effort was 90.1%. The Mississippi River Nature Festival's response rate was 100% (n=20) in 2006 and 92.3% (n=12) in 2007. The Hummingbird Migration Celebration had a 77.8% (n=49) response rate in 2006 and 100% (n=56) in 2008. Attendance from 2006 to 2008 for the Mississippi River Nature Festival was 145; 241; and 130 individuals, respectively. The 2008 attendance was much lower due to weather conditions (remnants of Hurricane Fay were moving through the area). For the same time period, the Hummingbird Migration Celebration had an attendance of 7,970; 8,100; and 8,500, respectively.

In 2006, participants at the Mississippi River Nature Festival incurred overall and in-state expenditures of $47.29 and $44.72/birdwatcher/activity day, respectively while in 2007 the expenditures were $67.56/birdwatcher/activity day for both overall and in-state (table 1). Participants reported in 2006 they would go birdwatching 101 days/year with 98 days being in Mississippi while in 2007 it was 133 days/year with 128 in Mississippi. Overall and in-state expenditures for the participants at the Hummingbird Migration Celebration in 2006 were $11.73 and $7.95/birdwatcher/activity day, respectively while in 2008 the expenditures were $13.89 and $12.28/birdwatcher/activity day, respectively (table 2). On average in 2006, they would go birdwatching 93 days/year with 53 days in Mississippi while in 2008 it was 91 days/year with 23 in Mississippi. A large percentage of respondents for the Hummingbird Migration Celebration participants were non-residents explaining the lower number of days spent birdwatching in Mississippi. The Mississippi River Nature Festival expenditures were much higher than those for the Hummingbird Migration Celebration. This can be explained by the fact that the Mississippi River Nature Festival has on-site lodging and many participants were birdwatching the entire weekend while the Hummingbird Migration Celebration does not provide on-site lodging and most participants were only birdwatching one day. According to the classification scheme of Hvenegaard (2002), the majority of Mississippi River Nature Festival participants would be classified as advanced-experienced while the majority of Hummingbird Migration Celebration participants would be novice birders. This can also explain the difference in expenditures as the advanced-experienced birders were willing to spend more money on birdwatching activities.

Table 1. Expenditures ($/birder/activity day) incurred by participants at the Mississippi River Nature Festival near Vicksburg, Mississippi in 2006 and 2007

	2006		2007	
Expenditure	Overall $	MS $	Overall $	MS $
Access fees	1.57	1.57	0.00	0.00
Entertainment	6.33	6.33	0.11	0.11
Equipment	5.61	3.04	8.35	8.35
Lodging	1.09	1.09	0.21	0.21
Package deal*	19.59	19.59	39.29	39.29
Restaurants/groceries	2.80	2.80	3.40	3.40
Retail	1.98	1.98	13.09	13.09
Transportation	8.32	8.32	3.11	3.11
Total	47.29	44.72	67.56	67.56

* Package deal included on-site access fees, lodging, and meals.

Economic impacts resulting from the birdwatching festivals were reported for direct, indirect, induced, and total impacts for aggregated sectors within the economy. The 2006 Mississippi River Nature Festival resulted in total direct impacts of $6,743 stimulating secondary impacts (i.e., indirect and induced) of $3,604 while supporting 0.1 full- and part-time jobs. In 2007, total direct impacts were $16,599 stimulating secondary impacts of $9,013 while supporting 0.3 full- and part-time jobs. The 2008 festival had total direct impacts of $8,954 with secondary impacts totaling $4,862 and supporting 0.2 full- and part-time jobs. Table 3 lists the combined total economic impacts for the Mississippi River Nature Festival in 2008 dollars.

Table 2. Expenditures ($/birder/activity day) incurred by participants at the Hummingbird Migration Celebration near Holly Springs, Mississippi in 2006 and 2008

	2006		2008	
Expenditure	Overall $	MS $	Overall $	MS $
Access fees	1.98	1.98	3.40	3.40
Entertainment	0.05	0.05	0.14	0.14
Equipment	3.26	0.13	0.07	0.04
Lodging	0.24	0.24	0.14	0.14
Package deal*	0.00	0.00	0.00	0.00
Restaurants/groceries	2.01	1.81	2.14	1.96
Retail	2.58	2.57	4.77	4.77
Transportation	1.61	1.17	3.23	1.83
Total	11.73	7.95	13.89	12.28

* Package deal included on-site access fees, lodging, and meals.

The 2006 Hummingbird Migration Celebration resulted in total direct impacts of $66,411 stimulating secondary impacts of $37,342 while supporting 1.4 full- and part-time jobs. The total direct impacts in 2007 were $67,494 stimulating secondary impacts of $37,951 while supporting 1.4 full- and part-time jobs. The 2008 event had total direct impacts of $104,380 with secondary impacts totaling $55,906 and supporting 1.8 full- and part-time jobs. Table 4 lists the combined total economic impacts for the Hummingbird Migration Celebration in 2008 dollars.

Combined, the two festivals had a direct impact of $270,581 over the three years. This resulted in $148,678 of secondary impacts while total impacts totaled $419,257. The major beneficiaries of birdwatching expenditures and impacts for both events were the aggregated sectors of manufacturing and services.

Table 3. Total economic impacts of birdwatchers at the Mississippi River Nature Festival, 2006 through 2008 (2008 dollars)

Industry	Direct ($)	Indirect ($)	Induced ($)	Total ($)
Ag., For., Fish.[a]	0	146	175	321
Mining	0	764	332	1,096
Construction	0	0	3	4
Manufacturing	8,898	4,015	4,729	17,641
TCPU[b]	0	819	254	1,073
Trade	0	798	187	986
FIRE[c]	0	198	130	328
Services	23,398	731	4,199	28,329
Total	32,296	7,471	10,008	49,774

[a] Agriculture, forestry, and fisheries.
[b] Transportation, communication, and public utilities.
[c] Finance, insurance, and real estate.

Table 4. Total economic impacts of birdwatchers at the Hummingbird Migration Celebration, 2006 through 2008 (2008 dollars)

Industry	Direct ($)	Indirect ($)	Induced ($)	Total ($)
Ag., For., Fish.[a]	0	1,368	1,282	2,650
Mining	0	5,458	2,435	7,893
Construction	0	5	20	26
Manufacturing	90,811	31,904	34,677	157,390
TCPU[b]	0	5,726	1,857	7,582
Trade	0	6,458	1,373	7,830
FIRE[c]	0	1,477	953	2,431
Services	147,474	5,418	30,789	183,680
Total	238,285	57,814	73,385	369,483

[a] Agriculture, forestry, and fisheries.
[b] Transportation, communication, and public utilities.
[c] Finance, insurance, and real estate.

As this research project continues, additional surveys will be conducted at other birdwatching festivals and events. This information will be combined with birdwatching expenditures gathered from a set of public and private sites, already established or under construction along the GRBT, as well as with operator site expenditures. In fact, a Web-based effort, working with Audubon Mississippi, is near completion to capture birdwatching expenditures along the GRBT throughout the year for all southern states affiliated with the Trail. This additional information will allow for a more detailed and accurate determination of the overall economic impact of all birdwatching activities along the GRBT. Results can be used by Audubon Mississippi and other public agencies (e.g., USDI, FWS, MSU Extension) to promote the further development of birdwatching sites on both public and privately held lands along the GRBT. State and federal agencies, conservation groups, private businesses, and landowners can improve local economies by marketing and planning improvements and developments that will enhance birding resources (e.g., bird habitat, birdwatching facilities, bird species sustainability) supported in large part by economic impact analysis, which will compliment the ecological considerations. An improved marketing strategy will increase awareness among birders nationwide about unique birding opportunities available in Mississippi and help increase the number of birding activities and events in the State, resulting in enhanced economic impacts.

4. Conclusion

As the demand for nature-based tourism and recreation continues to increase, many local and state governments and agencies are changing policies and focusing on multiple-use management. This information will assist natural resource (e.g., USDI, FWS) and tourism agencies (e.g., Mississippi Division of Tourism) and non-governmental organizations (e.g., Audubon Mississippi, The Nature Conservancy) in their efforts to complete the GRBT in the Lower Mississippi River area. Also, by connecting to the work already accomplished in the

Upper Mississippi River area and the Mississippi Coastal Birding Trail, already being established in the southern part of the State, it will be an invaluable asset for Mississippi and other states in the Lower Mississippi River area.

Quantifying total employment, income, value added, taxes, and total economic impacts will allow natural resource and tourism agencies, land use planners, and policy makers to estimate benefits accrued from various land management options related to birdwatching both on areas related to the GRBT and beyond. On this basis, funding for birdwatching area maintenance and restoration, species sustainability, and tourism promotion can be justified from both a biological and economic standpoint. The information will also be useful for establishing marketing and policy strategies related to eco-tourism and resource management oriented toward birdwatching and to garner legislative support for funding initiatives to address specific study areas for the GRBT.

ACKNOWLEDGMENTS

Special thanks to the staff at Audubon Mississippi, Strawberry Plains Audubon Center, and TARA Wildlife for allowing us to conduct this study.

REFERENCES

Bird Conservation. (1997). What's a bird worth? *Bird Conservation, Spring,* 6-8.

Crandall, K., Leones, J., & Colby, B. G. (1992). *Nature-based tourism and the economy of southeastern Arizona: Economic impacts of visitation to Ramsey Canyon Preserve and San Pedro Riparian National Conservation Area*. Tucson, AZ: University of Arizona.

Dillman, D. A. (2000). *Mail and Internet surveys: The Tailored Design Method* (2nd edition). New York, NY: Wiley and Sons, Inc.

Grado, S. C., Kaminski, R. M., Munn, I. A., & Tullos, T. A. (2001). Economic impact of waterfowl hunting on public lands and at private lodges in the Mississippi Delta. *Wildlife Society Bulletin, 29*(3), 846-855.

Hvenegaard, G. T. (2002). Birder specialization differences in conservation involvement, demographics, and motivations. *Human Dimensions of Wildlife*, *7*, 21-36.

Johnson, R. L. & Moore, E. (1993). Tourism impact estimation. *Annals of Tourism Research*, *20*, 279-288.

Kerlinger, P. (1992). Birding economics and birder demographics studies as conservation tools. In D. M. Finch & P. W. Stangel (Eds.), *Status and management of neotropical migratory birds*. (pp. 32-38). U.S. Forest Service General Technical Report, RM 229.

Kerlinger, P., Payne, R. H., & Eubanks, T. (1997). Not just for the birds: New studies demonstrate that wildlife refuges can be a financial boon for nearby communities. *Bird Conservation, Spring,* 12-13.

Loomis, J. B. & Walsh, R. G. (1997). *Recreation economic decisions: Comparing benefits and costs*. State College, PA: Venture Publishing, Inc.

Olson, D. & Lindall, S. (2000). *IMPLAN Professional* (2nd edition). Stillwater, MN: MIG, Co.

Scott, D. & Thigpen, J. (2003). Understanding the birder as tourist: Segmenting visitors to the Texas hummer/bird celebration. *Human Dimensions of Wildlife*, *8*, 199-218.

South Carolina Department of Parks, Recreation, and Tourism. (2001). *Birding and other wildlife watching: An overview of the demographics, economics, and trends of wildlife watching, including birdwatching, in South Carolina and the United States*. Columbia, SC: South Carolina Department of Parks, Recreation, and Tourism.

Strauss, C. H., Lord, B. E., & Grado, S. C. (1995). *Economic impact of travel and tourism in Southwestern Pennsylvania during 1994*. University Park, PA: School of Forest Resources, Pennsylvania State University.

U.S. Department of the Interior, Fish and Wildlife Service, and U.S. Department of Commerce, U.S. Census Bureau, (USDI, FWS, and USDC). (2007). *2006 National Survey of Fishing, Hunting, and Wildlife-Associated Recreation*. Washington, DC.: U.S. Gov. Printing Office.

U.S. Department of the Interior, Fish and Wildlife Service, and U.S. Department of Commerce, U.S. Census Bureau, (USDI, FWS, and USDC). (2008). *2006 National Survey of Fishing, Hunting, and Wildlife-Associated Recreation-Mississippi*. Washington, DC.: U.S. Gov. Printing Office.

Waldrop, J. (1994). For the birds. *American Demographics*, *16*(1), 4.

In: Global Change and Forestry
Editors: J. Gan, S. Grado and I. A. Munn

ISBN: 978-1-60876-262-0

Chapter 21

RECREATIONAL VISITATION AND SPENDING PATTERNS ON LAKE IMPOUNDMENTS IN EAST-CENTRAL MISSISSIPPI

Jon P. Rezek[*1] ***and Stephen C. Grado***[2]
[1] Department of Finance and Economics, Mississippi State University, MS, USA
[2] Department of Forestry, Mississippi State University, MS, USA

ABSTRACT

The southeastern United States is uniquely positioned to offer its residents and visitors a variety of recreational opportunities year-round. A favorable climate and an abundance of natural and impoundment water acreage provide long seasons for the region's anglers, boaters, campers, and other recreationists. Natural amenities such as water bodies and warm climates have been shown to stimulate economic growth in rural locations and provide substantial quality of life improvements to residents. While economic injections provided by surface water impoundments have been frequently studied, relatively little research has been conducted assessing how visitation patterns to such facilities vary by season. In this chapter, we discussed results of two on-site surveys carried out at lakes in east-central Mississippi. One survey was conducted at the peak of the recreation season and the other during the fall. Differences were analyzed along four dimensions: length of stay, party size, travel time, and local expenditures. Our findings indicated that significant seasonal variation existed for most visitor types across these dimensions. Since total new visitor spending in the local economy varied by length of stay, party size, and travel time, these results have important implications for the potential economic stimulus of water-based recreational facilities in the South.

Keywords: Lakes; Recreation spending; Visitation patterns; Visitor surveys; Water-based recreational activities.

---

[*] E-mail: jrezek@cobilan.msstate.edu, Department of Finance & Economics, College of Business and Industry, Mississippi State University, Box 9580, Mississippi State, MS 39762.

1. Introduction

The southeastern United States is uniquely positioned to offer its residents and visitors access to a variety of recreational opportunities on a year-round basis. A favorable climate and an abundance of natural and impoundment water acreage provide long seasons for the region's anglers, boaters, campers, picnickers, and other recreationists (USDI and USDC 2008). Environmental attributes such as water bodies and mild climates have been shown to stimulate economic growth in rural locations (Deller et al. 2001), which can increase real estate values, enhance the tax base, and provide substantial quality of life improvements to residents. Recently, several local governments in the state of Mississippi have entertained the possibility of creating surface water impoundments as a source of economic development. These projects were envisioned to serve as engines to stimulate rural development and enhance the quality of life for the State's citizens.

Economic injections provided by surface water impoundments have been frequently studied in Mississippi (Grado et al. 2002, Grado et al. 2004, Rezek, et al. 2006a, Rezek, et al. 2006b); however, relatively little research has been conducted on visitors to these sites. Past studies had to rely, for the most part, on secondary data sources to determine visitor characteristics and develop visitor expenditure profiles (Grado et al. 2002). For this chapter we acquired data from existing lakes in Mississippi, similar to those currently in the early stages of development. While survey dates were limited by funding and time constraints, an effort was made to acquire as much data as possible during peak and off-peak recreational seasons. Our objective was to acquire a realistic database of visitor spending profiles relative to Mississippi and its recreational water resources. We focused on the collection of data concerning local and off-site spending, length of stay, distance traveled, and party size. This data will be helpful in determining the economic impacts and financial feasibility of new lake impoundments as a source of rural development in Mississippi and surrounding states. More specifically, they will be useful in fine-tuning estimates of the size of lake-driven injections and in identifying the extent to which these injections generate permanent, as opposed to seasonal, income and employment impacts.

2. Study Area and Survey Implementation

During 2006, we surveyed recreationists at two U.S. Army Corps of Engineer lakes in the Mobile District. These lake sites were Okatibbee Lake, a 3,800 acre lake near Meridian, Mississippi, and Columbus Lake, an 8,900 acre lake on the Tennessee-Tombigbee Waterway in Columbus, Mississippi. Face-to-face surveys were conducted by two teams of Mississippi State University students enrolled in College of Business and Industry's Master of Business Administration program as part of their capstone Strategic Business Consulting course. Summer surveys were conducted from July 1st through July 7th and fall surveys were conducted from September 16th through September 24th[1].

[1] While few local monthly visitation records were available, statistics collected at national parks indicate a substantial decline in visitation from August to September. Visitation at the Grand Canyon, for instance, typically declines by 50% in September (National Parks Service, 2005). Given this evidence, mid to late September is assumed to be off-peak for the purposes of this study.

Okatibbee Lake and Columbus Lake were selected for several reasons. First, they offered a full range of recreational opportunities including: fishing, boating, waterskiing, picnicking, hunting, sightseeing, and camping. These impoundments also had substantial recreational infrastructure including: boat ramps, a marina, public docks, rustic and developed campgrounds, trails, parks, picnic areas, swimming beaches, and other land-based facilities. Furthermore, these sites were within close proximity to both Mississippi State University and new impoundments under consideration in several central Mississippi counties. Travel times for interviewers were relatively short, increasing the time when students could conduct surveys at these sites.

An on-site survey instrument was selected to eliminate mailing costs, enable interviewers to clearly explain the data being requested, and maximize response rates. Each recreationist encountered at boat landings, picnic sites and other areas was approached by a team of student canvassers and asked if they would be willing to participate in a research study. Response rates were exceptionally high as only a few visitors declined to be interviewed.[2] We included questions eliciting the location of recreationists' residence, their trip duration, number in their party, their primary recreational activity at the lake, and number of days they would recreate at the lake site during 2006. They were also asked open-ended questions about what other activities they would like to see introduced at the lake, and how they found out about the site.

The main focus of our survey, however, was to determine recreation-related expenditure patterns of lake visitors. Interviewers asked each individual to detail their short-term on-site and trip-related expenditures, and their long-term equipment expenditures. For on-site and local area expenses, we asked recreationists to provide their current 24-hour trip expenditures rather than total trip expenditures. This strategy was employed to minimize recall errors for recreationist expenses. In instances where recreationists were making a day trip, we asked them to estimate additional trip expenses for the balance of the day. Long-term expenditures were limited to equipment brought to the site and purchased within the previous year. In-state expenses were cataloged by amount and location, preferably by county. For out-of-state expenses, only the state location of the purchase was documented. Long-term expenditures, however, were not addressed in this chapter.

3. RESULTS

We focused on differences in visitation and spending patterns by visitor type and season. Specifically we analyzed party size, length of stay, distance traveled, local expenditures, and off-site expenditures of recreationists participating in the surveys. For the purposes of statistical comparison we assumed respondents were drawn from the same population – recreationists in Mississippi. Results were organized in tables 1-7. Table 1 detailed the mean, median, and standard deviation of party size at the two lakes surveyed. The first panel showed that the mean and median party size for all recreationists in the summer sample was approximately four persons per party. However, in the fall sample groups were smaller,

[2] It is unknown if this survey methodology introduced any bias into our results. However, every effort was made to minimize potential bias by selecting as many respondents as possible at varying locations and varying times throughout the day.

averaging 2.73 per party with a median of only two persons per group. This result implied that recreation at these facilities were based more on family activities, with larger groups, in the summer and based more on individual type activities in the fall. Results were consistent across the three visitor types; however, they were only statistically significant (at the 1% level) for anglers and all recreationists generally.[3] The average fishing party in the fall was about 2/3 the size experienced at the peak of the summer season. Boating and water sports parties were about 1/4 smaller in the fall than in the peak season. Point estimates for average party size were only slightly smaller in the fall for the 'other' category, which included picnickers, swimmers, sightseers, and campers.

Table 1. Median and Mean Party Size by Visitor Type and Season Derived from On-site Surveys at Okatibbee Lake and Columbus Lake in Mississippi during 2006

Summer (# in Party)			
Type of Recreationist	Median	Mean	St Dev
Anglers (43)	3.00	3.72	2.40
Boaters[1] (64)	4.00	5.09	4.10
Picnickers[2] (64)	3.00	3.30	1.85
All	4.00	4.06	3.08
Fall (# in Party)			
Type of Recreationist	Median	Mean	St Dev
Anglers (37)***	2.00	2.22	1.29
Boaters (7)	3.00	3.57	2.57
Picnickers (96)	2.00	2.87	2.49
All***	2.00	2.73	2.26

[1] includes water-skiers and others engaged in water sports

[2] includes sightseers, swimmers, campers and other recreationists

*** seasonal difference in means significant at 1% level.

A few differences in length of stay were also evident across seasons as detailed in table 2.[4] The median visit was only one day in both periods and the average length of stay was between 2 and 2 1/2 days. This difference was not statistically significant for recreationists as a whole or for anglers or picnickers. Length of stay for boaters and water-skiers was longer for those participating in the summer. According to the point estimates, differences in average stay were most pronounced among anglers. For these visitors, the median length of stay dropped by 2/3 in the fall and the average visit decreased by nearly a full day, but the length of stay did not differ substantially by season for picnickers and other recreationists. Patterns presented in this table were probably due largely to weather and school-related events in the case of boaters which is reinforced by the sample size. Boaters, water-skiers, and others participating in water sports were few and far between by the end of September as children

---

[3] Standard t-tests were conducted comparing the two sample means – one from the summer survey and one from the fall survey.

[4] For comparison purposes we eliminated five observations from the sample which heavily skewed the results. Each of these outliers resided at the lakes for more than a month and as long as six months. Two stayed for one month, two stayed for two months, and one stayed for six months. Four of these long-term visitors were interviewed in the summer; one was interviewed in the fall.

returned to school and temperatures dropped. In the case of fishermen, declining length of stay may also be the result of diminishing returns. After a long fishing season, anglers may be less interested in pursuing their pastime than they would be earlier in the season.

Table 2. Median and Mean Length of Stay by Visitor Type and Season Derived from On-site Surveys at Okatibbee Lake and Columbus Lake in Mississippi during 2006

Summer (Days)			
Type of Recreationist	Median	Mean	St Dev
Anglers (43)	3.00	3.07	2.55
Boaters[1] (64)	1.00	1.41	1.50
Picnickers[2] (64)	1.00	2.10	2.39
All	1.00	2.08	2.23
Fall (Days)			
Type of Recreationist	Median	Mean	St Dev
Anglers (37)	1.00	2.19	2.59
Boaters (6)*	1.00	1.00	0.00
Picnickers (96)	1.00	2.56	2.29
All	1.00	2.39	2.34

[1] includes water-skiers and others engaged in water sports
[2] includes sightseers, swimmers, campers and other recreationists
* seasonal difference in means significant at 10% level.

Travel times by visitor type and season were reported in table 3. These data were computed for each visitor using Mapquest.com to calculate travel time between the respondent's reported county of residence and lake location. Descriptive statistics listed were only for in-state residents, which could be measured more precisely from collected data. While not much difference occurred across seasons for recreationists as a whole, anglers and boaters and water-skiers tended to travel longer distances in the fall than in the summer. Point estimates for picnickers, swimmers, sightseers, and campers also indicated longer travel in the fall. However, none of these results were conclusive as they were not significant at 5 or even 10% levels.

Table 4 provided some additional documentation regarding recreational travel patterns across seasons. In the summer, 15.5% of visitors came from other U.S. states. While many originated in neighboring Alabama, some traveled from as far away as New Jersey and Montana. In the fall, only 10.6% of visitors come from out-of-state. The proportion of local visitors also differed seasonally. Forty-eight percent of fall visitors came from the county in which the lake was located, but only 42.5% of summer visitors were local residents. The lower incidence of out-of- state visitors and the higher the incidence of local visitors in the fall have important ramifications for gauging the economic impact of recreational facilities. Debate continues on just how much of local visitor spending should be included in economic impact models but this data suggested a higher percentage of off-season spending was not new money injected into the local economy but rather local money. However, these figures do identify spending that was not escaping to other recreation sites or other entertainment options.

Table 3. Median and Mean Travel Times by Visitor Type and Season Derived from On-site Surveys at Okatibbee Lake and Columbus Lake in Mississippi during 2006

Summer (Minutes)			
Type of Recreationist	Median	Mean	St Dev
Anglers (43)	35.00	42.56	30.51
Boaters[1] (52)	24.00	39.10	42.22
Picnickers[2] (61)	35.00	40.51	38.96
All	24.00	40.48	38.19
Fall (Minutes)			
Type of Recreationist	Median	Mean	St Dev
Anglers (37)	51.00	55.70	43.88
Boaters (6)	24.00	48.43	34.45
Picnickers (82)	24.00	34.44	31.67
All	24.00	41.46	36.83

[1] includes water-skiers and others engaged in water sports
[2] includes sightseers, swimmers, campers and other recreationists
Note: In-state visitors only.

Table 4. Origin of Recreational Visitors by Season Derived from On-site Surveys at Okatibbee and Columbus Lake in Mississippi during 2006

Season	Within County (%)	Other MS Counties (%)	Other U.S. States (%)
Summer	42.5	41.9	15.5
Fall	48.2	41.1	10.6

While results depicted in tables 1-4 were interesting, perhaps the most important implications of this research were shown in subsequent tables. As part of the recreational survey, respondents were asked to estimate the per person on-site and local area expenditures they made during their trip. Table 5 reported on the median, mean, and standard deviation for these types of spending. These numbers included such items as access fees, food, drinks, ice, lodging, souvenirs, bait, equipment rents, parking, and other expenses incurred at the site itself. They also included transportation, off-site lodging, off-site food, miscellaneous shopping, but did not include other long-term expenditures such as equipment or registration fees.

Table 5 indicated that overall per visitor expenditures rose in the fall compared to the summer season. Anglers spent about $35 locally per visitor day in the summer and about $40 in the fall. However, this difference was not statistically significant. Similarly, boater spending increased from about $28 per visitor day to about $36.50. The largest increase in spending occurred among picnickers, swimmers, sightseers, and other recreationists who spent about $15 more per visitor day in the fall than in the summer season. Both these categories showed statistically significant increases in per visitor spending in the fall. It was likely that these results were driven by the type of activity in which recreationists were engaged. For instance, in the fall, when temperatures were cooler, more visitors were camping and fewer swimming. Camping required greater expenditures than swimming but also usually entailed a lengthier stay. Our surveys reinforced the common result that day-users typically spend less than overnighters.

Table 5. Median and Mean Local Expenditures by Visitor Type and Season Derived from On-site Surveys at Okatibbee Lake and Columbus Lake in Mississippi during 2006

Summer ($)			
Type of Recreationist	Median	Mean	St Dev
Anglers (43)	26.00	35.17	41.80
Boaters[1] (64)	19.75	27.67	33.34
Picnickers[2] (67)	13.50	24.21	31.71
All	19.75	28.22	35.08
Fall ($)			
Type of Recreationist	Median	Mean	St Dev
Anglers (37)	29.00	40.19	28.40
Boaters (7)**	38.00	36.54	13.75
Picnickers (96)***	22.50	39.73	44.03
All	25.00	39.69	39.31

[1] – includes water-skiers and others engaged in water sports
[2] – includes sightseers, swimmers, campers and other recreationists
** seasonal difference in means significant at 5% level.
*** seasonal difference in means significant at 1% level.

Table 6 detailed expenditure patterns by length of stay for both summer and fall visitors. In all cases, the point estimates indicated that per visitor day expenditures were greater in the fall than in the summer. Day users spent in the neighborhood of $22 to $30 on average, while overnighters spent in the range of $40 to $56 on average, depending on length of stay and season. In the summer, those making 2-3 day stays spent about $49 per day locally and in the fall they spent about $56. The numbers fell slightly for visitors making longer stays, down to $40 and $52 for summer and fall visitors, respectively. These results suggested that more casual visitors – those staying for the day or those visiting in the summer – spend less than those who were presumably more engaged recreationists – fall visitors and multi-day users. However, due to wide variation in spending patterns across individuals, larger expenditures in the fall were not statistically significant, even at the 10% level. Not surprisingly, t-tests showed that those staying four or more days spend larger sums per day than day users in both the summer and fall. There was also a statistically significant increase in per day expenditures for those staying 2-3 days relative to day users in the fall.

More detailed data regarding the breakdown of these local expenditures is of significant interest to economists and public officials interested in developing expenditure profiles of recreationists. Such data can be used to determine economic impacts and ultimately the financial feasibility of such public sector projects. Expenditure data was presented in disaggregated form in table 7. Reported were the averages across five spending categories - on-site expenditures, and four types of local spending, including: transportation, lodging, food, and other retail. On-site expenditures referred to expenditures at the lake site itself. These may include access fees, parking, camping fees, equipment rentals, and similar expenditures. Transportation included only gasoline and other automotive expenditures spent in the local economy. Similarly, lodging included spending at local hotels, motels, campgrounds, or recreational vehicle parks not associated with the lake facilities. The food category included groceries, beverages, and ice sales as well as restaurant meals purchased in the local economy. All other spending was incorporated into the other retail category. Such items included bait, sunscreen, bug spray, equipment rental or repair, and entertainment expenditures.

Table 6. Median and Mean Local Expenditures by Length of Stay and Season Derived from On-site Surveys at Okatibbee Lake and Columbus Lake in Mississippi during 2006

Summer ($)			
Length of Stay (n)	Median	Mean	St Dev
1 day (123)	14.00	22.39	24.70
2-3 days (15)	28.25	48.84	65.55
4 or more days (36)**	29.10	39.56	42.48
Fall ($)			
Length of Stay (n)	Median	Mean	St Dev
1 day (83)	22.38	29.50	36.42
2-3 days (25)***	47.00	55.61	35.10
4 or more days (33)***	40.00	53.28	41.70

** difference in mean spending by length of stay significant at 5% level.

*** difference in mean spending by length of stay significant at 1% level.

The largest increase from the summer recreationists to those recreating in the fall was among the food and retail categories. This was particularly true of picnickers, sightseers, hikers, and other miscellaneous recreationists. These individuals were more likely to be camping in the fall period as opposed to day sightseers, hikers, or those engaging in other less expensive activities, which were more prevalent in the summer. It was also noteworthy that on-site expenditures appeared to be higher during the summer months. This was in all likelihood due to the fact that some businesses immediately surrounding an impoundment were highly seasonal, and they likely were closed at the end of the peak visitor season. However, it appeared that these purchases were picked up by the surrounding economy in the form of increased sales in the 'other retail' category during the fall period.

Table 7. Visitor Spending by Expenditure Type, Visitor Type, and Season Derived from On-site Surveys at Okatibbee Lake and Columbus Lake in Mississippi during 2006

Summer($)			
Expenditure	Anglers (43)	Boaters (64)	Picnickers (67)
On-site	8.61	7.92	8.84
Local			
Transportation	9.36	6.93	7.82
Lodging	1.34	2.16	2.77
Food	8.75	8.98	3.89
Other Retail	7.11	1.77	0.88
TOTAL	35.17	27.76	24.21

Table 7. (Continued)

Fall ($)			
Expenditure	Anglers (37)	Boaters (7)	Picnickers (96)
On-site	7.36	2.61	5.99
Local			
Transportation	8.38	11.79	7.99
Lodging	2.36	0.00	2.28
Food	10.69	9.58	15.00
Other Retail	11.38	12.56	8.46
TOTAL	40.17	36.54	39.73

4. CONCLUSIONS

Our study reported on results of a local recreational survey conducted at existing U.S. Army Corps of Engineers lakes in Mississippi. Our findings suggested that seasonal fluctuations occurred not only in visitor numbers, but also in length of stay, party size, distance traveled, and local expenditures. These results suggested that researchers interested in determining the economic or fiscal benefits of recreational surface water impoundments should proceed with caution when incorporating data collected in one visitor season to make yearly projections. Visitors in off-peak periods appeared to incur greater expenditures than those in the peak periods. Our survey showed that visitors came from further away and spent a longer time at the facility during the milder part of the year. The off-peak visitors may be more avid recreationists and, therefore, willing to spend greater amounts at the lake. Similarly, different types of recreationists appeared to have slightly different spending patterns, which should be acknowledged in economic impact and feasibility studies. Spending profiles reported provided reasonable estimates on the per visitor economic injection that can be expected from surface water impoundments in central Mississippi and the rural South generally.

On the basis of our study, we felt that surface water impoundments can generate substantial quality-of-life improvements and economic impacts for rural areas in Mississippi and elsewhere. Assessments can also lead to recommendations for additional low impact production of goods and services in the economy. This can compliment the natural resource base by further supporting the contention that there is an economic stimulus from water-based recreational facilities. As stated, these impacts do depend critically on how many recreationists frequent the lake, where they come from, how long they stay, what activities they were engaged in by season, and most importantly how much they spend. Beyond the immediate economic benefits we have outlined there are also potential opportunities for expanding upon nature-based tourism, improving real estate values, partnering with agencies and organizations to achieve project goals, and enhancing community services to name a few.

Research results can also be used to promote or market impoundment projects or specific recreational or service activities associated with these lakes. They can also be used to justify adjusting the size of the impoundment or their influencing their location. In general, economic analyses such as ours provide information that becomes part of the argument for a project and associated activities that can compliment the social and environmental aspects. The end result can often lead to policy initiatives or incentives to make these types of projects more attractive to rural areas. There also are other benefits to doing this type of study, beyond an economic analysis. In the case of federally supported impoundments, there is a requirement that this type of study be undertaken prior to a more involved environmental impact assessment. Last, we also feel our study methods will not only serve to assist in a determination of the economic success supporting a lake location, but can be used for other similar developments.

ACKNOWLEDGMENTS

Funding for this project was provided by the Choctaw County Board of Supervisors via Pickering, Inc. Special thanks to Marcus Measells, Darren Hudson, Garen Evans, Brent Doolittle, Andrew Hite, Justin McDaniel, Shu-mei Lin, Anna-Haskins Patton, and Wayne Lagrone, for their participation in this project.

REFERENCES

Deller, S.C., Tsai, T., Marcouiller, D.W., and English, D.B.K. (2001). The Role of Amenities and Quality of Life in Rural Economic Growth. *American Journal of Agricultural Economics* 83 (2), 352-365.

Grado, S.C., Grebner, D.L., Munn, I.A., and Drier, R.O. (2002). *Economic Feasibility Study for Recreational Development on the Bienville National Forest in Mississippi* (Final Project Report to the USDA Forest Service) Mississippi State, MS: Forest and Wildlife Research Center.

Grado, S.C., Grebner, D.L., Drier, R.O., and Munn, I.A. (2004). Economic feasibility study for recreational development on the Bienville National Forest in Mississippi. *Proceedings of the 2003 Southern Forest Economics Workshop.* 211-220.

Rezek, J., Evans G., Grado, S.C., and Hudson, M.D. (2006a). *Choctaw County Lake Impoundment: Economic Impact Report*. (Final Report to the USDA Natural Resources and Conservation Service). Mississippi State, MS: Mississippi State University.

Rezek, J. Evans, G., Grado, S.C., and Hudson, M.D. (2006b). *Madison County Economic Development Report.* (Final Report to the Madison County Economic Development Authority). Mississippi State, MS: Mississippi State University.

U.S. Department of the Interior, Fish and Wildlife Service, and U.S. Department of Commerce, U.S. Census Bureau. (2008). *2006 National Survey of Fishing, Hunting, and Wildlife-Associated Recreation-Mississippi (*Publication FHW/06-MS). Washington, DC: U.S. Government Printing Office.

United States Department of the Interior. (2005). *Passenger Statistics by Month.* Retrieved January 25, 2009 from National Park Service Web site, http://www.nps.gov/grca/ parkmgmt/ upload/VTS-Month-Stats.pdf

In: Global Change and Forestry
Editors: J. Gan, S. Grado and I. A. Munn
ISBN: 978-1-60876-262-0

Chapter 22

HEDONIC ANALYSIS OF AUCTIONED HUNTING LEASES: A CASE STUDY OF MISSISSIPPI 16TH SECTION LANDS[1]

Jacob D. Rhyne, Ian A. Munn* and Anwar Hussain
Mississippi State University, MS, USA

ABSTRACT

This chapter evaluates factors affecting the value of hunting leases on Sixteenth Section Lands in Mississippi using hedonic analysis. These lands are owned by the state and hunting rights are auctioned to the public. Because this market data are generated in a competitive setting, this analysis has certain advantages over previous investigations of hunting lease markets, which relied on surveys of non-industrial private landowners to elicit information about their hunting leases or contingent valuation methods to infer lease values from landowners or hunters. Due to the competitive nature of the issuance of Sixteenth Section leases, their prices better reflect the full impact of lease characteristics such as cover type, game quality, distance to urban areas, and location on hunter valuation of hunting access. Estimates of the implicit prices of these characteristics suggest that land managers should adopt shorter lease lengths, smaller lease sizes and improve habitat to increase lease revenue.

Keywords: Auctioned leases; Hedonic analysis; Hunting leases; Public lands.

---

[1] Originally published in *Human Dimensions of Wildlife*, 14(4): 227-239. Rhyne, J.D., I.A. Munn, A. Hussain. (2009). Hedonic analysis of auctioned hunting leases: A case study of Mississippi 16th sections lands. Reprinted by permission of the publisher (Taylor & Francis, http://www.informaworld.com)

* Email: imuun@cfr.msstate.edu, phone: (662) 325-4546, Department of Forestry, Mississippi State University, Box 9681, Mississippi State, MS 39762.

1. Introduction

To provide landowners with useful information about hunting lease markets, a thorough and accurate understanding of the values that hunters place on lease characteristics is required. Previous studies have relied on non-market data obtained by using the contingent valuation method or data from hunting leases on non-industrial private lands to draw conclusions about values that hunters place on leases and their characteristics. Hypothetical data generated via contingent valuation has limitations and drawbacks and are less reliable than data generated from actual market transactions. Information from hunting leases on non-industrial lands is typically priced in a non-competitive manner and do not capture the full market value of hunting access. Competitively issued hunting leases, in contrast, more likely capture full market value and provide more reliable information for land owners engaged in the lease market.

Data issues have prevented a clear understanding of the value that hunters place on hunting lease characteristics. Most leases are issued on private lands, resulting in no public record of the market transactions available for study. To gather information for empirical studies, researchers have relied on surveys to (a) infer lease values from landowners or hunters using contingent valuation (CV) (e.g., Goodwin, Offenbach, Cable, & Cook 1993; Hussain, Zhang, & Armstrong, 2004; Stribling, Caulfied, Lockaby, Thompson, Quicke, & Clonts, 1992), or (b) elicit transaction information from NIPF landowners about their hunting leases (e.g., Hussain, Munn, Grado, West, Jones & Jones, 2007; Messonier & Luzar, 1990; Munn, Loden, Grado, Jones, & Jones, 2005; Shrestha & Alavalapati, 2004; Standiford & Howitt, 1993; Zhang, Hussain, & Armstrong, 2006). The resultant data are the basis for conclusions about the hunting lease market and the valuation of lease characteristics (Buller, Hudson, Parkhurst, & Whittington, 2006). Hypothetical data generated via CV, however, do not necessarily represent actual market conditions and there are concerns about the reliability and accuracy (Freeman, 1993; Mitchell & Carson, 1989). Data based on actual market transactions (revealed preferences) are preferable. Even market data are not ideal for valuating lease characteristics, however, if the market is not competitive. Hunting lease research has focused on leases on non-industrial private lands. Evidence suggests that NIPF hunting leases are not necessarily issued in a fully competitive manner. Lease prices are typically negotiated. Few landowners advertise to the public, either on the internet or in print media. Most rely on word-of-mouth or family and friends to develop their customer base (Miller & West, 2007; Munn, et al., 2007). Munn, Hussain, West, Grado, and Jones (2007) found that landowners experienced in the hunting lease market generated higher lease rates than their less experienced, less knowledgeable counterparts. Given that less than 12% of NIPF landowners in the study were classified as experienced, a large majority of landowners were not capturing full market value. Finally, landowners who lease to family and friends may derive non-pecuniary benefits from such arrangements and the stated lease price would not reflect the full value of the lease. In such cases, using the stated lease price in a hedonic analysis would undervalue the lease components.

Additional case studies of the hunting lease market under different institutional arrangements are needed (Mozumber, Starbuck, Berrens, & Alexander, 2007). Studies of auctioned leases have been limited but would provide valuable insight into the actual value hunters place on leases. Auctioning goods or services generally results in greater revenue than if the price is negotiated (McAfee & McMillan, 1987; Milgrom, 1989).

Hunting leases on Sixteenth Section Lands in Mississippi present an excellent opportunity for study because these leases are advertised and auctioned to the public. These tracts were set aside to benefit public education in the Land Ordinance of 1785 (Public Lands Division, 2005). The sixteenth section, an approximately 640 acre block, of every township was reserved for public school districts to use in support of education. In Mississippi, the Board of Directors of each school district decides how these lands will be utilized. Revenue is generated from these lands through the sale of timber and from various leases such as oil, gas, mineral, farming, and hunting. Hunting leases are allowed only on forested land and are awarded to the highest bidder in a sealed bid auction. School districts advertise lease sales in local newspapers for two consecutive weeks prior to the lease auction.

Greater competition vying for the leases coupled with a competitive bidding format can be expected to push the lease price upward. Hunters bid based upon what they think the lease is worth to acquire the lease. Comparing the average lease price on hunting leases on non-industrial lands in Mississippi to the average lease price on Sixteenth Section Lands would reveal if studies of hunting leases on non-industrial lands have captured the full value that hunters placed on lease characteristics.

In this chapter we examined hunting leases on Sixteenth Section Lands in Mississippi to determine what role lease length, size of the lease area, habitat quality, market segmentation, game quality, and distance to the closest urban area have on hunting lease prices. We first compared lease prices for NIPF and Sixteenth Section lands to ascertain if lease prices vary significantly between these market segments. The study also added to previous research on the impact that game quality and cover type has on hunting lease prices because we used quantifiable measures of these attributes. Our goal was to provide information on factors affecting hunting lease prices to non-industrial private landowners, companies and corporations that lease, and the Board of Directors of Public Schools in Mississippi. Analysis of competitively issued hunting leases will provide more accurate estimates of the value hunters place on leases.

2. Conceptual Background

Analyzing hunting leases is challenging and complex. Previous studies have improved our knowledge of hunting leases and have determined that game quality, cover type, proximity to urban areas, lease length and size, and market segmentation influence hunting lease prices. These factors and the likely direction of their impact on lease rates are described below.

2.1. Game Quality

Game quality is one of the most important site attribute that influences hunters' valuation. In a study of hunting leases on private lands in California, Loomis and Fitzhugh (1989) found that hunters were willing to pay \$106 more per hunter for a 10% increase in trophy quality deer in the total deer harvest. Likewise, Standiford and Howitt (1993) observed a positive correlation between lease price and trophy size of deer in California hardwood rangeland. Both studies classified game quality as the number or percent of trophy deer; trophy size was not defined and was subjective. We hypothesized hunting sites that have greater potential of harvesting quality game to command higher lease rates than otherwise similar sites.

2.2. Cover Type

Cover type is a significant determinant of food availability and defines the amount of game an area can sustain. Hardwoods, in particular, provide year-round food sources and are ideal habitat for deer and other game (Dickson, 2004; Harris, Sullivan, & Badger, 1984; Hazel, 1995). Hardwoods also provide diverse food sources such as browse, fruits, mushrooms, and soft and hard mast that are relished by deer and other game (Hazel, 1995). More food sources equate to larger game populations and game density is positively related to hunting lease revenue (Livengood, 1983). In contrast, mature pine cover provide deer and other game with relatively little food (Thill, 1990). Gigliotti (2000), Hussain et al. (2004), and Messmer et al. (1998) showed that hunting success related variables influence hunter willingness to pay for hunting access. Stribling et al. (1992) found that hunters in Alabama preferred a mix of pine-hardwood age classes with less than 50% in regenerated stands of young pines.

2.3. Impact of Urban/Rural Issues

Distance to the closest major metropolitan centers negatively impacted hunting lease prices in Texas (Pope & Stoll, 1985). In Kansas, urban residents were more likely to purchase a hunting lease than rural residents (Goodwin et al., 1993). These two finding combine to suggest that urban hunters are driving lease prices, not their rural counterparts. As the distance and associated travel cost for urban hunters increases, aggregate demand for leases decreases as evidenced by decreasing prices with distance.

2.4. Length of Lease and Number of Acres Leased

Findings concerning the impacts of lease size on lease price have been contradictory. Shrestha and Alavalapati (2004) found a positive correlation between hunting lease revenue and lease length in a study of recreational hunting on ranches in Florida. Likewise, Messonier and Luzar (1990), Pope and Stoll (1985), and Standiford and Howitt (1993) found a positive relationship between lease revenue per acre and lease size. Zhang et al. (2006), however, reported a negative relationship, citing diminishing returns to size due to management limitations as a possible explanation.

2.5. Market Segmentation

Studies evaluating hunting leases should account for different hunting lease markets within the study area. Separate markets result from differing supply and demand structures coupled with barriers that prevent market integration (Freeman, 1993). If market segmentation exists and only one hedonic price function is calculated, the study will provide inaccurate estimates of model coefficients (Freeman, 1993). In a study of hunting leases in the coastal and Delta regions of Mississippi, Munn et al. (2005) found that lease characteristics impacted hunting lease prices differently in each region. In the coastal region, lease prices were significantly impacted by the number of forested acres and wildlife management expenditures. In the Delta region, the number of agricultural acres, number of forested acres, percent increase in amount of wetlands of the total acres leased significantly influenced hunting lease prices.

3. METHODS

To support our argument that NIPF lease prices do not represent full market value, we first compare lease prices collected in this study to lease prices collected from Mississippi NIPF landowners in a 2003 statewide survey (Munn et al., 2007). NIPF lease prices were adjusted for inflation to 2005 levels using GNP deflator values. The mean values were then compared using a two-sample *t*-test with unequal variances.

3.1. The Hedonic Method

The hedonic pricing method is a market evaluation technique used to estimate the economic value consumers place on non-market characteristics of a good or service. It is often used with goods that have varying attributes or characteristics that can only be sold as a collective unit (Rosen, 1974). Hunting leases can be considered composite goods with varying attributes. Hedonic analysis enabled lease prices for heterogeneous sites made up of a bundle of characteristics to be decomposed into a specific price for each characteristic (Brown & Mendelssohn, 1984). Other studies have used the hedonic method to evaluate hunting leases (Buller et al., 2006; Hussain et al., 2007; Munn et al., 2005; Pope & Stoll, 1985; Zhang et al., 2006). Based on previous research, we specified a hedonic price equation for hunting leases as:

Lease price per acre = F [lease size, lease length, habitat quality, market segment, game quality, and distance to urban area]

3.2. Data Sources and Construction of Variables

Lease data were obtained for hunting leases on Mississippi's Sixteenth Section Lands from the Public Lands Division of the Mississippi Secretary of State's Office. In 2005, there were 875 hunting leases on Sixteenth Section Lands (figure 1). Revenue generated from these leases was over 2.5 million dollars.

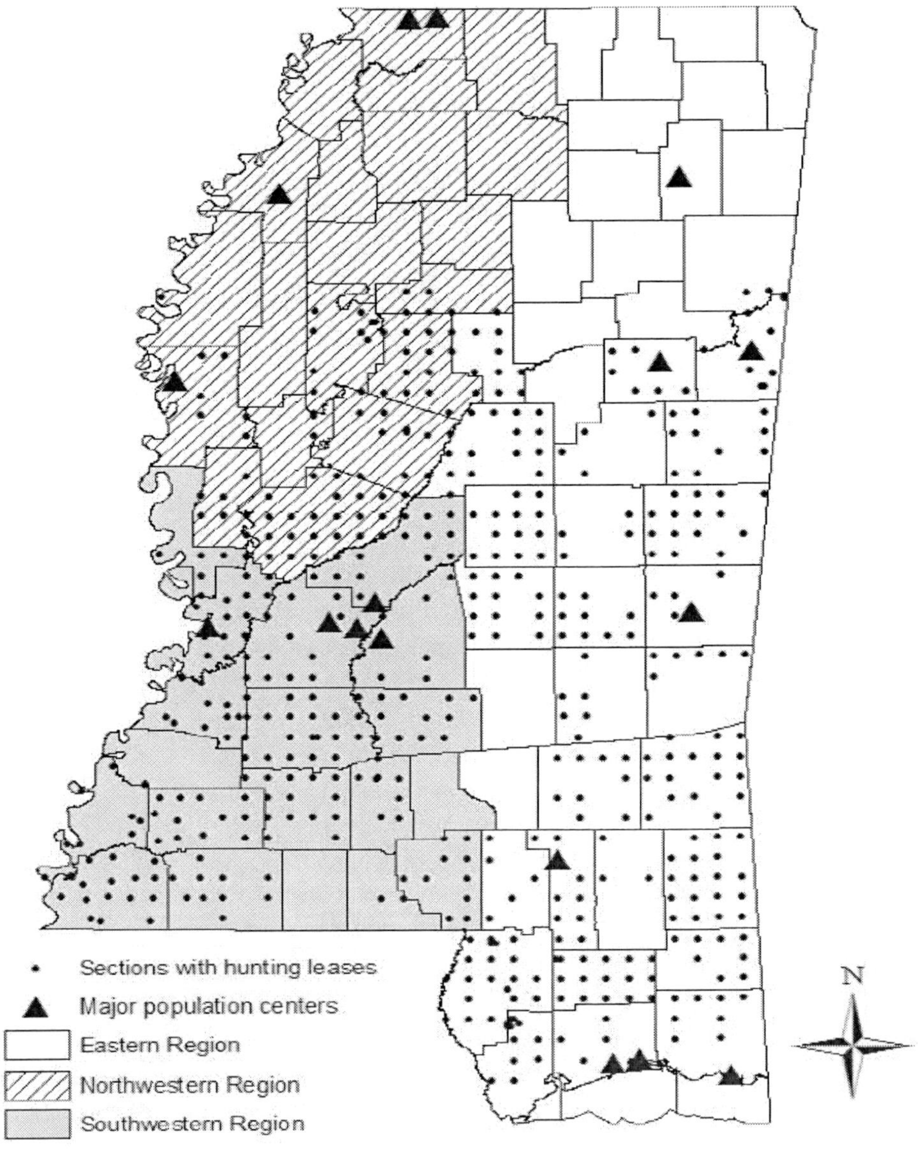

Figure 1. Map of Sixteenth Sections with hunting leases in Mississippi in 2005.

Not all Mississippi counties have auctioned lease data. There are fifteen counties in northern Mississippi that do not have Sixteenth Section Lands set aside for public education. When Mississippi became a state in 1817, a large portion of northern Mississippi was still owned and inhabited by the Chickasaw and was not subject to the Land Ordinance of 1785. The U.S. government subsequently sold this land occupied by the Chickasaw and failed to set aside the Sixteenth Section Lands for public schools. A small number of other counties throughout the state do not lease hunting rights for a variety of reasons.

The usable dataset had 715 observations, where each observation was composed of one hunting lease on Sixteenth Section Lands in 2005. A section may have multiple hunting leases. Data collected for each hunting lease included the per acre revenue for 2005, lease acres, lease length, cover type information, average county-level Boone and Crocket score (B&C), and the distance from each hunting lease to the nearest urban area (table 1). Specific details about data sources and construction of variables used in the analysis were:

Lease Price. Hunting lease information was provided by the Public Lands Division of the Mississippi Secretary of State's Office. The annual lease price per acre for 2005 was the dependent variable for the hedonic price function.

Lease length and acres leased. This information was also provided by the Public Lands Division of the Mississippi Secretary of State's Office. Given the clustered nature of lease lengths, three dummy variables were generated to represent lease lengths of one to four years, five years, and over five years. The over five years category served as the base (omitted) category in the regression analysis.

Cover type. Cover type information was provided by the Mississippi Institute of Forest Inventory (Parker et al., 2005). The information included the number of acres in the following cover types for each sixteenth section: pine, hardwoods, mixed pine-hardwoods, water, regenerated, and open. Acreages by cover type were converted to percentages of each section. We theorized that the cover type information for each section was representative of the cover type for each hunting lease on that section. Percent pine cover type was used as the base category.

Geographic regions of the state. Three regions were delineated as distinct market segments based on the major population centers in the state (figure 1). These segments were southwest Mississippi, northwest Mississippi, and east Mississippi. To model market segmentation, dummy variables for each of the three regions were employed. The eastern region served as the base category.

Projected average Boone and Crocket Scores. Projected average B&C scores for four+-year-old bucks for each county were obtained from Strickland and Demarais (2000). Their technique effectively approximates B&C scores from a subset of the antler measurements required for a direct B&C score. Strickland and Demaris (2000) calculated projected B&C scores from deer harvest data collected by the Mississippi Department of Wildlife Fisheries and Parks through the Deer Management Assistance Program (DMAP). DMAP monitors the deer population in Mississippi by taking biological samples from harvested game on wildlife management areas and from participating landowners and hunters.

Distance to urban area. The straight line distance in miles from each lease to the closest urban area was determined using ArcGIS. Five areas in Mississippi are classified as urban areas in the 2000 U.S Census. The areas are Biloxi, Pascagoula, Hattiesburg, Jackson, and Olive Branch, a suburb of Memphis.

Table 1. Definitions of explanatory variables used to predict hunting lease prices on Sixteenth Section lands in Mississippi in 2005

Name	Description	Expected Sign
Size of Lease		
log (leased, acres)	Logarithm of number of acres leased	+
Length of Lease		
One year to four	1 if lease is less than five years, otherwise 0;	+
Five years	1 if lease length is five years, otherwise 0;	+
Six year and over	1 if lease length is greater than five years; otherwise 0;	-
Cover Type		
% pine	Percentage of land in pine	-
% mixed pine-hardwoods	Percentage of land in mixed-pine hardwoods	+
% water	Percentage of land under permanent/temporary lakes, streams, ponds	+
% regenerated	Percentage of land that has been recently regenerated	-
% open	Percentage of land that is open	-
% hardwoods	Percentage of land that is in hardwoods	+
Market Segmentation		
Northwest	1 if land is located in northwest, MS, otherwise 0;	+
Southwest	1 if land is located in southeast, MS, otherwise 0;	+
East	1 if land is located in east, MS, otherwise 0;	-
Game Quality	Projected average Boone and Crocket Score by county	+
Distance to urban area	Miles from hunting lease to closest urban area	-

4. RESULTS

4.1. Descriptive Statistics

The average annual lease price was \$2,959.73 or \$8.73 per acre; the average lease size was 349 acres. However, there were broad departures from these averages. Annual lease price ranged from \$10 to \$32,000 and lease size varied from 3 to 3,059 acres. Pine stands constituted 35% of the forest cover types covering the sections containing leases, hardwoods 29%, mixed pine-hardwoods 10%, regenerated forests 9%, and open land and water accounted for the residual (table 2).

Of the 715 leases, 80% were of a five year length, 13% were of 1 to 4 year duration, and the remaining leases were of six or more year length. The leases were not evenly distributed between regions. East Mississippi accounted for 48% of the leases. Southwest Mississippi region accounted for 36% of the leases and northwest Mississippi accounted for the remaining 16%. The average county Boone and Crocket Score was 114. The average distance to the closest urban area was 54 miles.

Table 2. Descriptive statistics related to hunting leases on Sixteenth Section Lands in Mississippi in 2005 ($N = 715$)

Name	Mean	Std Dev	Min	Max
Dependent Variable				
Annual lease price	2,959.73	3,501.36	10.00	32,000.00
Annual lease price/acre	8.73	6.97	0.82	52.41
Log-annual lease price/acre	0.84	0.28	-0.09	1.72
Independent Variables				
Size of Lease				
Average lease size (acres)	348.73	261.14	3.00	3,059.00*
Log-acres leased	5.45	1.08	1.10	8.03
Length of Lease				
One year to four	0.13	0.34	0.00	1.00
Five years	0.80	0.40	0.00	1.00
Six year and over	0.07	0.26	0.00	1.00
Cover Type				
% pine	0.35	0.25	0.00	0.95
% mixed pine-hardwoods	0.11	0.07	0.00	0.40
% water	0.01	0.04	0.00	0.64
% regenerated	0.09	0.10	0.00	0.81
% open	0.15	0.16	0.00	0.90
% hardwoods	0.29	0.22	0.01	0.98
Market Segmentation				
Northwest	0.16	0.37	0.00	1.00
Southwest	0.36	0.48	0.00	1.00
East	0.48	0.50	0.00	1.00
Game Quality	113.77	11.62	81.50	133.50
Distance to urban area	54.45	29.82	1.80	261.66

* Giles Island is 3059 acres entrusted to the Mississippi School Board.

4.2. Comparison to NIPF Lease Prices

The mean price per acre for NIPF leases in 2003 reported by Munn et al. (2007) was \$6.50, which was significantly lower (p-value < 0.01) than the mean lease price per acre for Sixteenth Section leases reported above even after adjusting for inflation.

4.3. Factors Influencing Hunting Lease Revenue

Ordinary least squares regression was used to estimate the hedonic price equation relating the dependent variable (lease revenue per acre) to the independent variables (lease acres, lease length, cover type, average county-level Boone and Crocket scores, and distance to the nearest urban area). As hedonic price theory does not specify which functional form to use, a variety of models were estimated involving the Box-Cox procedure. The log-log functional form provided the best fit for the data. This agreed with Cropper, Leland, and McConnell (1988) who report that when proxies were used, simpler data forms such as linear, semi-log, and log-log work best. A variety of diagnostic tests were performed to determine whether assumptions underlying the ordinary least square method were satisfied. In particular, diagnostics related to heteroskedasticity, multicollinearity and model specification were conducted. Based on the Breusch-Pagan test, the hypothesis of nonconstant variance was rejected; thus heteroskedasticity was not a problem. Based on the Ramsey specification test, the null hypothesis of model misspecification was also rejected. To examine multicollinearity, variance inflation factors (VIF) and correlations between explanatory variables were estimated. The correlation matrix did not indicate signification correlations between any pair of explanatory variables. The VIF statistics did not indicate that multicollinearity was an issue.

Of the 12 variables in the model, nine were significant at the 10% level of confidence (table 3). The coefficient for lease size was negative and significant. Thus, a 1% increase in the size of the lease caused the average lease price per acre to decrease by 0.027%.[2] Variables representing lease length were significant and positive. Hunting leases less than five years in length generated 18% more per acre revenue than hunting leases greater than five years in duration.[3] Five year hunting leases generated 11% more per acre revenue than hunting leases greater than five years.

Table 3. Estimated coefficients of hedonic price model for hunting leases on Sixteenth Section Lands in Mississippi in 2005 ($N = 715$)

Variable	Coefficient	*p*-value	Elasticity
Dependent Variable			
Log-annual lease price/acre			
Independent Variables			
Size of Lease			
Log-acres leased	-0.027	0.003	-0.027
Length of Lease			
One year to four	0.225	0.000	18.252
Five years	0.182	0.000	10.649
Cover Type			
% pine	-0.283	0.000	-0.100
% mixed pine-hardwoods	-0.021	0.875	-0.002
% water	-0.094	0.695	-0.001
% regenerated	-0.420	0.000	-0.036
% open	-0.096	0.162	-0.014
Market Segmentation			
Northwest	0.226	0.000	17.241
Southwest	0.171	0.000	5.722
Game Quality	0.003	0.002	0.364
Distance to urban area	0.001	0.066	0.029

---

2 Elasticities, evaluated at means, for explanatory variables were derived by using: $\partial \log price / \partial x_k = \hat{B}_k \overline{X}_k$.

Elasticities for log-acres leased was based on $\partial \log price / \partial \log leaseacres = \hat{B}acres$. For details, see Johnson et al. (1987), p. 251.

3 Calculated using Halvorsen and Palmquist (1980) and Kennedy (1981) elasticity effects for dummy variables :$\{\text{Exp}[\hat{\beta}_k - 1/2V(\hat{\beta}_k)] - 1\} * 100$.

Table 3. (Continued)

Variable	Coefficient	*p*-value	Elasticity
Market Segmentation			
Intercept	0.477	*0.000*	
Adjusted R-squared	0.340		

Of the variables representing habitat quality, percent pine and regenerated lands were significant and had negative coefficients. In line with expectations that hardwoods are preferred habitat for deer and other game, results indicated hunters valued pine and regenerated stands significantly less than hardwoods. Thus, a 1% increase in the percent share of land in pines and regenerated areas with a corresponding decrease in the share in hardwoods caused lease prices to decrease by 0.10% and 0.04%, respectively. Collectively, the coefficients representing mixed pine-hardwoods, water, and open areas on hunting lease property were all negative, albeit not significant in the model, reinforcing the notion that hunters perceive hardwoods as preferred habitat for game. The lack of significance for these variables individually, however, indicates that any hunter preference against these cover types is weak at best.

Consistent with our expectations, there were different hunting lease markets in Mississippi; the coefficients for variables representing northwest Mississippi and southwest Mississippi were significant and positive. Accordingly, hunting leases in the northwestern and southwestern region generated approximately 17% and 6% higher revenue per acre than hunting leases in the eastern portion of the state.

The estimated coefficient on the projected average Boone and Crocket Score by county was positive and significant. A 1% increase in the projected average Boone and Crocket score increased the average lease price by 0.36%. Finally, the coefficient for distance to an urban area was significant and positive. A 1% increase in distance to urban areas caused the lease price to increase by 0.03%.

5. Conclusion

Evaluating hunting leases that are awarded competitively in an open market provides valuable data in assessing the "going rate" of hunting lease characteristics. Hunting leases on Sixteenth Section Lands in Mississippi are different from a vast majority of issued hunting leases because they are competitively awarded. Goods and services sold by auction generate more revenue than if the price is negotiated (McAfee & McMillan, 1987; Milgrom, 1989). Compared to negotiated hunting leases, competitively issued leases more accurately reflect the value that hunters place on hunting lease characteristics. In theory, hunters' bid prices approach the full market value when leases are auctioned in a competitive manner, otherwise they might lose out on the lease. Studying these leases can shed light on the value hunters place on hunting locations.

To effectively analyze hunting leases on Sixteenth Section Lands in Mississippi, we used hunting lease data from market transactions instead of hypothetical data as generated by contingent valuation methods typically used to study this market. Information collected included: the amount of revenue generated, lease length, lease acres, cover type of the lease area, county-level Boone and Crocket Scores to proxy game quality, and distance of the hunting lease to nearest urban area. A functioning hunting lease market has not been examined in previous studies and provided more conclusive results than previous studies that relied on hypothetical methods to evaluate the hunting lease market. Results of this study provide land managers with a better understanding of how to generate more financial benefit from their hunting leases.

Our findings regarding specific hunting lease attributes largely mirrored that reported in previous research. Cover type composition, as proxy for habitat quality, was strongly linked to lease prices, which agreed with results by Stribling et al. (1992). Pine and regenerated areas did not generate as much lease revenue as hardwoods and clearly indicated that habitat quality matters in the hunting lease market – as would be expected. Given that game quality is in large part a function of habitat quality, this result is also reflected in the higher prices for leases in counties with higher mean Boone and Crocket scores for 4.5-year-old deer. Hunters in this study paid significantly more for hunting leases as the quality of game increased. This objective result corroborated subjective findings by Loomis and Fitzhugh (1989) and Standiford and Howitt (1993).

In addition to game and habitat characteristics, this study supported previous results regarding lease location. The result that hunting lease prices increased with distance from urban areas agreed with Pope and Stoll (1985) who found that lease prices increased when distances to metropolitan areas exceeded 89 miles. Lease location was also important in that supply and demand for leases could vary dramatically from one location to the next. The existence of multiple hunting lease markets within Mississippi was consistent with Munn et al. (2005) who found different hunting lease characteristics impacted lease prices differently in different parts of Mississippi. Future studies that examine hunting leases must also take into consideration the possibility of distinct hunting lease markets within the study area.

Some of our findings regarding specific hunting lease attributes conflicted with previous research. Estimation results indicated that shorter duration leases generated greater revenue than longer term leases. These results are contrary to Shrestha and Alavalapati (2004) who reported that longer term leases generate greater per acre revenue than leases of a shorter duration. Similarly, in this study, smaller leases brought higher per acre prices than larger leases, while agreeing with Zhang et al. (2006), contrasted with most of the literature, which found a positive relationship between per acre revenue generated and lease size (Messonier & Luzar, 1990; Pope & Stoll, 1985; Standiford & Howitt, 1993). Our study indicated that there was a market for smaller size leases. This is likely due to a number of individual and small groups of hunters who can effectively compete for small leases but cannot match the financial resources of large clubs in bidding on large leases. A limitation of our study is that we did not consider the possibility of interaction effects between variables. While this is consistent with most of the literature, there are exceptions (e.g., Little & Berrens, 2008). One approach is to include specific interaction terms based on *a priori* expectations.

There are several practical implications from this study that benefit land managers and future hunting lease studies. First, land managers can increase the amount of revenue generated from leasing by offering hunting leases five years or less in duration. Of course, trade-offs between increased revenues from shorter leases and increased costs of more frequent auctions must be considered. Second, if access is available, and there are logical and enforceable boundaries such as roads, rights-of-way, or major streams between adjacent blocks, landowners should consider dividing up large hunting leases into multiple smaller hunting leases to generate more lease income. Third, providing better habitat for game by leaving hardwoods in regenerated stands is an alternative that can also lead to increased revenues. Future studies, however, will have to determine whether habitat improvements of this magnitude will be feasible and financially advantageous. Fourth, many hunters prefer hunting leases in remote locations away from urban areas, so advertising rural hunting leases in urban newspapers might be worthwhile in attracting prospective lessees. Fifth, advertising and auctioning hunting leases to the public is worth considering. The disparity between the average lease price on Sixteenth Section Lands and non-industrial lands in Mississippi suggests that studies that have examined hunting leases on non-industrial lands have not captured the full value that hunters place on lease characteristics. Competitively issued hunting leases more accurately reflect these values.

REFERENCES

Brown, G., & Mendelssohn, R. (1984). The hedonic travel cost method. *The Review of Economics and Statistics. 66*(3), 427-433.

Buller, V. M., Hudson, M. D., Parkhurst, G., & Whittington, A. F. (2006). The impact of hunting package attributes on hunting package prices in Mississippi. *Mississippi State University Department of Agricultural Economics. Research Report.* 2006-01. January

Cropper, M. L., Leland, B. D., & Mconnell, K. E. (1988). On the choice of functional form for the hedonic price functions. *The Review of Economics and Statistics. 70*(4), 668-675.

Dickson, J. G. (2004). Wildlife and Upland Oak Forests. Gen. Tech. Rep. SRS-73. Asheville, NC: *U.S. Department of Agriculture, Forest Service, Southern Research Station.* pages 106-115.

Freeman, A. M. III. (1993). *The measurement of environmental resource values: Theory and methods.* Resources for the Future, Washington, D.C.

Gigliotti, L. M. (2000). A classification scheme to better understand satisfaction of Black Hills hunters: the role of harvest success. *Human Dimensions of Wildlife. 5*(1), 32-51.

Goodwin, B. K., Offenbach, L. A., Cable, T. T., & Cook, P. S. (1993). Discrete/continuous contingent valuation of private hunting access in Kansas. *Journal of Environmental Management. 39*, 1-12.

Halvorsen, R., & Palmquist, R. (1980). The interpretation of dummy variables in semi-logarithmic equations. *American Economic Review. 70*(3), 474-475.

Harris, L. D., Sullivan, R., & Badger, L. (1984). Bottomland Hardwoods: Valuable, Vanishing, Vulnerable. Florida Cooperative Fish and Wildlife Research Unit and the National Coastal Ecosystem Team of the Fish and Wildlife Service

Hazel, B. R. (1995). Deer management. *North Carolina Cooperative Extension Service. Retrieved September 11, 2005, from* http:www.ces.nscu.edu.nreos/forest/woodland/won-12htm

Hussain, A., Zhang, D., & Armstrong, J. B. (2004). Willingness to pay for hunting leases in Alabama. *Southern Journal of Applied Forestry. 28*(1), 21-27.

Hussain, A., Munn, I. A., Grado, S. C., West, B. C., Jones, W. D, & Jones, J. C. (2007). Hedonic analysis of hunting lease revenue and landowner willingness to allow hunting access. *Forest Science. 53*(14), 493-506.

Johnson, A. A., Johnson, M. B., & Buse, R. C. (1987). *Econometrics: basic and applied.* New York: Macmillan Publishing, Inc.

Kennedy, E. P. (1981). Estimation with correctly interpreted dummy variables in semi-logarithmic equations. *American Economic Review. 71*(4), 801.

Little, J. P., & Berrens, R. P. (2008). The southwestern market for big-game hunting permits and services: A hedonic pricing analysis. *Human Dimensions of Wildlife. 13*,143-157.

Livengood, K. R. (1983). Value of big game from markets for hunting leases: The hedonic approach. *Land Economics. 59*(3), 287-291.

Loomis, J. B., & Fitzhugh, L. (1989). Financial returns to California landowners for providing hunting access: Analysis and determinants of returns and implications to wildlife management. *Trans of North American Wildlife Natural Resource Conference.* 197-201.

McAfee, R. P., & McMillan, J. (1987). Auctions and bidding. *Journal of Economic Literature. 25*(June), 699-738.

Messmer, T. A., Dixon, C. E., Shields, W. Barras, S. C., & Schroeder, S. A. (1998). Cooperative wildlife management units: Achieving hunter, landowner, and wildlife management agency objectives. *Wildlife Society Bulletin. 26*(20), 325-332.

Messonnier, M. L., & Luzar, E. J. (1990). A hedonic analysis of private hunting land attributes using an alternative functional form. *Southern Journal Agricultural Economics. 22*(2), 129-135.

Milgrom, P. (1989). Auctions and bidding: A primer. *Journal Economic Perspectives. 3*(3), 3-22.

Mississippi Secretary of State's Office. 2005. A brief history of sixteenth section public school trust lands. Public Lands Division Record.

Mitchell, R. C., & Carson, R.T. (1989). *Using surveys to value public goods: The contingent valuation method.* Washington D.C: Resources for the Future, pp. 122-124.

Mozumder, P., Starbuck, C.M., Berrens, R.P. & Alexander, S. (2007). Lease and fee hunting on private lands in the U.S.: A review of the economic and legal issues. *Human Dimensions of Wildlife. 12*,1-14.

Munn, I. A., Loden, E. K., Grado, S. C., Jones, J. C., & Jones, W. D. (2005). Comparing hunting lease prices: A price decomposition approach. (pp. 193-200). In: *Proceedings of the 2004 Annual Southern Forest Economics Workshop*. Alavalapati, R.R. and D.R. Carter, eds. March 14-16, 2004. St. Augustine, FL.

Munn, I. A., Hussain, A., Byrd, J.D., Grado, S. C., Jones, J. C., Jones, W. D., Loden, E. K., Miller, J. E., & West, B. C. (2007). Landowners involvement and attitudes: Fee access wildlife and fisheries recreation. *Forest and Wildlife Research Center, Research Bulletin. Fo332,* Mississippi State University. 27pp.

Munn, I. A., Hussain, A., West, B., Grado, S.C., & Jones, W.D. 2007. Analyzing landowner demand for wildlife and forest management information. *Journal of Agricultural and Applied Economics. 39*(3),557-569.

Parker, R. C., Glass, P. A., Londo, H. A., Evans, D. L., Belli, K. L. Matney, T. G., & Schultz, E. B. (2005). Mississippi's forest inventory pilot program. Use of computer and spatial technologies in large area inventories. *Forest and Wildlife Research Center, Bulletin FO. 274*, Mississippi State University. 43 pp.

Pope, C. A., & Stoll, J. R. (1985). The market value of ingress rights for white-tailed deer hunting in Texas. *Southern Journal of Agricultural Economics. 17*(1), 177-182.

Rosen, S. (1974). Hedonic prices and implicit markets: Product differentiation in pure competition. *Journal of Political Economy. 82*, 34-555.

Shrestha, R. K., & Alavalapati, J. R. (2004). Effect of ranchland attributes on recreational hunting in Florida: A hedonic price analysis. *Journal of Agricultural and Applied Economics. 36*(3), 763-772.

Standiford, R. B., &. Howitt, R. E. (1993). Multiple use management of California's hardwood rangelands. *Journal of Range Management. 46*(2), 176-182.

Stribling, H. L., Caulfied, J. P., Lockaby, B. G., Thompson, D. P., Quicke, H. E., & Clonts, H. A. (1992). Factors influencing willingness to pay for deer hunting in the Alabama piedmont. *Southern Journal of American Forestry. 16*(3), 125-129.

Strickland, B. K., & Demarais, S. (2000). Age and regional differences in antlers and mass of white-tailed deer. *Journal of Wildlife Management. 64*, 903-911.

Thill, R. (1990). Managing southern pine plantations for wildlife. Proceedings of XIXth IUFRO World Congress; Canadian International Union of Forestry Research Organizations: Vol 1: 58-68.

U.S. Department of Commerce. (2000). 2000 census of population and housing. Population and housing unit counts. Bureau of the Census, Washington D.C.

Zhang, D., Hussain, A., & Armstrong, J. B. (2006). Supply of hunting leases from non-industrial private forest lands in Alabama. *Human Dimensions of Wildlife. 11*(1), 1-14.

INDEX

A

B

C

D

E

F

G

H

I

J

K

L

M

P

Q

R

S